Advances in Experimental Medicine and Biology

Volume 1033

More information about this series at http://www.springer.com/series/5584

Laura R. McCabe • Narayanan Parameswaran
Editors

Understanding the Gut-Bone Signaling Axis

Mechanisms and Therapeutic Implications

Springer

Editors
Laura R. McCabe
Department of Physiology and
Department of Radiology
Biomedical Imaging Research Centre
Michigan State University
East Lansing, MI, USA

Narayanan Parameswaran
Comparative Medicine and Integrative
Biology Program
East Lansing, MI, USA

Department of Physiology
Michigan State University
East Lansing, MI, USA

ISSN 0065-2598 ISSN 2214-8019 (electronic)
Advances in Experimental Medicine and Biology
ISBN 978-3-319-66651-8 ISBN 978-3-319-66653-2 (eBook)
DOI 10.1007/978-3-319-66653-2

Library of Congress Control Number: 2017955552

Printed on acid-free paper

This Springer imprint is published by Springer Nature
The registered company is Springer International Publishing AG
The registered company address is: Gewerbestrasse 11, 6330 Cham, Switzerland

Contents

Contributors

L.R. Bielke The Ohio State University, Animal Sciences, Columbus, OH, USA

Robert A. Britton Baylor College of Medicine, Department of Molecular Virology and Microbiology, Alkek Center for Metagenomics and Microbiome Research, Houston, TX, USA

Glen Camfield Department of Physiology, Michigan State University, East Lansing, MI, USA

Saima Chishti Department of Physiology, Michigan State University, East Lansing, MI, USA

Sylvia Christakos Department of Microbiology, Biochemistry and Molecular Genetics, Rutgers the State University of New Jersey, New Jersey Medical School, Newark, NJ, USA

Fraser L. Collins Department of Physiology, Michigan State University, East Lansing, MI, USA

Kevin Dagenais Department of Physiology, Michigan State University, East Lansing, MI, USA

John P. Geibel Department of Surgery, Yale School of Medicine, New Haven, CT, USA

B.M. Hargis University of Arkansas, Department of Poultry Science, Fayetteville, AR, USA

C.J. Hernandez Sibley School of Mechanical and Aerospace Engineering, Cornell University, Ithaca, NY, USA

Meinig School of Biomedical Engineering, Cornell University, Ithaca, NY, USA

Hospital for Special Surgery, New York, NY, USA

Regina Irwin Department of Physiology, Michigan State University, East Lansing, MI, USA

Carlos M. Isales Institute for Regenerative and Reparative Medicine, Augusta University, Augusta, GA, USA

Department of Neuroscience and Regenerative Medicine, Augusta University, Augusta, GA, USA

Department of Orthopaedic Surgery, Augusta University, Augusta, GA, USA,

Department of Medicine, Augusta University, Augusta, GA, USA

Department of Cellular Biology and Anatomy, Augusta University, Augusta, GA, USA

Jun Kang Department of Physiology, Michigan State University, East Lansing, MI, USA

Ho Jun Kang Department of Physiology, Michigan State University, East Lansing, MI, USA

Alice M. Kitay Department of Surgery, Yale School of Medicine, New Haven, CT, USA

J. D. Latorre University of Arkansas, Department of Poultry Science, Fayetteville, AR, USA

Brigitte Lavoie Department of Neurological Sciences, The University of Vermont, Burlington, VT, USA

The University of Vermont, Burlington, VT, USA

Tristan Lemon Department of Physiology, Michigan State University, East Lansing, MI, USA

Jane B. Lian Department of Biochemistry, The University of Vermont, Burlington, VT, USA

Heather Mallin Department of Physiology, Michigan State University, East Lansing, MI, USA

Gary M. Mawe Department of Neurological Sciences, The University of Vermont, Burlington, VT, USA

Laura R. McCabe Department of Physiology and Department of Radiology, Biomedical Imaging Research Centre, Michigan State University, East Lansing, MI, USA

Narayanan Parameswaran Comparative Medicine and Integrative Biology Program, East Lansing, MI, USA

Department of Physiology, Michigan State University, East Lansing, MI, USA

Nishant Patel Department of Microbiology, Biochemistry and Molecular Genetics, Rutgers the State University of New Jersey, New Jersey Medical School, Newark, NJ, USA

Darin Quach Department of Microbiology and Molecular Genetics, Michigan State University, East Lansing, MI, USA

Sandi Raehtz Department of Physiology, Michigan State University, East Lansing, MI, USA

Walter Ramsey Institute for Regenerative and Reparative Medicine, Augusta University, Augusta, GA, USA

Naiomy Deliz Rios-Arce Comparative Medicine and Integrative Biology Program, East Lansing, MI, USA

Department of Physiology, Michigan State University, East Lansing, MI, USA

Jonathan D. Schepper Department of Physiology, Michigan State University, East Lansing, MI, USA

Daniel Schoenherr Department of Physiology, Michigan State University, East Lansing, MI, USA

Danny T. Schoenherr Department of Physiology, Michigan State University, East Lansing, MI, USA

Ally Shinouskis Department of Physiology, Michigan State University, East Lansing, MI, USA

Michael D. Steury Department of Physiology, Michigan State University, East Lansing, MI, USA

Francisco A. Sylvester Division Chief of Pediatric Gastroenterology, The University of North Carolina at Chapel Hil, Chapel Hill, NC, USA

Vaishali Veldurthy Department of Microbiology, Biochemistry and Molecular Genetics, Rutgers the State University of New Jersey, New Jersey Medical School, Newark, NJ, USA

Connie M. Weaver Purdue University, Nutrition Science, West Lafayette, IN, USA

Ran Wei Department of Microbiology, Biochemistry and Molecular Genetics, Rutgers the State University of New Jersey, New Jersey Medical School, Newark, NJ, USA

Corrie M. Whisner Arizona State University, School of Nutrition and Health Promotion, Arizona Biomedical Collaborative, Phoenix, AZ, USA

Introduction

The skeleton and the gastrointestinal tract are intricate, highly regulated organ systems. Each system contains vasculature, nerve supply, lymphatics, immune cells, and resident specialized cells (i.e., osteoblasts and enterocytes) as well as a complex signaling network that involves hormones, nerves, immune cells, and paracrine factors. While much research has focused on understanding the mechanisms defining organ function and health, the involvement of interactions between organs has received less attention.

During the past decade, however, the role of organ-organ communication has become more apparent. For example, the skeleton actively regulates other systems through its secretion of osteocalcin, lipocalin, and FGF23. Osteocalcin, made by osteoblasts and osteocytes within bone, regulates pancreatic beta cell insulin secretion, insulin sensitivity, muscle and brain function, and testosterone production [1–3]. Lipocalin 2, a glucoprotein released by osteoblasts in response to food, binds receptors in the brain to suppress appetite [4]. FGF23, produced by osteocytes, suppresses kidney phosphate absorption [5]. Similarly, studies are now demonstrating links between the gastrointestinal tract and the regulation of a multitude of distant organ functions/systems including the brain, metabolism, liver, muscle, and cardiovascular and immune systems [6–8].

Similar to the regulation of other organs by gut or bone, it has become apparent that these two organs communicate with each other through a gut-bone axis that is far more complex and powerful than originally anticipated (Fig. 1). Bone has long been linked with the GI system because of the requirement for calcium absorption to promote bone mineralization; however, the GI tract communicates with bone through a variety of additional mechanisms that utilize blood, nerves, and immune cells. This book provides a comprehensive look at our current understanding of various factors that are involved in gut-bone signaling axis. In this context, chapters in Part I will address the role of intestinal regulation of calcium absorption, incretins (GIP, GLPs), serotonin, microbiota, and immune cells as mediators of gut-to-bone signaling. Part II will examine GI pathologies that impact bone density and health. These will include effects of modulating gastric acid secretion, intestinal inflammation (IBD, celiac disease), conditions associated with epithelial barrier changes such

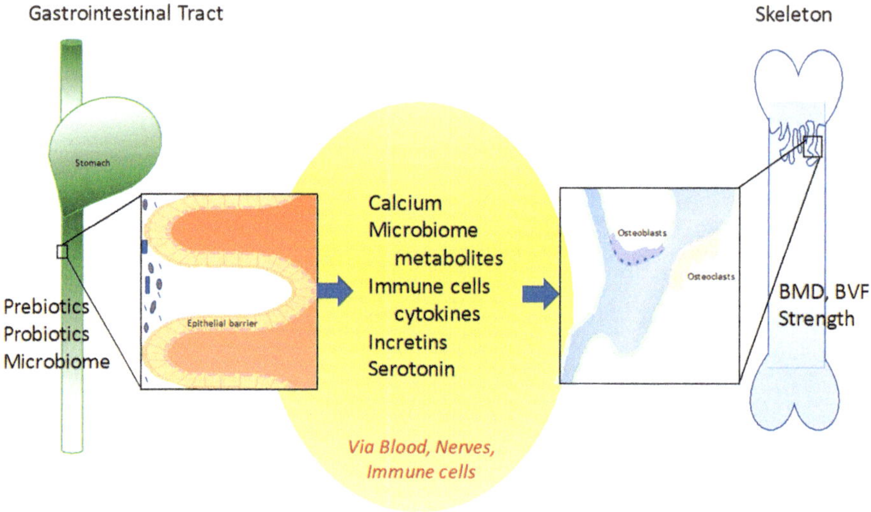

Fig. 1 Gastrointestinal-bone signaling axis. It is known that the intestine is critical for calcium absorption, but additional factors such as the microbiome and its metabolites, immune cells, and the cytokines and other factors such as incretins and serotonin contribute to the regulation of bone mineral density (BMD), bone volume fraction (BVF), and bone strength. Prebiotics, probiotics, and the microbiome itself can influence gut-bone axis signaling. Signals likely move to the bone through the blood, nerves, and movement of immune cells from the gut to the bone

as diabetes, menopause, and models of barrier breaks, as well as the role of dysbiosis. The last part, Part III, will focus on how manipulation of the intestinal environment, through pre- and/or probiotics, can benefit bone.

References

1. Suchacki KJ, Roberts F, Lovdel A, Farquharson C, Morton NM, MacRae VE, et al. Skeletal energy homeostasis: a paradigm of endocrine discovery. J Endocrinol. 2017;234(1):R67–79. doi:10.1530/JOE-17-0147.
2. Lee NK, Sowa H, Hinoi E, Ferron M, Ahn JD, Confavreux C, et al. Endocrine regulation of energy metabolism by the skeleton. Cell. 2007;130(3):456–69.
3. Oury F, Ferron M, Huizhen W, Confavreux C, Xu L, Lacombe J, et al. Osteocalcin regulates murine and human fertility through a pancreas-bone-testis axis. J Clin Invest. 2013;123(6): 2421–33. doi:10.1172/JCI65952.
4. Mosialou I, Shikhel S, Liu JM, Maurizi A, Luo N, He Z et al. MC4R-dependent suppression of appetite by bone-derived lipocalin 2. Nature. 2017;543(7645):385–90. doi:10.1038/nature21697.
5. Martin A, Quarles LD. Evidence for FGF23 involvement in a bone-kidney axis regulating bone mineralization and systemic phosphate and vitamin D homeostasis. Adv Exp Med Biol. 2012;728:65–83. doi:10.1007/978–1–4614-0887-1_4.

6. Powell N, Walker MM, Talley NJ. The mucosal immune system: master regulator of bidirectional gut-brain communications. Nat Rev. Gastroenterol Hepatol. 2017;14(3):143–59. doi:10.1038/nrgastro.2016.191.
7. Geurts L, Neyrinck AM, Delzenne NM, Knauf C, Cani PD. Gut microbiota controls adipose tissue expansion, gut barrier and glucose metabolism: novel insights into molecular targets and interventions using prebiotics. Benef Microbes. 2014;5(1):3–17. doi:10.3920/BM2012.0065.
8. Cani PD, Delzenne NM. The gut microbiome as therapeutic target. Pharmacol Ther. 2011;130(2):202–12. doi:10.1016/j.pharmthera.2011.01.012.

Part I
Gut-Bone Signaling Pathways

Intestinal Regulation of Calcium: Vitamin D and Bone Physiology

Sylvia Christakos, Vaishali Veldurthy, Nishant Patel, and Ran Wei

Introduction

Calcium, an essential ion for numerous physiological processes, is a major constituent of bone [1]. The intestine is the only source of new calcium. Absorption of dietary calcium, a process dependent on vitamin D, is essential for calcium homeostasis. The importance of vitamin D in this process is emphasized by the consequences of vitamin D deficiency which includes rickets in children and osteomalacia in adults [2–4]. Nearly a century ago, McCollum et al. identified vitamin D as the factor that cured rickets [5]. Solar or UVB irradiation is needed to convert 7-dehydrocholesterol in the skin to pre-vitamin D3 that ultimately is converted to vitamin D_3 (cholecalciferol) by thermo-isomerization [3, 6]. Since the synthesis of vitamin D in the skin depends on the intensity of ultraviolet irradiation, geographical location and season play an important role in contributing to vitamin D sufficiency in man [3, 7]. Although cutaneous production of vitamin D remains an important source, the fortification of foods (including fortification of dairy products) largely contributed to the marked decrease in the incidence of vitamin D-dependent rickets in the Western world by the mid-twentieth century [5]. However, vitamin D deficiency in children and adults is still prevalent worldwide due, in part, to lack of exposure to sunlight and low vitamin D intake [3]. In this chapter the vitamin D endocrine system as well as the mechanisms by which 1,25-dihydroxyvitamin D_3 ($1,25(OH)_2D_3$) acts to maintain calcium homeostasis will be briefly reviewed followed by an emphasis on the intestinal actions of $1,25(OH)_2D_3$ and effects of bone.

S. Christakos (✉) • V. Veldurthy • N. Patel • R. Wei
Department of Microbiology, Biochemistry and Molecular Genetics, Rutgers the State University of New Jersey, New Jersey Medical School,
185 South Orange Ave, Newark, NJ 07103, USA
e-mail: christak@njms.rutgers.edu

L.R. McCabe, N. Parameswaran (eds.), *Understanding the Gut-Bone Signaling Axis*, Advances in Experimental Medicine and Biology 1033,
DOI 10.1007/978-3-319-66653-2_1

The Vitamin D Endocrine System: Bioactivation and Mechanism of Action

Vitamin D, which is taken in the diet or which is synthesized in the skin by UV irradiation, must be metabolized to its active form in order to regulate calcium homeostasis. Vitamin D is transported by vitamin D binding protein (DBP, which binds and transports vitamin D and its metabolites in the serum) to the liver where 25-hydroxyvitamin D_3 (25(OH)D_3), the major circulating form of vitamin D and an important biomarker for vitamin D status, is generated [6, 8, 9]. It has been suggested that the cytochrome P450 (CYP) enzyme CYP2R1 is the key enzyme involved in the conversion of vitamin D to 25(OH)D_3 [10]. Patients with mutations in CYP2R1 are deficient in 25(OH)D_3 and develop vitamin D-dependent rickets [11]. Studies in *Cyp2r1* null mice which show that levels of 25(OH)D_3 are diminished but not eliminated suggest the presence of additional vitamin D 25-hydroxylases is yet to be identified [12]. In the proximal renal tubule, CYP27B1 (25(OH)D_3 1α hydroxylase) converts 25(OH)D_3 to 1,25-dihydroxyvitamin D_3 (1,25(OH)$_2$$D_3$), the hormonally active form of vitamin D which is responsible for the biological actions of vitamin D [6, 8, 9]. Mutations in CYP27B1 cause vitamin D-dependent rickets type 1 (VDDR1), characterized by hypocalcemia, hypophosphatemia, decreased mineralization, and low circulating 1,25(OH)$_2$$D_3$ levels [13]. With regard to the regulation of vitamin D metabolism, parathyroid hormone (PTH), whose synthesis is increased in response to hypocalcemia, induces CYP27B1 and is the major stimulator of 1,25(OH)$_2$$D_3$ production [8, 9]. 1,25(OH)$_2$$D_3$ and FGF23 (which promotes phosphate excretion) and its co-receptor α klotho negatively regulate CYP27B1. As an autoregulatory mechanism, 1,25(OH)$_2$$D_3$ induces CYP24A1 (25-hdyroxyvitamin D_3 24-hydroxylase), the enzyme that accelerates the catabolism of 1,25(OH)$_2$$D_3$ preventing hypercalcemia resulting from high circulating 1,25(OH)$_2$$D_3$ (see ref. [9] for review of the regulation of vitamin D metabolism). Thus, vitamin D, FGF23/α klotho, and serum calcium and phosphate act together to regulate calcium homeostasis.

The actions of 1,25(OH)$_2$$D_3$, similar to other steroid hormones, are mediated by the vitamin D receptor (VDR). 1,25(OH)$_2$$D_3$-occupied VDR heterodimerizes with the retinoid X receptor and together with chromatin active co-regulatory proteins interacts with vitamin D response elements in and around target genes resulting in the induction or suppression of gene expression [14].

Vitamin D and Intestinal Calcium Absorption

The principal function of 1,25(OH)$_2$$D_3$ in the maintenance of calcium homeostasis is to increase calcium absorption from the intestine. This conclusion was made from studies in VDR null mice which showed that rickets, osteomalacia, hypocalcemia, and hyperparathyroidism were prevented when VDR null mice were fed a rescue diet which included high calcium (2%), indicating that the skeletal abnormalities of VDR ablation are primarily the result of impaired intestinal calcium absorption (and resultant hyperparathyroidism and hypophosphatemia) [15, 16]. The abnormalities in the VDR null mice were reported to develop only after weaning [17] consistent with previous studies showing that intestinal VDR and the calcium binding protein calbindin-D_{9k} are induced at weaning, the time of onset of active intestinal calcium absorption [18, 19]. Although there is an increase in PTH and in the number of osteoblasts, osteoclast number is not increased in VDR null mice, suggesting that $1,25(OH)_2D_3$/VDR is needed for PTH-induced osteoclastogenesis [15]. Similar to VDR null mice, serum calcium and PTH were normalized in CYP27B1 null mice and in CYP27B1/VDR double null mice fed the high calcium rescue diet, confirming the importance of both $1,25(OH)_2D_3$ and VDR in intestinal calcium absorption [20, 21]. In addition, in CYP27B1 null mice, in spite of markedly elevated PTH levels, osteoclast numbers were also not increased above levels in normal wild-type (WT) mice, indicating that both VDR and $1,25(OH)_2D_3$ are necessary for PTH-mediated production of osteoclasts [21].

Direct evidence for the critical role of $1,25(OH)_2D_3$-mediated intestinal calcium absorption in bone homeostasis was noted in studies in transgenic mice with VDR expression limited to the intestine [22]. Transgenic expression of VDR in the intestine of VDR null mice restored calcium homeostasis and prevented the rachitic phenotype of the VDR null mice [22]. Thus, intestinal VDR is essential for controlling bone formation. In addition, when VDR is deleted specifically from the intestine (Vdr^{int-}), there is a decrease in intestinal calcium absorption, an inhibition of bone mineralization, and an increase in bone fractures in the Vdr^{int-}mice [23]. In these mice serum calcium is normal indicating that in the absence of VDR-mediated intestinal calcium absorption normal serum calcium will be maintained at the expense of skeletal integrity.

When there is an increased need for calcium (under low dietary calcium conditions, during growth, pregnancy, or lactation), the synthesis of $1,25(OH)_2D_3$ is increased and $1,25(OH)_2D_3$ acts at the intestine to increase active calcium absorption [6, 8]. The major defect from the loss of VDR is decreased intestinal calcium absorption resulting in decreased bone mineralization [15, 16]. If normal serum calcium cannot be maintained by intestinal calcium absorption, then $1,25(OH)_2D_3$ acts together with PTH to stimulate osteoclastogenesis resulting in the removal of calcium from the bone and to increase calcium reabsorption from the distal tubules of the kidney [6, 8].

Mechanisms Involved in $1,25(OH)_2D_3$ Regulation of Intestinal Calcium Absorption

Intestinal calcium absorption occurs by an active, saturable, transcellular mechanism or by a nonsaturable passive process which occurs through tight junctions and structures within intercellular spaces and requires a high luminal calcium concentration (>2–6 mM) [24]. $1,25(OH)_2D_3$ regulates the transcellular process by inducing TRPV6 (an apical membrane calcium channel), the calcium binding protein calbindin-D_{9k}, and the basolateral membrane calcium ATPase (PMCA1b) [8, 25, 26] (Fig. 1). Although it has been suggested that calbindin-D_{9k} mediates intracellular calcium diffusion, other studies suggest that a principal function of calbindin-D_{9k} is to buffer calcium preventing toxic levels from accumulating in the cell [27, 28]. When calcium is low, the $1,25(OH)_2D_3$-mediated transcellular calcium transport process is the predominant mechanism of calcium absorption [24]. Although TRPV6 and calbindin-D_{9k} are induced by $1,25(OH)_2D_3$, TRPV6 or calbindin-D_{9k} null mice have normal serum calcium and show no change in active intestinal calcium absorption compared to WT mice [29–31]. However, studies in TRPV6/calbindin-D_{9k} double null mice under conditions of low dietary calcium have shown that intestinal calcium absorption is least efficient in the absence of both proteins (compared to single null mice and WT mice), suggesting that TRPV6 and calbindin can act together in certain aspects of the absorptive process [29]. Findings in the single null mice suggest that in the absence of calbindin or TRPV6, there is compensation by other channels or proteins yet to be identified. Although other apical membrane calcium transporters may compensate for the loss of TRPV6, intestine-specific transgenic expression of TRPV6 has been shown to result in a marked increase in intestinal calcium absorption and bone density in VDR null mice, indicating a direct role for TRPV6 in the calcium absorptive process and that a primary defect in the VDR null mouse is low apical membrane calcium uptake [28].

The duodenum has been a focus of research related to $1,25(OH)_2D_3$ regulation of calcium absorption. However, it is the distal intestine where most of the ingested calcium is absorbed [24]. VDR, TRPV6, and calbindin-D_{9k} are expressed in all segments of the intestine and $1,25(OH)_2D_3$-regulated active calcium absorption occurs in the ileum, cecum, and colon [22, 32–36]. Recent studies have shown that transgenic expression of VDR specifically in the ileum, cecum, and colon can prevent abnormal calcium homeostasis and rickets in VDR null mice [37]. In addition, when VDR is deleted specifically from the distal region of the intestine, altered calcium metabolism is observed [38]. These findings indicate that the distal as well as the proximal segments of the intestine are important in vitamin D-mediated calcium homoeostasis and bone mineralization. Future studies related to mechanisms involved in $1,25(OH)_2D_3$-mediated regulation of calcium absorption in the distal intestine may suggest new strategies to increase the efficiency of calcium absorption in individuals at risk for bone loss including those with reduced calcium absorption due to small bowel resection or following menopause.

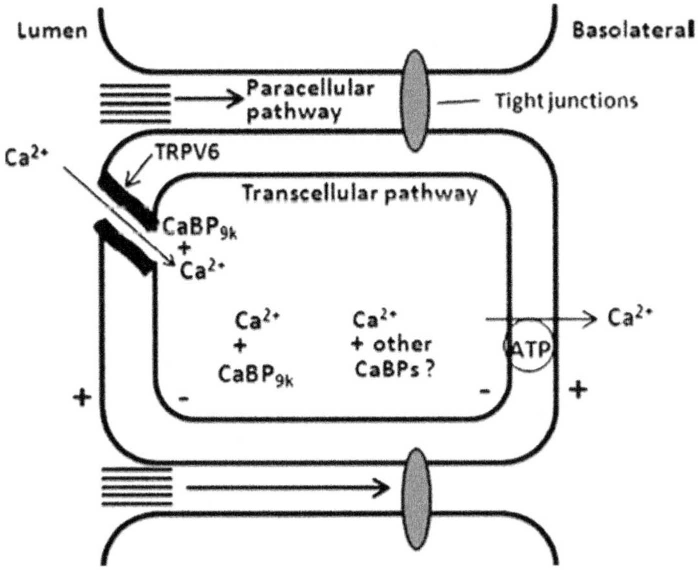

Fig. 1 Model of vitamin D-mediated intestinal calcium absorption. The transcellular pathway consists of influx through the apical calcium channel TRPV6, diffusion through the cytosol, and active extrusion at the basolateral membrane by the plasma membrane calcium ATPase (PMCA1b). Studies using TRPV6 and calbindin-D_{9k} ($CaBP_{9k}$) null mice indicate that our understanding of the vitamin D-mediated calcium transport process remains incomplete. In the absence of TRPV6 or calbindin-D_{9k} $1,25(OH)_2D_3$-mediated active transport still occurs, suggesting compensation by another channel or protein. Intracellular calcium transfer may involve calcium bound to calbindin as well as other calcium binding proteins. Calbindin as well as other calcium binding proteins may also act to prevent toxic levels of calcium from accumulating in the cell. It has been suggested that $1,25(OH)_2D_3$ can regulate the paracellular pathway by regulating tight junction proteins

In addition to regulation by vitamin D of the active transcellular process, early studies suggested that the passive, nonsaturable process of intestinal calcium absorption can also be enhanced by vitamin D [39]. More recent studies have shown that $1,25(OH)_2D_3$ can regulate intestinal proteins that are involved in tight junctions or cell adhesion. Claudin-2 and claudin-12 are induced and cadherin-17 is inhibited by $1,25(OH)_2D_3$ [40, 41]. These findings suggest that $1,25(OH)_2D_3$, by regulating these proteins, can facilitate calcium absorption through the paracellular path (Fig. 1). The identification of multiple mechanisms and novel vitamin D targets involved in $1,25(OH)_2D_3$-mediated calcium absorption in different segments of the intestine is needed in order to identify new approaches to maximize calcium absorption and minimize bone loss.

Vitamin D: Direct Effects on Bone

Although the primary role of the vitamin D endocrine system on bone is indirect (providing calcium to bone by stimulating intestinal calcium absorption), direct effects of $1{,}25(OH)_2D_3$ on bone cells have also been demonstrated (see [42] for review). As indicated above in studies in VDR and CYP27B1 null mice, VDR and $1{,}25(OH)_2D_3$ are needed for PTH-mediated osteoclastogenesis [15, 21]. Osteoclastogenesis mediated by $1{,}25(OH)_2D_3$ as well as by PTH involves upregulation of receptor activator of nuclear kB ligand (RANKL) in osteoblastic cells and requires cell to cell contact between osteoblasts and osteoclast precursors [43]. In addition $1{,}25(OH)_2D_3$ can also stimulate the production in osteoblasts of the calcium binding proteins osteocalcin and osteopontin (OPN) [8]. OPN has been reported to inhibit bone matrix mineralization [44]. Thus, during a negative calcium balance, $1{,}25(OH)_2D_3$ action can promote increased bone resorption and reduced bone matrix mineralization in order to maintain normal serum calcium levels [23]. $1{,}25(OH)_2D_3$ has also been shown to induce LRP5 (low density lipoprotein receptor-related 5) which facilitates β catenin activation and exerts an anabolic effect on bone formation [45]. These findings indicate that the effects of $1{,}25(OH)_2D_3$ on osteogenic cells are complex and can result in either bone resorption or formation.

Vitamin D: The Kidney and Calcium Homeostasis

Vitamin D-mediated calcium homeostasis is regulated by an integrated system involving not only the intestine and bone but also the kidney. Although most of the filtered calcium is reabsorbed by a passive, paracellular path in the proximal renal tubule that is independent of $1{,}25(OH)_2D_3$, 10–15% of the filtered calcium is reabsorbed in the distal convoluted tubule and connecting tubule and is regulated by PTH and $1{,}25(OH)_2D_3$ [46]. Similar to studies in the intestine, $1{,}25(OH)_2D_3$ regulates an active, transcellular process in the distal portion of the nephron by inducing the apical calcium channel TRPV5 (which shares 75% sequence homology with TRPV6) and by inducing the calbindins [both calbindin-D_{9k} (9,000 M_r) and calbindin-D_{28k} (28,000 M_r) are present in the mouse kidney and only calbindin-D_{28k} is present in rat and human kidney] [8]. Calcium is extruded via PMCA1b and the Na^+Ca^{++} exchanger [8]. It has been shown that calbindin-D_{28k} binds to TRPV5 and modulates calcium influx [47]. PTH has been reported to activate TRPV5 via protein kinase A phosphorylation [48]. The kidney is also the major site of production of $1{,}25(OH)_2D_3$ and its regulation [9]. Thus, the kidney, by regulating transport processes and as a major site of synthesis of $1{,}25(OH)_2D_3$, plays an essential role in the maintenance of calcium homeostasis.

Vitamin D: The Intestine and Bone Health

In aging intestinal calcium absorption declines which results in secondary hyperparathyroidism and increased fracture risk [49]. Decreased intestinal calcium absorption with age has been shown to correlate with decreased expression of TRPV6 and calbindin-D_{9k}, the two major targets of 1,25$(OH)_2D_3$ in the intestine [50, 51] (Fig. 2). Increasing evidence indicates that the reason for disturbed calcium balance with age is that vitamin D status is often inadequate in the elderly [52]. With age there is a decline in the ability of the kidney to synthesize 1,25$(OH)_2D_3$ and an increase in CYP24A1 which would result in enhanced catabolism of 1,25$(OH)_2D_3$ [53–55] (Fig. 2). Intestinal resistance to 1,25$(OH)_2D_3$ with age has also been reported [56–58]. It has been suggested that this resistance is due to a decrease in the content of intestinal VDR with age [57]. However, this has been a matter of debate [58]. It is possible that the resistance to 1,25$(OH)_2D_3$ may also be due to altered recruitment by 1,25$(OH)_2D_3$ of VDR and VDR coactivators to intestinal vitamin D target genes and/or to epigenetic changes. To reduce fracture risk, a combination of calcium and vitamin D supplementation has been recommended [59]. The current recommended daily doses for vitamin D-sufficient individuals are 800 IU calcium and 1,000 mg calcium [60].

In addition to effects of aging, gastric bypass surgery has also been reported to result in calcium malabsorption and decreased bone mineral density in patients [61–63]. Animal studies have also noted gastric bypass-associated bone resorption which is due to several factors including vitamin D and calcium malabsorption and acid/base dysregulation [64, 65]. Patients with inflammatory bowel disease are also at risk for bone disease due in part to impaired intestinal calcium absorption as well as to proinflammatory cytokines which are involved in the intestinal immune response but can also enhance bone resorption [66]. Future studies related to mechanisms involved in VDR-mediated activation of intestinal calcium absorption may suggest new mechanisms to compensate for calcium malabsorption in order to minimize bone loss due to aging, bariatric surgery, or inflammatory bowel disease.

Aging

Effects on the Vitamin D Endocrine System

↓Intestinal calcium absorption (↓TRPV6 and ↓Calbindin)

Vitamin D resistance

↓Synthesis of 1,25$(OH)_2D_3$

↑24(OH)ase expression (↑Catabolism of 1,25$(OH)_2D_3$)

(↑Age related bone loss)

Fig. 2 Age-related effects on the vitamin D endocrine system

Acknowledgements SC receives funding from the National Institutes of Health grants AG044552, DK112365 and AI121621.

Competing Interests The authors declare no conflict of interest.

References

1. Peacock M. Calcium metabolism in health and disease. Clin J Am Soc Nephrol. 2010;5(Suppl 1): S23–30.
2. Bouillon R, Bischoff-Ferrari H, Willett W. Vitamin D and health: perspectives from mice and man. J Bone Miner Res. 2008;23(7):974–9.
3. Holick MF. Resurrection of vitamin D deficiency and rickets. J Clin Invest. 2006;116(8):2062–72.
4. Lips P. Vitamin D deficiency and secondary hyperparathyroidism in the elderly: consequences for bone loss and fractures and therapeutic implications. Endocr Rev. 2001;22(4):477–501.
5. Deluca HF. History of the discovery of vitamin D and its active metabolites. Bonekey Rep. 2014;3:479.
6. DeLuca HF. Evolution of our understanding of vitamin D. Nutr Rev. 2008;66(10 Suppl 2):S73–87.
7. Webb AR, Kline L, Holick MF. Influence of season and latitude on the cutaneous synthesis of vitamin D3: exposure to winter sunlight in Boston and Edmonton will not promote vitamin D3 synthesis in human skin. J Clin Endocrinol Metab. 1988;67(2):373–8.
8. Christakos S, et al. Vitamin D: metabolism, molecular mechanism of action, and pleiotropic effects. Physiol Rev. 2016;96(1):365–408.
9. Christakos S, et al. Vitamin D: metabolism. Endocrinol Metab Clin N Am. 2010;39(2):243–53. table of contents
10. Cheng JB, et al. Genetic evidence that the human CYP2R1 enzyme is a key vitamin D 25-hydroxylase. Proc Natl Acad Sci U S A. 2004;101(20):7711–5.
11. Thacher TD, et al. CYP2R1 mutations impair generation of 25-hydroxyvitamin D and cause an atypical form of vitamin D deficiency. J Clin Endocrinol Metab. 2015;100(7):E1005–13.
12. Zhu JG, et al. CYP2R1 is a major, but not exclusive, contributor to 25-hydroxyvitamin D production in vivo. Proc Natl Acad Sci U S A. 2013;110(39):15650–5.
13. Kitanaka S, et al. Inactivating mutations in the 25-hydroxyvitamin D3 1alpha-hydroxylase gene in patients with pseudovitamin D-deficiency rickets. N Engl J Med. 1998;338(10):653–61.
14. Pike JW, Meyer MB. Fundamentals of vitamin D hormone-regulated gene expression. J Steroid Biochem Mol Biol. 2014;144(Pt A):5–11.
15. Amling M, et al. Rescue of the skeletal phenotype of vitamin D receptor-ablated mice in the setting of normal mineral ion homeostasis: formal histomorphometric and biomechanical analyses. Endocrinology. 1999;140(11):4982–7.
16. Masuyama R, et al. Dietary calcium and phosphorus ratio regulates bone mineralization and turnover in vitamin D receptor knockout mice by affecting intestinal calcium and phosphorus absorption. J Bone Miner Res. 2003;18(7):1217–26.
17. Yoshizawa T, et al. Mice lacking the vitamin D receptor exhibit impaired bone formation, uterine hypoplasia and growth retardation after weaning. Nat Genet. 1997;16(4):391–6.
18. Huang YC, et al. Effect of hormones and development on the expression of the rat 1,25-dihydroxyvitamin D3 receptor gene. Comparison with calbindin gene expression. J Biol Chem. 1989;264(29):17454–61.
19. Halloran BP, DeLuca HF. Calcium transport in small intestine during early development: role of vitamin D. Am J Phys. 1980;239(6):G473–9.
20. Dardenne O, et al. Correction of the abnormal mineral ion homeostasis with a high-calcium, high-phosphorus, high-lactose diet rescues the PDDR phenotype of mice deficient for the 25-hydroxyvitamin D-1alpha-hydroxylase (CYP27B1). Bone. 2003;32(4):332–40.

21. Panda DK, et al. Inactivation of the 25-hydroxyvitamin D 1alpha-hydroxylase and vitamin D receptor demonstrates independent and interdependent effects of calcium and vitamin D on skeletal and mineral homeostasis. J Biol Chem. 2004;279(16):16754–66.
22. Xue Y, Fleet JC. Intestinal vitamin D receptor is required for normal calcium and bone metabolism in mice. Gastroenterology. 2009;136(4):1317–27. e1-2
23. Lieben L, et al. Normocalcemia is maintained in mice under conditions of calcium malabsorption by vitamin D-induced inhibition of bone mineralization. J Clin Invest. 2012;122(5):1803–15.
24. Wasserman RH. Vitamin D and the dual processes of intestinal calcium absorption. J Nutr. 2004;134(11):3137–9.
25. Christakos S. Recent advances in our understanding of 1,25-dihydroxyvitamin D(3) regulation of intestinal calcium absorption. Arch Biochem Biophys. 2012;523(1):73–6.
26. Christakos S, et al. Vitamin D endocrine system and the intestine. Bonekey Rep. 2014;3:496.
27. Schroder B, et al. Role of calbindin-D9k in buffering cytosolic free Ca2+ ions in pig duodenal enterocytes. J Physiol. 1996;492(Pt 3):715–22.
28. Cui M, et al. Villin promoter-mediated transgenic expression of transient receptor potential cation channel, subfamily V, member 6 (TRPV6) increases intestinal calcium absorption in wild-type and vitamin D receptor knockout mice. J Bone Miner Res. 2012;27(10):2097–107.
29. Benn BS, et al. Active intestinal calcium transport in the absence of transient receptor potential vanilloid type 6 and calbindin-D9k. Endocrinology. 2008;149(6):3196–205.
30. Kutuzova GD, et al. TRPV6 is not required for 1alpha,25-dihydroxyvitamin D3-induced intestinal calcium absorption in vivo. Proc Natl Acad Sci U S A. 2008;105(50):19655–9.
31. Akhter S, et al. Calbindin D9k is not required for 1,25-dihydroxyvitamin D3-mediated Ca2+ absorption in small intestine. Arch Biochem Biophys. 2007;460(2):227–32.
32. Hirst MA, Feldman D. 1,25-Dihydroxyvitamin D3 receptors in mouse colon. J Steroid Biochem. 1981;14(4):315–9.
33. Stumpf WE, et al. Target cells for 1,25-dihydroxyvitamin D3 in intestinal tract, stomach, kidney, skin, pituitary, and parathyroid. Science. 1979;206(4423):1188–90.
34. Lee DB, et al. Intestinal and metabolic effect of 1,25-dihydroxyvitamin D3 in normal adult rat. Am J Phys. 1981;240(1):G90–6.
35. Favus MJ, Angeid-Backman E. Effects of 1,25(OH)2D3 and calcium channel blockers on cecal calcium transport in the rat. Am J Phys. 1985;248(6 Pt 1):G676–81.
36. Favus MJ, et al. Effects of diet calcium and 1,25-dihydroxyvitamin D3 on colon calcium active transport. Am J Phys. 1980;238(2):G75–8.
37. Christakos S, et al. Vitamin D biology revealed through the study of knockout and transgenic mouse models. Annu Rev Nutr. 2013;33:71–85.
38. Reyes-Fernandez PC, Fleet JC. Compensatory changes in calcium metabolism accompany the loss of vitamin D receptor (VDR) from the distal intestine and kidney of mice. J Bone Miner Res. 2016;31(1):143–51.
39. Wasserman RH, Kallfelz FA. Vitamin D3 and unidirectional calcium fluxes across the rachitic chick duodenum. Am J Phys. 1962;203:221–4.
40. Fujita H, et al. Tight junction proteins claudin-2 and -12 are critical for vitamin D-dependent Ca2+ absorption between enterocytes. Mol Biol Cell. 2008;19(5):1912–21.
41. Kutuzova GD, Deluca HF. Gene expression profiles in rat intestine identify pathways for 1,25-dihydroxyvitamin D(3) stimulated calcium absorption and clarify its immunomodulatory properties. Arch Biochem Biophys. 2004;432(2):152–66.
42. Bikle DD. Vitamin D and bone. Curr Osteoporos Rep. 2012;10(2):151–9.
43. Yasuda H, et al. Osteoclast differentiation factor is a ligand for osteoprotegerin/osteoclastogenesis-inhibitory factor and is identical to TRANCE/RANKL. Proc Natl Acad Sci U S A. 1998;95(7):3597–602.
44. Prince CW, Butler WT. 1,25-Dihydroxyvitamin D3 regulates the biosynthesis of osteopontin, a bone-derived cell attachment protein, in clonal osteoblast-like osteosarcoma cells. Coll Relat Res. 1987;7(4):305–13.

45. Fretz JA, et al. 1,25-Dihydroxyvitamin D3 regulates the expression of low-density lipoprotein receptor-related protein 5 via deoxyribonucleic acid sequence elements located downstream of the start site of transcription. Mol Endocrinol. 2006;20(9):2215–30.
46. Boros S, Bindels RJ, Hoenderop JG. Active Ca(2+) reabsorption in the connecting tubule. Pflugers Arch. 2009;458(1):99–109.
47. Lambers TT, et al. Calbindin-D28K dynamically controls TRPV5-mediated Ca2+ transport. EMBO J. 2006;25(13):2978–88.
48. de Groot T, et al. Parathyroid hormone activates TRPV5 via PKA-dependent phosphorylation. J Am Soc Nephrol. 2009;20(8):1693–704.
49. Ensrud KE, et al. Low fractional calcium absorption increases the risk for hip fracture in women with low calcium intake. Study of Osteoporotic Fractures Research Group. Ann Intern Med. 2000;132(5):345–53.
50. van Abel M, et al. Age-dependent alterations in Ca2+ homeostasis: role of TRPV5 and TRPV6. Am J Physiol Ren Physiol. 2006;291(6):F1177–83.
51. Brown AJ, Krits I, Armbrecht HJ. Effect of age, vitamin D, and calcium on the regulation of rat intestinal epithelial calcium channels. Arch Biochem Biophys. 2005;437(1):51–8.
52. Cauley JA, et al. Serum 25-hydroxyvitamin D concentrations and risk for hip fractures. Ann Intern Med. 2008;149(4):242–50.
53. Armbrecht HJ, Zenser TV, Davis BB. Effect of age on the conversion of 25-hydroxyvitamin D3 to 1,25-dihydroxyvitamin D3 by kidney of rat. J Clin Invest. 1980;66(5):1118–23.
54. Matkovits T, Christakos S. Variable in vivo regulation of rat vitamin D-dependent genes (osteopontin, Ca,Mg-adenosine triphosphatase, and 25-hydroxyvitamin D3 24-hydroxylase): implications for differing mechanisms of regulation and involvement of multiple factors. Endocrinology. 1995;136(9):3971–82.
55. Johnson JA, et al. Age and gender effects on 1,25-dihydroxyvitamin D3-regulated gene expression. Exp Gerontol. 1995;30(6):631–43.
56. Francis RM, et al. Calcium malabsorption in elderly women with vertebral fractures: evidence for resistance to the action of vitamin D metabolites on the bowel. Clin Sci (Lond). 1984;66(1):103–7.
57. Ebeling PR, et al. Evidence of an age-related decrease in intestinal responsiveness to vitamin D: relationship between serum 1,25-dihydroxyvitamin D3 and intestinal vitamin D receptor concentrations in normal women. J Clin Endocrinol Metab. 1992;75(1):176–82.
58. Wood RJ, et al. Intestinal calcium absorption in the aged rat: evidence of intestinal resistance to 1,25(OH)2 vitamin D. Endocrinology. 1998;139(9):3843–8.
59. Ross AC, et al. The 2011 report on dietary reference intakes for calcium and vitamin D from the Institute of Medicine: what clinicians need to know. J Clin Endocrinol Metab. 2011;96(1):53–8.
60. Gallagher JC. Vitamin D and aging. Endocrinol Metab Clin N Am. 2013;42(2):319–32.
61. Casagrande DS, et al. Changes in bone mineral density in women following 1-year gastric bypass surgery. Obes Surg. 2012;22(8):1287–92.
62. Fleischer J, et al. The decline in hip bone density after gastric bypass surgery is associated with extent of weight loss. J Clin Endocrinol Metab. 2008;93(10):3735–40.
63. Schafer AL, et al. Intestinal calcium absorption decreases dramatically after gastric bypass surgery despite optimization of vitamin D status. J Bone Miner Res. 2015;30(8):1377–85.
64. Abegg K, et al. Roux-en-Y gastric bypass surgery reduces bone mineral density and induces metabolic acidosis in rats. Am J Phys Regul Integr Comp Phys. 2013;305(9):R999–R1009.
65. Canales BK, et al. Gastric bypass in obese rats causes bone loss, vitamin D deficiency, metabolic acidosis, and elevated peptide YY. Surg Obes Relat Dis. 2014;10(5):878–84.
66. Bernstein CN, Leslie WD. The pathophysiology of bone disease in gastrointestinal disease. Eur J Gastroenterol Hepatol. 2003;15(8):857–64.

Intestinal Incretins and the Regulation of Bone Physiology

Walter Ramsey and Carlos M. Isales

Abbreviations

DPP-IV	Dipeptidyl peptidase-IV
EEH	Enteroendocrine hormones
EOH	Entero-osseous hormones
GIP	Gastric inhibitory polypeptide/glucose-dependent insulinotropic polypeptide
GLP-1	Glucagon-like peptide-1
GLP-2	Glucagon-like peptide-2
PC	Prohormone convertase
RA	Receptor agonist
SCFA	Short-chain fatty acids

Role of Nutrition in Bone Turnover

The process of bone remodeling serves a twofold purpose in that it renews the structural integrity of bone while playing a key role in calcium homeostasis. Serum calcium is maintained within a narrow range based on tightly regulating calcium

W. Ramsey
Institute for Regenerative and Reparative Medicine, Augusta University, Augusta, GA 30912, USA

C.M. Isales, MD (✉)
Institute for Regenerative and Reparative Medicine, Augusta University, Augusta, GA 30912, USA

Department of Neuroscience and Regenerative Medicine, Augusta University, Augusta, GA 30912, USA

Department of Orthopaedic Surgery, Augusta University, Augusta, GA 30912, USA

Department of Medicine, Augusta University, Augusta, GA 30912, USA

Department of Cellular Biology and Anatomy, Augusta University, Augusta, GA 30912, USA
e-mail: cisales@augusta.edu

L.R. McCabe, N. Parameswaran (eds.), *Understanding the Gut-Bone Signaling Axis*, Advances in Experimental Medicine and Biology 1033,
DOI 10.1007/978-3-319-66653-2_2

absorption or excretion in three different pools: gut, bone, and urine. Dietary calcium intake is normally the main source of daily calcium requirements, and this seems to be mainly a threshold dose rather than dose dependent with intake values below about 700 mg associated with a higher incidence of fractures [1]. During fasting the skeleton is an important calcium/mineral reservoir, and bone resorption helps maintain serum calcium levels. Although homeostatic mechanisms can compensate for even prolonged fasts, these processes are activated to some degree even during an overnight fast.

Because humans evolved under conditions where intermittent or prolonged fasting occurred frequently, the intestine is primed to maximize nutrient absorption and utilization when these become available. In the postprandial setting, the skeleton switches from catabolic to anabolic processes. Dietary calcium levels, calcitonin, and parathyroid hormone have been extensively investigated as modulators of this process. However, it has become clear that as the nutrients transit through the intestine, this organ serves an important endocrine function where a series of gut hormones are released to promote anabolic functions including on the skeleton, a connection we have named the entero-osseous axis [2]. Although it was initially believed that the rate of bone turnover varied in a circadian manner, it was later reported that fasting modulates bone turnover independent of circadian hormones [3, 4].

This chapter focuses on reviewing the mechanism of incretin hormone (such as GIP and GLP-1 as well as GLP-2) effects on bone turnover. The term "entero-osseous hormones" (EOH) will be used herein to distinguish GIP, GLP-1, and GLP-2 from other enteroendocrine hormones (EEH).

Entero-osseous Hormone Processing, Signaling, and Metabolism

Introduction to Entero-osseous Hormones

In the early 1900s, it was discovered that signals originating in the gut could impact a variety of endocrine processes. One such process, the "incretin effect," was coined to describe the difference between the observed greater increase in pancreatic insulin secretion in response to oral glucose administration as compared to intravenous glucose administration. The gut hormones responsible for this process became known as incretins [5].

The incretins are a group of intestinal peptide hormones with known insulinotropic action. The most widely recognized incretins are gastric inhibitory polypeptide (also known as glucose-dependent insulinotropic polypeptide or GIP) and glucagon-like peptide-1 (GLP-1). These incretins are secreted by the endocrine cells of the intestine in response to oral glucose, as well as a number of other biological signals discussed later in this chapter. GIP is secreted by the "K-cells" which line the mucosa of the duodenum and jejunum [6], while GLP-1 and its counterpart GLP-2 are secreted by the "L-cells," which are primarily located in the distal jejunum, ileum, and colon [7]. Unlike GIP and GLP-1, GLP-2 lacks classical insulinotropic action.

All three of these hormones belong to the glucagon-related superfamily [7]. They are classified as members of this superfamily because of the amino acid sequence homology they share with glucagon, but they do not necessarily act in a similar manner to glucagon physiologically.

Entero-osseous Hormone Synthesis

Prior to their release by the enteroendocrine K- and L-cells, the entero-osseous hormones must be processed into their active forms from their inactive prohormone precursors. GLP-1 and GLP-2 are synthesized from the peptide hormone pre-proglucagon. Pre-proglucagon is a 180 amino acid molecule which contains, in addition to the sequence of glucagon, the sequences of GLP-1 and GLP-2 [8]. The gene that encodes pancreatic pre-proglucagon contains six exons. The first four exons encode proteins corresponding to the signal peptide, glucagon, GLP-1, and GLP-2, and the final two are characteristically cleaved during the posttranslational processing of prohormones [9]. The mRNA that encodes the proglucagon gene is the same in the various tissues; however, the form of proglucagon varies, due to alternative posttranslational modification [10]. This differential processing is tissue dependent [11].

The enzymes involved in alternative posttranslational processing of proglucagon are known as prohormone convertases (PCs) . Proteolytic cleavage by these enzymes ensures that peptide hormones like GLP-1 and GLP-2 become activated at the appropriate site [12]. In the alpha cells of the pancreas, the predominant PC2 cleaves active glucagon from residues 33–61 of proglucagon, leaving a major fragment from the C-terminal of the molecule unprocessed. However, in the L-cells of the intestine, glucagon is handled by the protein convertase PC1/3. This proteolytic cleavage yields GLP-1 and GLP-2, rather than glucagon [13]. The preponderance of PC1/3 compared to the paucity of PC2 expression in the L-cells is responsible for this.

Unlike its EOH counterparts, GIP is not cleaved from proglucagon. Rather, GIP is uniquely produced from the precursor transcript proGIP. The cleavage of proGIP to the active form of GIP takes place in the enteroendocrine K-cells. Much like GLP-1 and -2 posttranslational processing, this cleavage requires PC1/3 but not PC2 [14, 15].

Incretin	Precursor	Released from	Size
GIP	ProGIP	Enteroendocrine K-cells	42 amino acids
GLP-1	Proglucagon	Enteroendocrine L-cells	30 amino acids
GLP-2	Proglucagon	Enteroendocrine L-cells	33 amino acids

Regulation of Entero-osseous Hormone Secretion

The most common stimulus for enteric hormone secretion is glucose, but other nutrient/stimuli can also induce their secretion as well. Ingestion of other macronutrients including fat, protein, and non-glucose carbohydrates can also stimulate EOH secretion. In vivo studies have shown fat and protein consumption both significantly increase the serum concentrations of GIP and GLP-2 compared to fasting [15]. In fact, fat ingestion induces GIP concentrations approximately twice as high as those measured after glucose or protein ingestion. Oral fructose induces GLP-2 secretion in quantities comparable to glucose, but it has no significant impact on GIP. Further, because postprandial GLP-1 and GLP-2 secretion have been shown to occur in parallel [16] and in equimolar fashion, it is likely that GLP-1 secretion also increases in response to fat, protein, and fructose ingestion.

Hormone	Fat	Protein	Fructose	Glucose
GIP	+ +	+	No change	+
GLP-1	+	+	+	+
GLP-2	+	+	+	+

Secondly, it has been shown that a variety of non-nutrient mechanisms contribute to the regulation of EOH-secreting cells. These processes include bile acid signaling, microbiotic fermentation, neural regulation, and hormonal control [17]. While bile acids have been shown to act directly on basolateral L-cell receptors to induce GLP-1 [18], other mechanisms involve more complex processes.

Microbiota in the colon produce short-chain fatty acids (SCFAs) by fermenting fiber. In vivo data shows SCFAs in turn increase the secretion of GLP-1 [19]. This observable phenomenon is supported by PCR studies that confirm the expression of SCFA receptors on human L-cells [20].

Neural control of incretin-secreting cells occurs through a variety of pathways. In fact, it is likely that both the sympathetic and parasympathetic systems play a role. Excitatory signals are transmitted by muscarinic cholinergic, beta-adrenergic, and peptidergic fibers. EOH secretion is diminished by adrenergic and somatostatin innervation [21].

Hormonal control of human EOH secretion has long been debated as well. Contrary to animal research studies, it has been shown that humans do not respond to GIP infusion by increasing secretion of GLP-1 [22]. However, it is still believed that the pancreatic hormone somatostatin may negatively control both GIP and GLP-1 secretion by suppressing cAMP levels in human enteroendocrine cells, as this mechanism has been confirmed in porcine and rodent models [23, 24]. In fact, somatostatin has been demonstrated to downregulate GIP in humans [25]. As the incretins GIP and GLP-1 have been shown to significantly increase secretion of somatostatin in vivo [26], this relationship has received consideration as a local negative feedback mechanism. Although these appear to be the most significant hormonal controls of enteroendocrine secretion, a variety of other hormonal pathways have been studied [17].

Entero-osseous Hormone Receptors

The biological actions of entero-osseous hormones are mediated by G protein-coupled receptors, which bind their respective ligands with extremely high specificity. GIP, GLP-1, and GLP-2 receptors all belong to the secretin-glucagon receptor seven-transmembrane G protein-coupled family and are widely distributed in the body (Fig. 1). Their binding events activate signal transduction cascades that involve both changes in cAMP and intracellular calcium. GLP-1 signal transduction events have been mainly studied in pancreatic beta cells. In these cells GLP-1 primarily increases cAMP production to potentiate glucose-induced insulin secretion. However, in response to increasing glucose concentrations, GLP-1 also increases calcium influx through voltage-dependent calcium channels [27]. Some investigators have reported that GLP-1 also activates phospholipase C and mobilizes intracellular calcium pools [28]. These signaling pathways may serve different physiological functions with cAMP important in glucose-stimulated insulin secretion [29], and intracellular calcium/protein kinase C pathway may mediate the viability-enhancing, antiapoptotic properties of GLP-1 [30] (Fig. 2).

The GLP-2 receptor is widely expressed in enteric tissues including stomach, duodenum, jejunum, ileum, and colon [31]. GLP-2 receptor expression in bone is less widespread although it is present in immature osteoblasts. Binding of GLP-2 to its receptors expressed in BHK fibroblast cells leads to an increase in cAMP and activation of PKA and AP-1 pathways [32]. It would appear likely though that GLP-2 activates additional signaling pathways by either direct or indirect mechanisms. GLP-2 receptors are also expressed in bone cell lines, and GLP-2 binding resulted in an increase in osteocalcin in MG-63 cells [33].

GIP receptors are widely expressed throughout the intestine including the gut, pancreas, adrenal cortex, and adipose tissue, and in addition it is present in vascular tissue, brain, and bone cells [34–36]. Activation of the GIP receptor results in an increase in both cAMP and an increase in intracellular calcium [34, 35]. Downstream these second messengers activate the ERK1/2/MAPK pathway and in a cell-dependent manner modulate cell survival or proliferation [37]. The presence of GIP receptor splice variants has been suggested in some tissues [38], although these might not be functional variants [39].

Biological actions that are activated upon receptor binding are covered in more detail later in this chapter.

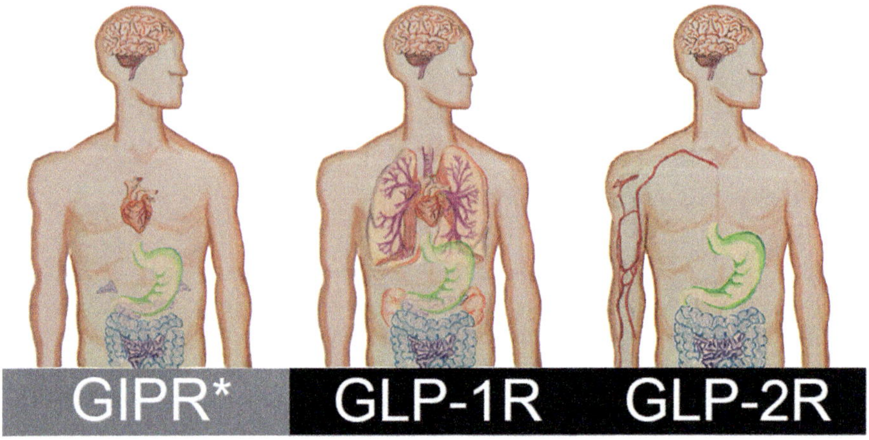

Fig. 1 Incretin hormone receptor expression: GIP receptors are present in the pancreas, gut, adipose, heart, adrenal cortex, and numerous brain regions, including the cerebral cortex, hippocampus, olfactory bulb, and pituitary. GLP-1 receptors are expressed in the pancreas, duodenum, kidney, heart, lung, large intestine, stomach, breast, and brain. The GLP-2 receptor is widely expressed in enteric tissues including stomach, duodenum, jejunum, ileum, and colon

Entero-osseous Hormone Metabolism

The main degradation pathway for GIP, GLP-1, and GLP-2 is aminopeptidase dipeptidyl peptidase IV (DPP-IV). This ubiquitous enzyme exists in both a membrane-bound and soluble form. Immediately following enteroendocrine secretion, inactivation by DPP-IV enzymes embedded in the endothelium of intestinal capillary beds begins [40]. DPP-IV proteolytically cleaves two N-terminal amino acids, dramatically reducing, and in some cases completely eliminating, the biological activity of these hormones [41]. In GLP-1 and GLP-2, the dipeptide His-Ala is truncated, whereas in active GIP the dipeptide Tyr-Ala gets removed [42].

The metabolic action of DPP-IV explains the often-observed discrepancy between serum GLP-1 and GLP-2 concentrations despite their equimolar secretion. An in vitro study by Lambier et al. demonstrates that DPP-IV preferentially truncates GLP-1 over GLP-2 at equivalent concentrations [43]. In fact, in vivo human studies have shown that the half-life of GLP-1 is 1–2 min, while the half-life of GLP-2 is closer to 7 min [44, 45].

Finally, clearance of enteroendocrine metabolites is believed to be chiefly handled by the kidney [40]. This is supported by evidence that GLP-2 markers are elevated in patients with chronic kidney failure [46].

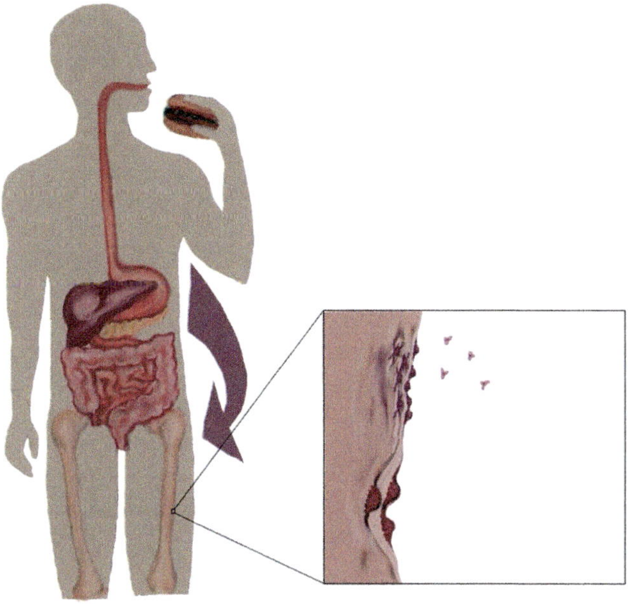

Fig. 2 Nutrients modulate bone turnover: Intestinal hormones serve to signal tissues in the body that nutrients are available for utilization. Incretin hormones in particular serve as anabolic signals to the bone leading to suppression of bone breakdown (osteoclastic activity) and increased bone formation (osteoblastic activity). Incretin hormones such as GIP appear to have predominant effects on osteoblasts, while GLP-1 and GLP-2 appear to mainly modulate bone breakdown through effects on the osteoclasts

Osseous Effects of GIP, GLP-1, and GLP-2

The Entero-osseous Signaling Axis

After discovery that bone metabolism was modulated, at least in part, by hormones that are released following nutritional intake, the role of gut-derived peptide hormones in bone turnover has been extensively evaluated in both in vitro and in vivo studies. The existence of the entero-osseous signaling axis was supported by the fact that receptors for enteroendocrine hormones were expressed on bone cells.

GIP receptors (GIPR) were reported to be present on osteoblast cells [34], and a subsequent study by Pacheco et al. examined the presence of GIP, GLP-1, and GLP-2 receptors in osteoblastic precursor cells, reporting that GIPR expression increased with increasing differentiation while GLP-1R and GLP-2R were only observed in less-differentiated osteoblastic precursors [33]. GIP was found to increase the viability of further-differentiated precursor cells, while GLP-1 and GLP-2 exposure significantly enhanced viability of less-differentiated precursors. Other studies examined osteoclasts for EEH receptors and found both GIPR expression in these cells [36].

	Immature osteoblasts	Mature osteoblasts	Mature osteoclasts
GIPR	No	Yes	Yes
GLP-1R	Yes	Yes[a]	No
GLP-2R	Yes	No	No

[a]Pacheco et al. do not report the presence of GLP-1R on osteoblasts [33]. However, GLP-1R was identified on osteoblastic cells in another study [47]. GLP-2 infusion decreases bone resorption [48]

Apart from studies that have focused on identifying EEH receptors on bone, a plethora of experiments have examined the entero-osseous axis by measuring the specific effects of EEH administration on bone turnover. In general, GIP, GLP-1, and GLP-2 achieve net bone formation by either enhancing osteoblast activity or attenuating osteoclast activity. The specific actions of these three peptide hormones, however, are unique, and as such they will be addressed individually.

Effects of GIP on Bone

The osseous effects of GIP are the first – and perhaps the best – documented of the enteroendocrine hormones. While osteoblastic and osteoclastic activity are coupled under most normal physiological conditions, it has been noted postprandially that GIP effectively uncouples these processes [49]. GIP operates in a twofold manner to increase bone quality: it has both pro-anabolic and anti-catabolic actions.

To date human studies involving GIP administration are limited with most of the in vivo studies being performed in murine models and in vitro models to a number of osteoprogenitor or osteoblastic cell lines.

In osteoblastic cell lines, addition of GIP results in increases in cAMP and intracellular calcium. GIP stimulation of these cells results in an elevation of alkaline phosphatase activity and increased expression of collagen type I mRNA [34], consistent with an anabolic effect of this hormone on bone. In the osteoblastic cell line, MC3T3-E1 GIP was also found to increase cAMP levels and promote collagen maturity and modulates collagen fiber diameter [50]. An additional function of GIP (and other incretins) seems to be the ability to protect cells from injury. In both osteoblasts and human bone marrow-derived mesenchymal stem cells, GIP reduced the apoptosis normally seen with tissue culture serum deprivation by inhibiting activation of caspase-3 and caspase-7 [51].

A number of murine models have been used to evaluate GIP actions on bone including GIP injections, transgenic GIP peptide overexpressing mice, and GIP receptor knockout mice. A daily GIP injection for 6 weeks was shown to prevent ovariectomy-induced bone loss in Sprague-Dawley rats [2]. In transgenic mice in which systemic GIP levels are between two- and threefold higher than control mice, bone density was increased by about 5% accompanied by an increase in the marker of bone formation, osteocalcin, and a decrease in the marker of bone breakdown, PYD. Finally, knockout of the GIP receptor in mice resulted in decreased bone mineral density (3.7% decrease in total BMD at 5 months of age compared to wild type), decreased bone strength, and a decrease in markers of bone formation, alkaline phosphatase, and osteocalcin [52]. Bone histomorphometry in GIPR -/- mice reveals decreased bone formation and increased osteoclast number. In addition postprandial plasma calcium was higher in the GIPR -/- mice suggesting a potential role for GIP in calcium deposition in bone [53]. GIPR knockout models have also demonstrated that GIP contributes to cortical thickness, load-bearing ability, and bone matrix mineralization [54]. GIPR-deficient mice have diminished bone mechanical properties despite increased total bone volume and trabecular bone mass [55]. To further confirm that these changes in bone mass were GIP related and not due to upregulation by other incretin hormones, Mieczkowska et al. [56] evaluated bone mass in double incretin receptor (GIPR and GLP-1) knockout mice (DIRKO) and found that cortical bone was still decreased in this mouse model consistent with a role for GIP as an intestinal hormone linking nutrient intake to bone formation.

Postprandial suppression of bone breakdown in human subjects has been well documented [57]. Suppression of bone breakdown by nutrients was initially narrowed to an enteric hormone as activity was suppressible by somatostatin infusion [58], but this nutrient-mediated antiresorptive effect was clearly not fully explained by food-induced increases in insulin levels [59]. A study by Henriksen et al. [15] evaluated the effects of GIP infusion on markers of bone turnover in postmenopausal women. These investigators found that GIP given as a bolus had no statistically significant effect (although a 16% drop was observed) on markers of bone turnover (either s-CTX or osteocalcin) over 48 min. In a follow-up study by the same group, in which GIP was co-infused during a euglycemic or hyperglycemic glucose clamp, the investigators found that GIP by itself decreased CTX (a maker of bone breakdown) but that this inhibition was more dramatic in the presence of hyperglycemia (49% decrease from basal) [60]. The authors discuss that differences between the two studies might relate to the fact that in the first study GIP was administered as a bolus rather than an infusion and also note that under physiological conditions GIP would be rising after a meal, when blood glucose would also be normally rising. Thus, it is possible that the second study was more physiologically relevant. This possibility was further supported by a study from Denmark involving 1,424 perimenopausal women followed longitudinally for 10 years. This study by Torekov et al. found that women with a functional polymorphism (Glu354Gln) of the GIP receptor had a significantly lower femoral neck bone mineral density and women homozygous for this variant were at increased risk for non-vertebral fractures (hazard ratio 1.6; CI: 1.0–2.6, $p < 0.05$) [61].

More recently with the emergence of inhibitors of the enzyme that breaks down incretin hormones (DPP-IV inhibitors) as a therapy for patients with diabetes mellitus, there has been increasing interest in the impact of these medications on bone health [62].

Glorie et al. [63] evaluated the impact of the DPP-IV inhibitor, sitaglipitin, on streptozotocin-treated male Wistar rats for up to 12 weeks. They found that sitagliptin improved bone parameters in the diabetic rat including improved trabecular number, decreased trabecular spacing, improved femoral bone strength, and decreased marker of bone resorption CTX. In a study in diabetic mice, the mice were treated for 8 weeks with either alendronate, pioglitazone (a drug known to negatively affect bone), or DPP-IV inhibitor MK-0626 [64]. The investigators reported that while alendronate improved cortical and trabecular bone microarchitecture and pioglitazone negatively impacted trabecular bone in the wild-type bone, the DPP-IV inhibitor had no impact on either trabecular or cortical bone in diabetic mice. Similarly in a study in Zucker Diabetic Fatty rats, treated with either the DPP-IV inhibitor vildagliptin, pioglitazone, or the combination of these two drugs for 5 weeks, the DPP-IV inhibitor protected the bones of diabetic mice from the negative effects of pioglitazone [65].

In a study in human diabetic patients [66] randomized to either control or therapy with the DPP-IV inhibitor vildagliptin for 1 year, the investigators found that at the end of the year there was no difference in the markers of bone breakdown (sCTx) or bone formation (alkaline phosphatase). Similarly in a meta-analysis involving 22,961 diabetic patients, the use of DPP-IV inhibitors was not associated with a reduction in fractures [67]. The same group performed a retrospective study involving 328,254 patients from the Clinical Practice Research Datalink (CPRD) database who receive a prescription for an antidiabetic drug [68]. The authors found that long-term use of a DPP-IV inhibitor (4–8.5 years) was not associated with any greater risk of osteoporotic or hip fracture. Thus, the consensus to far of these animal and limited human studies appears to be that DPP-IV inhibitors used to treat patients with diabetes do not weaken bone or increase fracture risk. Whether these drugs might be beneficial for increasing bone mass or quality in normal or diabetic patients remains unclear.

The anti-resorptive effects of GIP have also been well documented. Human studies have shown that GIP infusion significantly decreases markers for bone resorption in both the euglycemic and hyperglycemic states [60]. Rat studies show that GIP is capable of attenuating bone resorption both directly and indirectly. GIP infusion decreases the area of osteoclastic pits by acting directly on osteoclast cells in a dose-dependent manner. Meanwhile, GIP also dose dependently restricts bone resorption through modulation of parathyroid hormone-induced resorption [36].

GIP has a very short half-life, and thus any potential therapeutic use of this peptide would be limited by the need for multiple repeated administrations. A rapidly developing area is the synthesis of long-lived GIP analogues resistant to degradation. DPP-IV cleaves the amino terminal dipeptide; modifications of these amino acids by acetylation can result in analogs like N-AcGIP that has a half-life of over 24 h [69]. When this compound was administered to Copenhagen rats as a daily injection for 4 weeks, it resulted in improvement in the mechanical properties of cortical bone. Similarly, [D-AL2]GIP injected daily for 21 days improved bone material properties in diabetic mice (STZ treated) [70]. Hybrid incretin receptor agonists are also in development. [D-Ala^2]GIP-Oxm has the ability to bind the GLP-1, GIP, and glucagon receptors [71]. When this compound was injected daily into the diabetic mouse model (db/db) for 21 days, there was an improvement in bone strength in microarchitecture over that seen in saline-injected mice.

GIP's Osseous Effects in Aging Populations

GIP has been shown to continue to have anabolic effects on bone mass even in the setting of aging bone. Administration of GIP has been shown to induce osteoblastic differentiation and counter the effects of age-induced bone loss in mice. In fact, it has been observed that a reduction of GIPR expression in bone marrow stromal cells may partially explain the pathophysiology of age-induced bone loss [72]. Accordingly, an in vitro study of human osteoblasts and bone marrow mesenchymal stem cells has shown GIP has an antiapoptotic effect [51]. These exciting discoveries could pave the way for innovative anti-osteoporotic therapies.

Effects of GLP-1 on Bone

Mouse studies have shown that GLP-1 is also necessary for optimal bone metabolism and strength [73]. Similar to GIPR-KO models, GLP-1R-knockout rodent experiments have documented that a lack of GLP-1 action in bone significantly reduces maximal load-bearing, cortical thickness and bone matrix quality [74]. Conversely, GLP-1R agonist treatment has been shown to have osteogenic effects [75]. However, the rodent response appears to be mediated by increased calcitonin secretion from the C-cells of the thyroid. Corresponding action of GLP-1 in the human thyroid has not been observed [76]. Therefore, it is possible that GLP-1 affects human bone via another pathway, or not at all.

The efficacy of GLP-1R agonists to aid in fracture prevention has been documented in some, but not all, human studies [77]. As GLP-1R agonists are commonly used for the treatment of type II diabetes mellitus, this is a topic of considerable relevance. However, definitive proof linking GLP-1RA-class drugs to decreased fracture rates is lacking at this point [78].

Effects of GLP-2 on Bone

In contrast, the pro-osseous properties of GLP-2 have been noted in a variety of human studies, yet paradoxically, the mechanism underlying GLP-2 action on bone remains poorly understood. Initially it was discovered that parenteral administration of GLP-2 reduced markers for bone resorption without affecting bone deposition [15]. This led to the hypothesis that GLP-2 either acts directly on osteoclasts (rather than osteoblasts) or operates via secondary signaling factors. Later it was revealed that GLP-2 significantly attenuated the normally observed overnight rise in bone resorption, again without significantly stimulating bone formation [79]. Thus, a more prolonged 14-day study was performed to confirm these findings, and indeed bone resorption markers were significantly reduced while indicators of bone formation were unchanged [80].

A study by Henriksen et al. focused on GLP-2's effect on bone turnover – particularly hip BMD – in postmenopausal women and found that GLP-2 increased bone formation [81]. Another study showed that GLP-2 administration significantly increased spinal bone mineral density in short-bowel patients with no colon. While the mechanism mediating GLP-2 anabolic effects on bone mass remains unclear, this study suggests that GLP-2 promotes increased intestinal calcium absorption [82].

Extraosseous Effects of Entero-osseous Hormones

In addition to being present in bone, GIP, GLP-1, and GLP-2 receptors have been identified on the cells of many organs throughout the body. This suggests that these entero-osseous hormones cause direct effects in the corresponding organs and tissues. However, in many cases these functions have not been precisely identified.

The Incretin Effect

As previously mentioned, GIP and GLP-1 are best known for their insulinotropic action. In 1973, Dupre et al. first reported that GIP stimulates insulin secretion and improves glucose tolerance [17]. This led to the concept that gut hormones are responsible for the "incretin effect" observed in response to oral but not IV-administered glucose. Then, in 1987 Kerymann et al. reported that GLP-1 also has incretin action and even noted that GLP-1 induces a more potent insulin release than does GIP [83]. Accordingly, the presence of GIP and GLP-1 receptors in the pancreatic islets has been documented by numerous sources. However, as this aspect of GIP and GLP-1 physiology has already been reviewed extensively in the existing literature, this text will avoid focusing further on the insulinotropic actions of the incretin hormones.

Extraosseous GLP-1 Receptors

The GLP-1 receptor is widespread throughout humans. Obviously it is highly expressed in the pancreas, but a number of studies have also detected GLP-1R in the duodenum, kidney, heart, lung, large intestine, stomach, breast, and brain [47, 84, 85]. Although the biological importance of many of these receptors is unknown, several represent exciting opportunities for future research.

GLP-1 receptors are highly expressed in Brunner's glands of the duodenum. Recent studies show these receptors may be involved in inflammatory processes, and they are thus being investigated for their potential role in inflammatory bowel diseases [86]. The physiological significance of the dense distribution of GLP-1R in the human neurohypophysis is unclear [85], although GLP-1 has been shown to stimulate the neurohypophyseal secretion of oxytocin and vasopressin in rats [87].

In the kidney, GLP-1R is consistently found in the vascular wall of the arteries and arterioles as well as in the juxtaglomerular apparatus [88]. This suggests a potential role for GLP-1 in the regulation of kidney blood flow and renin secretion.

Controversy surrounds the action of GLP-1 and its metabolites in the heart. While some studies report a lack of GLP-1R in the heart, others have identified GLP-1R in myocytes of the sinoatrial node and even postulated they may mediate an increase in heart rate and blood pressure [85]. This may explain the observed cardiovascular changes following the administration of some GLP-1 targeting drugs. However, it has also been proposed that GLP-1R agonists may directly influence cardiac performance via signaling the GLP-1R of vascular smooth muscle cells, while endogenous GLP-1 may operate through a different pathway. Regardless, it has been noted that the cardiac effects of GLP-1 targeting drugs directly affect the atrial, rather than the ventricular, myocardium [89]. Because GLP-1 and GLP-1R agonists appear to produce cardioprotective effects but their mechanisms are not well explained, this field merits further investigation.

According to a GLP-1R mapping study by Körner et al., lesser quantities of GLP-1 receptors are found in the intestines, lung, and breast. While GLP-1 receptors pervade the small muscular vessels of the lung, they are not found in lung tissue. In the ileum and colon, GLP-1R expression is restricted to the myenteric nerve plexus. It is thought that GLP-1 elicits an inhibitory effect on large intestine motility through the neural release of nitric oxide [90]. GLP-1 receptors have also been reported in the stomach, but their purpose is unknown [91].

Expression of GLP-1 receptors in breast tissue suggests the possibility of GLP-1 signaling in the milk ducts. The same study failed to show GLP-1 receptor expression in multiple tissues including the liver, spleen, lymph nodes, thymus, adrenal gland, adenohypophysis, prostate, heart, skeletal muscle, and fat. Körner et al. also reported GLP-1R overexpression in gut and lung neuroendocrine tumors.

It is also noteworthy that murine studies have revealed GLP-1R presence on thyroid C-cells, suggesting a calcitonin-mediated role for GLP-1 in the regulation of bone turnover. However, despite extensive investigation, no such receptors have ever been identified in humans.

Extraosseous GLP-2 Receptors

In contrast to GLP-1R, GLP-2R expression is confined to the gastrointestinal tract. The pattern of GLP-2R distribution supports the notion that GLP-2 action is mostly limited to the gut. In a study mapping GLP-2 receptor expression in human organs, immunopositive results were predominantly linked to the endocrine cells of the stomach, small intestine, and colon [31]. In light of the fact that GLP-producing L-cells primarily occupy the distal ileum, jejunum, and colon, this receptor distribution suggests GLP-2's direct effects are due to endocrine signaling via the bloodstream. However, this does not rule out the possibility that GLP-2 signals adjacent endocrine cells in a paracrine manner. Furthermore, the lack of GLP-2R-positive findings outside of enteroendocrine cells suggests that indirect effects likely mediate the well-documented actions of GLP-2 in promoting mucosal growth in the intestines, enhancing nutrient absorption, and decreasing gastric acid secretion [92]. It has been shown that GLP-2 and GLP-1 work in concert to significantly inhibit secretion of gastric acid in humans [93].

Peripheral blood mononuclear cells expressing GLP-2R have been detected in small but significant quantities in the circulation. The implications of this finding are unclear, yet are believed that these cells may play a role in the differentiation of osteoblasts [94].

Murine analyses have revealed a potential role for GLP-2-mediated signaling as a central regulator of glucose homeostasis. When activated, GLP-2Rs in the mouse hypothalamus initiate a signaling cascade that stimulates insulin sensitivity and glucose tolerance while it simultaneously attenuates glucose production [95]. To date, however, no significant analyses of GLP-2R expression in human CNS tissues have been performed.

Finally, it is noteworthy that aberrant GLP-2R expression has been spotted in some human GI carcinoid tumors, within distinct foci. The presence of GLP-2R was not detected in the majority of cells, but anomalous clusters of GLP-2R-immunopositive cells were identified. The same study detected no GLP-2R-mRNA transcripts in cancers of the pancreatic islets, adrenal glands, or central nervous system [31].

Extraosseous GIP Receptors

Unlike the glucagon-like peptide receptors, whole body distribution of gastric inhibitory polypeptide receptors has not been reported in humans, although it has been reported in mouse. GIPR-mRNA is present in a variety of murine tissues including the pancreas, gut, adipose, heart, and adrenal cortex. In addition, it inhabits numerous brain regions, including the cerebral cortex, hippocampus, olfactory bulb, and pituitary. It does not appear to be present in the liver or spleen tissues [35].

GIP's functions in the pancreas and bone have been well described; a role for GIP in the brain has also been reported in studies by Nyberg et al. who found a correlation between increased neurogenesis and expression of the GIPR gene in the adult rat brain. Their findings also demonstrate that GIP administration stimulates the proliferation of progenitor cells [96] in the rat hippocampal dentate gyrus. Interestingly, the dentate gyrus is also an area of active neurogenesis in adult humans [97], symbolizing the importance of further research regarding GIP's effects in the brain.

Therapeutic Potential of Entero-osseous Hormones

As the entero-osseous hormones GIP, GLP-1, and GLP-2 have a diverse array of receptor expression throughout the body, they possess considerable potential as therapeutic targets. To date, most attempts at EOH-related therapies have aimed at the incretin actions of GLP-1 and GIP for the treatment of diabetes. Similarly, DPP-IV inhibitors (targeting EOH breakdown) have been successfully employed as treatment for hyperglycemia and to induce insulin secretion [98]. However, evidence has also arisen that may support the use of EOH-based therapies in other organs including the liver and the central nervous system. Finally, there is the aforementioned potential for EOH-related therapies in the treatment of osteoporosis and other diseases related to poor bone quality.

Diabetic Treatments and EOH-Associated Benefits

The most promising diabetic treatments with regard to bone status appear to be GLP-1 receptor agonists. Studies have found that liraglutide is a "safe weight-lowering agent" which can contribute to superior bone formation despite low-calorie dietary intake. Subjects not taking this GLP-1RA showed a fourfold decrease in bone mineral concentration when compared to those who did receive liraglutide [76].

Additionally, incretin-based therapies have demonstrated great potential to attenuate a number of the negative consequences of diabetes throughout the body. GLP-1 therapies reduce chronic hyperglycemia-related inflammation, thus benefitting the kidneys and the blood vessels. DPP-4 inhibitors may also aid in decreasing vascular tone by attenuating the metabolism of certain non-incretin substrates. Notably, these therapies may be of benefit to diabetic patients developing nephropathy [99].

Prospective EOH Therapies

Only recently have the entero-osseous hormones been investigated for their therapeutic potential in the brain and the peripheral organs. As such, most of the effects of EOH administration have been explored solely in rats and mice. Recent findings in these models suggest potential use in diseases such as degenerative brain diseases, inflammation, and liver regeneration.

GIP mimetics and GLP-1 receptor agonists have shown the ability to cross the blood-brain barrier and elicit an array of neuroprotective responses. Among these beneficial consequences, the reduction of beta amyloid plaques along with the induction of long-term potentiation pathways represent interesting prospects for the treatment of Alzheimer's disease [100, 101]. The neuroregenerative capacity of hormones has also been studied in the realm of epilepsy treatment [101–103].

GLP-2 has recently been linked to accelerated liver regeneration in murine models. In addition, GLP-2R mRNA has been reported in human liver samples, revealing a possible avenue for its use in treating human liver disease. Similarly, GLP-2 may be useful in the treatment of IBD patients. Samples from these individuals have revealed decreased expression of GLP-2R transcripts in many tissues, including inflamed segments of the colon and ileum [104]. Finally, studies have shown GLP-2 may modulate insulin sensitivity in mice, despite never having been discovered in humans [105].

References

1. Warensjo E, Byberg L, Melhus H, Gedeborg R, Mallmin H, Wolk A, et al. Dietary calcium intake and risk of fracture and osteoporosis: prospective longitudinal cohort study. BMJ. 2011;342:d1473.
2. Bollag RJ, Zhong Q, Ding KH, Phillips P, Zhong L, Qin F, et al. Glucose-dependent insulinotropic peptide is an integrative hormone with osteotropic effects. Mol Cell Endocrinol. 2001;177(1–2):35–41.
3. Bjarnason NH, Henriksen EE, Alexandersen P, Christgau S, Henriksen DB, Christiansen C. Mechanism of circadian variation in bone resorption. Bone. 2002;30(1):307–13.
4. Blumsohn A, Herrington K, Hannon RA, Shao P, Eyre DR, Eastell R. The effect of calcium supplementation on the circadian rhythm of bone resorption. J Clin Endocrinol Metab. 1994;79(3):730–5.
5. Wolffbrandt KH, Damm Jorgensen K, Moody AJ, Pedersen PC. The effects of porcine GIP on insulin secretion and glucose clearance in the pig. Horm Metab Res Hormon Stoffwechselforschung = Horm Metab. 1986;18(3):159–62.
6. Kieffer TJ. Gastro-intestinal hormones GIP and GLP-1. Ann Endocrinol. 2004;65(1):13–21.
7. Kieffer TJ, Francis HJ. The glucagon-like peptides. Endocr Rev. 1999;20(6):876–913.
8. Heinrich G, Gros P, Lund PK, Bentley RC, Habener JF. Pre-proglucagon messenger ribonucleic acid: nucleotide and encoded amino acid sequences of the rat pancreatic complementary deoxyribonucleic acid. Endocrinology. 1984;115(6):2176–81.
9. Bell GI, Sanchez-Pescador R, Laybourn PJ, Najarian RC. Exon duplication and divergence in the human preproglucagon gene. Nature. 1983;304(5924):368–71.
10. Mojsov S, Heinrich G, Wilson IB, Ravazzola M, Orci L, Habener JF. Preproglucagon gene expression in pancreas and intestine diversifies at the level of post-translational processing. J Biol Chem. 1986;261(25):11880–9.
11. Lee YC, Brubaker PL, Drucker DJ. Developmental and tissue-specific regulation of proglucagon gene expression. Endocrinology. 1990;127(5):2217–22.
12. Steiner DF. The proprotein convertases. Curr Opin Chem Biol. 1998;2(1):31–9.
13. Rouille Y, Martin S, Steiner DF. Differential processing of proglucagon by the subtilisin-like prohormone convertases PC2 and PC3 to generate either glucagon or glucagon-like peptide. J Biol Chem. 1995;270(44):26488–96.
14. Ugleholdt R. Glucose-dependent Insulinotropic Polypeptide (GIP): from prohormone to actions in endocrine pancreas and adipose tissue. Dan Med Bull. 2011;58(12):B4368.
15. Henriksen DB, Alexandersen P, Bjarnason NH, Vilsboll T, Hartmann B, Henriksen EE, et al. Role of gastrointestinal hormones in postprandial reduction of bone resorption. J Bone Mineral Res: Off J Am Soc Bone Mineral Res. 2003;18(12):2180–9.

16. Hartmann B, Johnsen AH, Orskov C, Adelhorst K, Thim L, Holst JJ. Structure, measurement, and secretion of human glucagon-like peptide-2. Peptides. 2000;21(1):73–80.
17. Pais R, Gribble FM, Reimann F. Stimulation of incretin secreting cells. Ther Adv Endocrinol Metab. 2016;7(1):24–42.
18. Brighton CA, Rievaj J, Kuhre RE, Glass LL, Schoonjans K, Holst JJ, et al. Bile acids trigger GLP-1 release predominantly by accessing basolaterally located G protein-coupled bile acid receptors. Endocrinology. 2015;156(11):3961–70.
19. Freeland KR, Wilson C, Wolever TM. Adaptation of colonic fermentation and glucagon-like peptide-1 secretion with increased wheat fibre intake for 1 year in hyperinsulinaemic human subjects. Br J Nutr. 2010;103(1):82–90.
20. Tolhurst G, Heffron H, Lam YS, Parker HE, Habib AM, Diakogiannaki E, et al. Short-chain fatty acids stimulate glucagon-like peptide-1 secretion via the G-protein-coupled receptor FFAR2. Diabetes. 2012;61(2):364–71.
21. Deacon CF. What do we know about the secretion and degradation of incretin hormones? Regul Pept. 2005;128(2):117–24.
22. Nauck MA, Heimesaat MM, Orskov C, Holst JJ, Ebert R, Creutzfeldt W. Preserved incretin activity of glucagon-like peptide 1 [7-36 amide] but not of synthetic human gastric inhibitory polypeptide in patients with type-2 diabetes mellitus. J Clin Invest. 1993;91(1):301–7.
23. Moss CE, Marsh WJ, Parker HE, Ogunnowo-Bada E, Riches CH, Habib AM, et al. Somatostatin receptor 5 and cannabinoid receptor 1 activation inhibit secretion of glucose-dependent insulinotropic polypeptide from intestinal K cells in rodents. Diabetologia. 2012;55(11):3094–103.
24. Hansen L, Hartmann B, Bisgaard T, Mineo H, Jorgensen PN, Holst JJ. Somatostatin restrains the secretion of glucagon-like peptide-1 and -2 from isolated perfused porcine ileum. Am J Physiol Endocrinol Metab. 2000;278(6):E1010–8.
25. Salera M, Pironi L, Giacomoni P, Venturi S, Capelli M, Miglioli M, et al. Effect of somatostatin on fasting and glucose-stimulated gastric inhibitory polypeptide release in man. Digestion. 1982;24(2):126–32.
26. Adriaenssens A, Lam BY, Billing L, Skeffington K, Sewing S, Reimann F, et al. A transcriptome-led exploration of molecular mechanisms regulating somatostatin-producing D-cells in the gastric epithelium. Endocrinology. 2015;156(11):3924–36.
27. Lu M, Wheeler MB, Leng XH, Boyd AE 3rd. The role of the free cytosolic calcium level in beta-cell signal transduction by gastric inhibitory polypeptide and glucagon-like peptide I(7-37). Endocrinology. 1993;132(1):94–100.
28. Wheeler MB, Lu M, Dillon JS, Leng XH, Chen C, Boyd AE 3rd. Functional expression of the rat glucagon-like peptide-I receptor, evidence for coupling to both adenylyl cyclase and phospholipase-C. Endocrinology. 1993;133(1):57–62.
29. Fridolf T, Ahren B. GLP-1(7–36) amide stimulates insulin secretion in rat islets: studies on the mode of action. Diabetes Res. 1991;16(4):185–91.
30. Zhang L, Wang Y, Wang J, Liu Y, Yin Y. Protein kinase C pathway mediates the protective effects of glucagon-like peptide-1 on the apoptosis of islet beta-cells. Mol Med Rep. 2015;12(5):7589–94.
31. Yusta B, Huang L, Munroe D, Wolff G, Fantaske R, Sharma S, et al. Enteroendocrine localization of GLP-2 receptor expression in humans and rodents. Gastroenterology. 2000;119(3):744–55.
32. Yusta B, Somwar R, Wang F, Munroe D, Grinstein S, Klip A, et al. Identification of glucagon-like peptide-2 (GLP-2)-activated signaling pathways in baby hamster kidney fibroblasts expressing the rat GLP-2 receptor. J Biol Chem. 1999;274(43):30459–67.
33. Pacheco-Pantoja EL, Ranganath LR, Gallagher JA, Wilson PJ, Fraser WD. Receptors and effects of gut hormones in three osteoblastic cell lines. BMC Physiol. 2011;11:12.
34. Bollag RJ, Zhong Q, Phillips P, Min L, Zhong L, Cameron R, et al. Osteoblast-derived cells express functional glucose-dependent insulinotropic peptide receptors. Endocrinology. 2000;141(3):1228–35.

35. Usdin TB, Mezey E, Button DC, Brownstein MJ, Bonner TI. Gastric inhibitory polypeptide receptor, a member of the secretin-vasoactive intestinal peptide receptor family, is widely distributed in peripheral organs and the brain. Endocrinology. 1993;133(6):2861–70.
36. Zhong Q, Itokawa T, Sridhar S, Ding KH, Xie D, Kang B, et al. Effects of glucose-dependent insulinotropic peptide on osteoclast function. Am J Physiol Endocrinol Metab. 2007;292(2):E543–8.
37. Ehses JA, Pelech SL, Pederson RA, McIntosh CH. Glucose-dependent insulinotropic polypeptide activates the Raf-Mek1/2-ERK1/2 module via a cyclic AMP/cAMP-dependent protein kinase/Rap1-mediated pathway. J Biol Chem. 2002;277(40):37088–97.
38. Zhong Q, Bollag RJ, Dransfield DT, Gasalla-Herraiz J, Ding KH, Min L, et al. Glucose-dependent insulinotropic peptide signaling pathways in endothelial cells. Peptides. 2000;21(9):1427–32.
39. Pujadas G, Drucker DJ. Vascular biology of glucagon receptor superfamily peptides: mechanistic and clinical relevance. Endocr Rev. 2016;37(6):554–83.
40. Deacon CF. Circulation and degradation of GIP and GLP-1. Horm Metab Res Horm Stoffwechselforschung Horm Metab. 2004;36(11–12):761–5.
41. Kieffer TJ, McIntosh CH, Pederson RA. Degradation of glucose-dependent insulinotropic polypeptide and truncated glucagon-like peptide 1 in vitro and in vivo by dipeptidyl peptidase IV. Endocrinology. 1995;136(8):3585–96.
42. Brubaker PL, Drucker DJ. Structure-function of the glucagon receptor family of G protein-coupled receptors: the glucagon, GIP, GLP-1, and GLP-2 receptors. Recept Channels. 2002;8(3–4):179–88.
43. Lambeir AM, Proost P, Scharpe S, De Meester I. A kinetic study of glucagon-like peptide-1 and glucagon-like peptide-2 truncation by dipeptidyl peptidase IV, in vitro. Biochem Pharmacol. 2002;64(12):1753–6.
44. Hartmann B, Harr MB, Jeppesen PB, Wojdemann M, Deacon CF, Mortensen PB, et al. In vivo and in vitro degradation of glucagon-like peptide-2 in humans. J Clin Endocrinol Metab. 2000;85(8):2884–8.
45. Deacon CF, Nauck MA, Toft-Nielsen M, Pridal L, Willms B, Holst JJ. Both subcutaneously and intravenously administered glucagon-like peptide I are rapidly degraded from the NH2-terminus in type II diabetic patients and in healthy subjects. Diabetes. 1995;44(9):1126–31.
46. Brubaker PL, Crivici A, Izzo A, Ehrlich P, Tsai CH, Drucker DJ. Circulating and tissue forms of the intestinal growth factor, glucagon-like peptide-2. Endocrinology. 1997;138(11):4837–43.
47. Nuche-Berenguer B, Portal-Nunez S, Moreno P, Gonzalez N, Acitores A, Lopez-Herradon A, et al. Presence of a functional receptor for GLP-1 in osteoblastic cells, independent of the cAMP-linked GLP-1 receptor. J Cell Physiol. 2010;225(2):585–92.
48. Askov-Hansen C, Jeppesen PB, Lund P, Hartmann B, Holst JJ, Henriksen DB. Effect of glucagon-like peptide-2 exposure on bone resorption: effectiveness of high concentration versus prolonged exposure. Regul Pept. 2013;181:4–8.
49. Yavropoulou MP, Yovos JG. Incretins and bone: evolving concepts in nutrient-dependent regulation of bone turnover. Hormones. 2013;12(2):214–23.
50. Mieczkowska A, Bouvard B, Chappard D, Mabilleau G. Glucose-dependent insulinotropic polypeptide (GIP) directly affects collagen fibril diameter and collagen cross-linking in osteoblast cultures. Bone. 2015;74:29–36.
51. Berlier JL, Kharroubi I, Zhang J, Dalla Valle A, Rigutto S, Mathieu M, et al. Glucose-dependent insulinotropic peptide prevents serum deprivation-induced apoptosis in human bone marrow-derived mesenchymal stem cells and osteoblastic cells. Stem Cell Rev. 2015;11(6):841–51.
52. Xie D, Cheng H, Hamrick M, Zhong Q, Ding KH, Correa D, et al. Glucose-dependent insulinotropic polypeptide receptor knockout mice have altered bone turnover. Bone. 2005;37(6):759–69.
53. Tsukiyama K, Yamada Y, Yamada C, Harada N, Kawasaki Y, Ogura M, et al. Gastric inhibitory polypeptide as an endogenous factor promoting new bone formation after food ingestion. Mol Endocrinol. 2006;20(7):1644–51.

54. Mieczkowska A, Irwin N, Flatt PR, Chappard D, Mabilleau G. Glucose-dependent insulinotropic polypeptide (GIP) receptor deletion leads to reduced bone strength and quality. Bone. 2013;56(2):337–42.
55. Gaudin-Audrain C, Irwin N, Mansur S, Flatt PR, Thorens B, Basle M, et al. Glucose-dependent insulinotropic polypeptide receptor deficiency leads to modifications of trabecular bone volume and quality in mice. Bone. 2013;53(1):221–30.
56. Mieczkowska A, Mansur S, Bouvard B, Flatt PR, Thorens B, Irwin N, et al. Double incretin receptor knock-out (DIRKO) mice present with alterations of trabecular and cortical micro-morphology and bone strength. Osteoporos Int: J Established Result Cooperation Eur Found Osteoporos Natl Osteoporos Found USA. 2015;26(1):209–18.
57. Clowes JA, Hannon RA, Yap TS, Hoyle NR, Blumsohn A, Eastell R. Effect of feeding on bone turnover markers and its impact on biological variability of measurements. Bone. 2002;30(6):886–90.
58. Clowes JA, Allen HC, Prentis DM, Eastell R, Blumsohn A. Octreotide abolishes the acute decrease in bone turnover in response to oral glucose. J Clin Endocrinol Metab. 2003;88(10):4867–73.
59. Clowes JA, Robinson RT, Heller SR, Eastell R, Blumsohn A. Acute changes of bone turnover and PTH induced by insulin and glucose: euglycemic and hypoglycemic hyperinsulinemic clamp studies. J Clin Endocrinol Metab. 2002;87(7):3324–9.
60. Nissen A, Christensen M, Knop FK, Vilsboll T, Holst JJ, Hartmann B. Glucose-dependent insulinotropic polypeptide inhibits bone resorption in humans. J Clin Endocrinol Metab. 2014;99(11):E2325–9.
61. Torekov SS, Harslof T, Rejnmark L, Eiken P, Jensen JB, Herman AP, et al. A functional amino acid substitution in the glucose-dependent insulinotropic polypeptide receptor (GIPR) gene is associated with lower bone mineral density and increased fracture risk. J Clin Endocrinol Metab. 2014;99(4):E729–33.
62. Meier C, Schwartz AV, Egger A, Lecka-Czernik B. Effects of diabetes drugs on the skeleton. Bone. 2016;82:93–100.
63. Glorie L, Behets GJ, Baerts L, De Meester I, D'Haese PC, Verhulst A. DPP IV inhibitor treatment attenuates bone loss and improves mechanical bone strength in male diabetic rats. Am J Physiol Endocrinol Metab. 2014;307(5):E447–55.
64. Gallagher EJ, Sun H, Kornhauser C, Tobin-Hess A, Epstein S, Yakar S, et al. The effect of dipeptidyl peptidase-IV inhibition on bone in a mouse model of type 2 diabetes. Diabetes Metab Res Rev. 2014;30(3):191–200.
65. Eom YS, Gwon AR, Kwak KM, Kim JY, Yu SH, Lee S, et al. Protective effects of vildagliptin against pioglitazone-induced bone loss in type 2 diabetic rats. PLoS One. 2016;11(12):e0168569.
66. Bunck MC, Poelma M, Eekhoff EM, Schweizer A, Heine RJ, Nijpels G, et al. Effects of vildagliptin on postprandial markers of bone resorption and calcium homeostasis in recently diagnosed, well-controlled type 2 diabetes patients. J Diabetes. 2012;4(2):181–5.
67. Driessen JH, de Vries F, van Onzenoort H, et al. The use of incretins and fractures - a meta-analysis on population-based real life data. Br J Clin Pharmacol. 2017;83(4):923–6.
68. Driessen JH, van den Bergh JP, van Onzenoort HA, Henry RM, Leufkens HG, de Vries F. Long-term use of dipeptidyl peptidase-4 inhibitors and risk of fracture: A retrospective population-based cohort study. Diabetes Obes Metab. 2017;19(3):421–8.
69. Mabilleau G, Mieczkowska A, Irwin N, Simon Y, Audran M, Flatt PR, et al. Beneficial effects of a N-terminally modified GIP agonist on tissue-level bone material properties. Bone. 2014;63:61–8.
70. Mansur SA, Mieczkowska A, Bouvard B, Flatt PR, Chappard D, Irwin N, et al. Stable incretin mimetics counter rapid deterioration of bone quality in type 1 diabetes mellitus. J Cell Physiol. 2015;230(12):3009–18.
71. Mansur SA, Mieczkowska A, Flatt PR, Bouvard B, Chappard D, Irwin N, et al. A new stable GIP-Oxyntomodulin hybrid peptide improved bone strength both at the organ and tissue levels in genetically-inherited type 2 diabetes mellitus. Bone. 2016;87:102–13.

72. Ding KH, Shi XM, Zhong Q, Kang B, Xie D, Bollag WB, et al. Impact of glucose-dependent insulinotropic peptide on age-induced bone loss. J Bone Mineral Res:Off J Am Soc Bone Mineral Res. 2008;23(4):536–43.
73. Yamada C, Yamada Y, Tsukiyama K, Yamada K, Udagawa N, Takahashi N, et al. The murine glucagon-like peptide-1 receptor is essential for control of bone resorption. Endocrinology. 2008;149(2):574–9.
74. Mabilleau G, Mieczkowska A, Irwin N, Flatt PR, Chappard D. Optimal bone mechanical and material properties require a functional glucagon-like peptide-1 receptor. J Endocrinol. 2013;219(1):59–68.
75. Sun HX, Lu N, Luo X, Zhao L, Liu JM. Liraglutide, the glucagon-like peptide-1 receptor agonist, has anabolic bone effects in diabetic Goto-Kakizaki rats. J Diabetes. 2015;7(4):584–8.
76. Irwin N, Flatt PR. New perspectives on exploitation of incretin peptides for the treatment of diabetes and related disorders. World J Diabetes. 2015;6(15):1285–95.
77. Mabilleau G, Mieczkowska A, Chappard D. Use of glucagon-like peptide-1 receptor agonists and bone fractures: a meta-analysis of randomized clinical trials. J Diabetes. 2014;6(3):260–6.
78. Su B, Sheng H, Zhang M, Bu L, Yang P, Li L, et al. Risk of bone fractures associated with glucagon-like peptide-1 receptor agonists' treatment: a meta-analysis of randomized controlled trials. Endocrine. 2015;48(1):107–15.
79. Henriksen DB, Alexandersen P, Byrjalsen I, Hartmann B, Bone HG, Christiansen C, et al. Reduction of nocturnal rise in bone resorption by subcutaneous GLP-2. Bone. 2004;34(1):140–7.
80. Henriksen DB, Alexandersen P, Hartmann B, Adrian CL, Byrjalsen I, Bone HG, et al. Disassociation of bone resorption and formation by GLP-2: a 14-day study in healthy postmenopausal women. Bone. 2007;40(3):723–9.
81. Henriksen DB, Alexandersen P, Hartmann B, Adrian CL, Byrjalsen I, Bone HG, et al. Four-month treatment with GLP-2 significantly increases hip BMD: a randomized, placebo-controlled, dose-ranging study in postmenopausal women with low BMD. Bone. 2009;45(5):833–42.
82. Haderslev KV, Jeppesen PB, Hartmann B, Thulesen J, Sorensen HA, Graff J, et al. Short-term administration of glucagon-like peptide-2. Effects on bone mineral density and markers of bone turnover in short-bowel patients with no colon. Scand J Gastroenterol. 2002;37(4):392–8.
83. Kreymann B, Ghatei M, Williams G, Bloom S. Glucagon-like peptide-1 7-36: a physiological incretin in man. Lancet. 1987;330(8571):1300–4.
84. Korner M, Stockli M, Waser B, Reubi JC. GLP-1 receptor expression in human tumors and human normal tissues: potential for in vivo targeting. J Nucl Med: Off Publ Soc Nucl Med. 2007;48(5):736–43.
85. Pyke C, Heller RS, Kirk RK, Orskov C, Reedtz-Runge S, Kaastrup P, et al. GLP-1 receptor localization in monkey and human tissue: novel distribution revealed with extensively validated monoclonal antibody. Endocrinology. 2014;155(4):1280–90.
86. Bang-Berthelsen CH, Holm TL, Pyke C, Simonsen L, Sokilde R, Pociot F, et al. GLP-1 induces barrier protective expression in Brunner's glands and regulates colonic inflammation. Inflamm Bowel Dis. 2016;22(9):2078–97.
87. Bojanowska E, Stempniak B. Effects of glucagon-like peptide-1 (7-36) amide on neurohypophysial hormone secretion induced by acute hyperosmotic challenge. Neuropeptides. 2003;37(1):45–50.
88. Mima A, Hiraoka-Yamomoto J, Li Q, Kitada M, Li C, Geraldes P, et al. Protective effects of GLP-1 on glomerular endothelium and its inhibition by PKCbeta activation in diabetes. Diabetes. 2012;61(11):2967–79.
89. Ussher JR, Drucker DJ. Cardiovascular actions of incretin-based therapies. Circ Res. 2014;114(11):1788–803.
90. Amato A, Baldassano S, Liotta R, Serio R, Mule F. Exogenous glucagon-like peptide 1 reduces contractions in human colon circular muscle. J Endocrinol. 2014;221(1):29–37.

91. Wei Y, Mojsov S. Tissue-specific expression of the human receptor for glucagon-like peptide-I: brain, heart and pancreatic forms have the same deduced amino acid sequences. FEBS Lett. 1995;358(3):219–24.
92. Meier JJ, Nauck MA, Pott A, Heinze K, Goetze O, Bulut K, et al. Glucagon-like peptide 2 stimulates glucagon secretion, enhances lipid absorption, and inhibits gastric acid secretion in humans. Gastroenterology. 2006;130(1):44–54.
93. Wojdemann M, Wettergren A, Hartmann B, Hilsted L, Holst JJ. Inhibition of sham feeding-stimulated human gastric acid secretion by glucagon-like peptide-2. J Clin Endocrinol Metab. 1999;84(7):2513–7.
94. Clowes J, Xiao J, Eastell R, Khosla S, editors. Evidence that gut hormone receptors are expressed on circulating peripheral blood cells. J Bone Mineral Res. 2005: Am Soc Bone Mineral Res 2025 M ST, NW, STE 800, Washington, DC 20036–3309 USA.
95. Guan X. The CNS glucagon-like peptide-2 receptor in the control of energy balance and glucose homeostasis. Am J Phys Regul Integr Comp Phys. 2014;307(6):R585–96.
96. Nyberg J, Anderson MF, Meister B, Alborn A-M, Ström A-K, Brederlau A, et al. Glucose-dependent insulinotropic polypeptide is expressed in adult hippocampus and induces progenitor cell proliferation. J Neurosci. 2005;25(7):1816–25.
97. Eriksson PS, Perfilieva E, Björk-Eriksson T, Alborn A-M, Nordborg C, Peterson DA, et al. Neurogenesis in the adult human hippocampus. Nat Med. 1998;4(11):1313–7.
98. McIntosh CH. Dipeptidyl peptidase IV inhibitors and diabetes therapy. Front Biosci. 2008;13:1753–73.
99. Kawanami D, Matoba K, Sango K, Utsunomiya K. Incretin-based therapies for diabetic complications: basic mechanisms and clinical evidence. Int J Mol Sci. 2016;17(8):1223.
100. Holscher C. The incretin hormones glucagonlike peptide 1 and glucose-dependent insulinotropic polypeptide are neuroprotective in mouse models of Alzheimer's disease. Alzheimers Dement: J Alzheimers Assoc. 2014;10(1 Suppl):S47–54.
101. Figueiredo CP, Pamplona FA, Mazzuco TL, Aguiar AS Jr, Walz R, Prediger RD. Role of the glucose-dependent insulinotropic polypeptide and its receptor in the central nervous system: therapeutic potential in neurological diseases. Behav Pharmacol. 2010;21(5–6):394–408.
102. During MJ, Cao L, Zuzga DS, Francis JS, Fitzsimons HL, Jiao X, et al. Glucagon-like peptide-1 receptor is involved in learning and neuroprotection. Nat Med. 2003;9(9):1173–9.
103. Figueiredo CP, Antunes VL, Moreira EL, de Mello N, Medeiros R, Di Giunta G, et al. Glucose-dependent insulinotropic peptide receptor expression in the hippocampus and neocortex of mesial temporal lobe epilepsy patients and rats undergoing pilocarpine induced status epilepticus. Peptides. 2011;32(4):781–9.
104. El-Jamal N, Erdual E, Neunlist M, Koriche D, Dubuquoy C, Maggiotto F, et al. Glucagon-like peptide-2: broad receptor expression, limited therapeutic effect on intestinal inflammation and novel role in liver regeneration. Am J Physiol Gastrointest Liver Physiol. 2014;307(3):G274–85.
105. Amato A, Baldassano S, Mule F. GLP2: an underestimated signal for improving glycaemic control and insulin sensitivity. J Endocrinol. 2016;229(2):R57–66.

Regulation of Bone Metabolism by Serotonin

Brigitte Lavoie, Jane B. Lian, and Gary M. Mawe

Introduction

The bone is a highly dynamic tissue that is influenced by many intrinsic and extrinsic signaling factors capable of modulating its growth and turnover. The balance between new bone formation by osteoblasts and resorption of calcified bone by osteoclasts is responsible for accurate bone modeling and remodeling. Impairment in this equilibrium can lead to skeletal disorders involving bone loss, such as osteoporosis, or the contrasting high bone mass syndromes that result from genetic mutations [1]. A variety of hormones, growth factors, and cytokines regulate the coordinated activity of osteoblasts and osteoclasts. Recently, serotonin (5-hydroxytryptamine; 5-HT) is becoming appreciated as one of the key players in bone tissue dynamic, and the mechanisms by which it acts are still being revealed [2]. Significantly, 5-HT can exert divergent effects on bone density, both through actions in the brain and through actions on cells within the bone [3].

B. Lavoie (✉)
Department of Neurological Sciences, The University of Vermont,
Burlington, VT 05405, USA

The University of Vermont, D406 Given Building, Burlington, VT 05405, USA
e-mail: blavoie@uvm.edu

J.B. Lian
Department of Biochemistry, The University of Vermont, Burlington, VT 05405, USA

G.M. Mawe
Department of Neurological Sciences, The University of Vermont,
Burlington, VT 05405, USA

L.R. McCabe, N. Parameswaran (eds.), *Understanding the Gut-Bone Signaling Axis*, Advances in Experimental Medicine and Biology 1033,
DOI 10.1007/978-3-319-66653-2_3

Serotonin as a Signaling Molecule in the Brain and Periphery

While serotonin is best known for the roles it plays as a neurotransmitter in the central nervous system (CNS), it was first discovered in the periphery during searches for signaling molecules that could contract smooth muscle. In 1937, the Italian physiologist Vittorio Erspamer isolated an indole molecule, highly abundant in the gastrointestinal (GI) tract of various species, which caused muscle contraction [4]. He named this compound "enteramine" and described it as the main secretory molecule of enterochromaffin (EC) cells. Nearly a decade later, Maurice Rapport in collaboration with Arda Green and Irvine Page (1948) identified a compound in bovine serum that caused blood vessel contraction, and they named it "serotonin" [5, 6]. Structural analysis of enteramine and serotonin proved that enteramine and serotonin were the same molecule: 5-hydrotryptamine (5-HT) [7]. It was Betty Twarog, who in 1953 demonstrated the presence of serotonin in the mammalian brain and suggested a role for the molecule as a neurotransmitter rather than a hormone [8]. Since the blood-brain barrier is quite impermeable to serotonin, serotonin pools in the CNS vs the periphery are considered as two distinct pools that are independently regulated.

Despite its widespread influences, serotonin is produced by relatively few cells in the body. Within the CNS, serotonin is synthesized by clusters of neurons, designated B1-B9, that are restricted to the brainstem raphe nuclei. These neurons, which use tryptophan hydroxylase 2 (TPH2) as their rate-limiting enzyme for serotonin synthesis, extend ascending and descending axonal projections that reach most regions of the CNS and mediate a wide variety of functions. The axon terminals of these neurons express the serotonin-selective reuptake transporter (SERT) in their membranes, so once serotonin is released from the nerve terminals and mediates its actions on nearby receptors, it is removed from the synaptic cleft. Within serotonergic nerve terminals, serotonin is deaminated by monoamine oxidase A (MAO-A) and converted to 5-hydroxyindole acetic acid (5-HIAA) by aldehyde dehydrogenase [9].

In the periphery, the major site of serotonin synthesis is the EC cells of the intestinal mucosa, which are a form of enteroendocrine cells, and in fact EC cells are the source of the vast majority of the serotonin in the body [10]. EC cells use TPH1 to synthesize serotonin, as do all other peripheral serotonin-synthesizing cells, with the exception of enteric serotonergic neurons, which like CNS neurons utilize TPH2. After EC cells release serotonin in response to chemical and/or mechanical stimuli, it acts as a paracrine signaling molecule on nearby nerve fibers, epithelial cells, and immune cells that express serotonin receptors, and it is then taken up by epithelial cells, all of which express SERT. The serotonin that is not transported into epithelial cells moves into the blood stream, where it is transported into platelets, which also express SERT. Until recently, it was presumed that intestinal EC cells represented the sole source of peripheral serotonin because 5-HIAA levels in the urine are undetectable within 24–48 h after removal of the entire GI tract from experimental animals [11]. However, the finding that non-neuronal serotonin is synthesized by TPH1 has led to the discovery of local sources of peripheral serotonin synthesis, including adipocytes [12], pancreatic islet β-cells [13, 14], and cells within the bone [15].

Serotonin and Bone Remodeling in Health

Brain-Derived Serotonin

Serotonin in the CNS has a positive effect on bone growth, since young TPH2-/- mice (12 weeks and younger) have low bone mass with decreased bone formation [16]. The osteogenic actions of serotonin in the CNS involve the inhibition of inhibitory inputs in the following sequence of events (Fig. 1a): (1) serotonergic raphe neurons provide excitatory input to neurons in the ventromedial hypothalamus; (2) hypothalamic neurons provide inhibitory input to sympathetic preganglionic neurons; and (3) the decrease in sympathetic outflow releases the bone from the resorption influence of β2 adrenoreceptor activation. Therefore, the CNS serotonergic system favors bone accrual. In older TPH2-/- mice (21–83 weeks), moderately elevated trabecular bone in the vertebrae has been reported in both sexes, and females have decreased femur cortical bone [17].

Central serotonin regulates bone mass accrual through stimulation of 5-HT_{2C} receptors on hypothalamic neurons, while appetite, which is also influenced by leptin acting on serotonergic neurons, is controlled by activation of 5-HT_{1A} and $_{2B}$ receptors in the arcuate nucleus [16, 18]. Activation of 5-HT_{2C} receptors leads to decrease sympathetic tone that is activated by calmodulin kinase (CaMK)-dependent signaling cascade. Decreased sympathetic tone releases bone cells from the impact of β_2 adrenoreceptor activation, which can promote bone growth through two opposite actions: by promoting the proliferation of osteoblasts and by inhibiting the proliferation and differentiation of osteoclasts.

The CNS serotonin bone pathway is negatively regulated by leptin arising from adipocytes. Both Ob/ob and db/db mouse models, which lack leptin or the leptin receptor, respectively, have increased bone mass, and this bone phenotype can be corrected by specific inactivation of central serotonin signaling [2, 16, 19]. Furthermore, deletion of the leptin receptor on brainstem neurons, which would release these neurons from the inhibitory influence of leptin, mimics the elevated bone mass phenotype seen in the Ob/ob and db/db mouse models. Leptin inhibits serotonergic raphe neurons both by reducing their serotonin synthesis and their level of excitability, which, in turn, results in decreased serotonergic input to the hypothalamic nuclei. It has been reported that leptin regulation of appetite does not involved CNS serotonin neurons, and relevant to this discussion, leptin receptors were only found on non-serotonin neurons in the raphe nuclei [20].

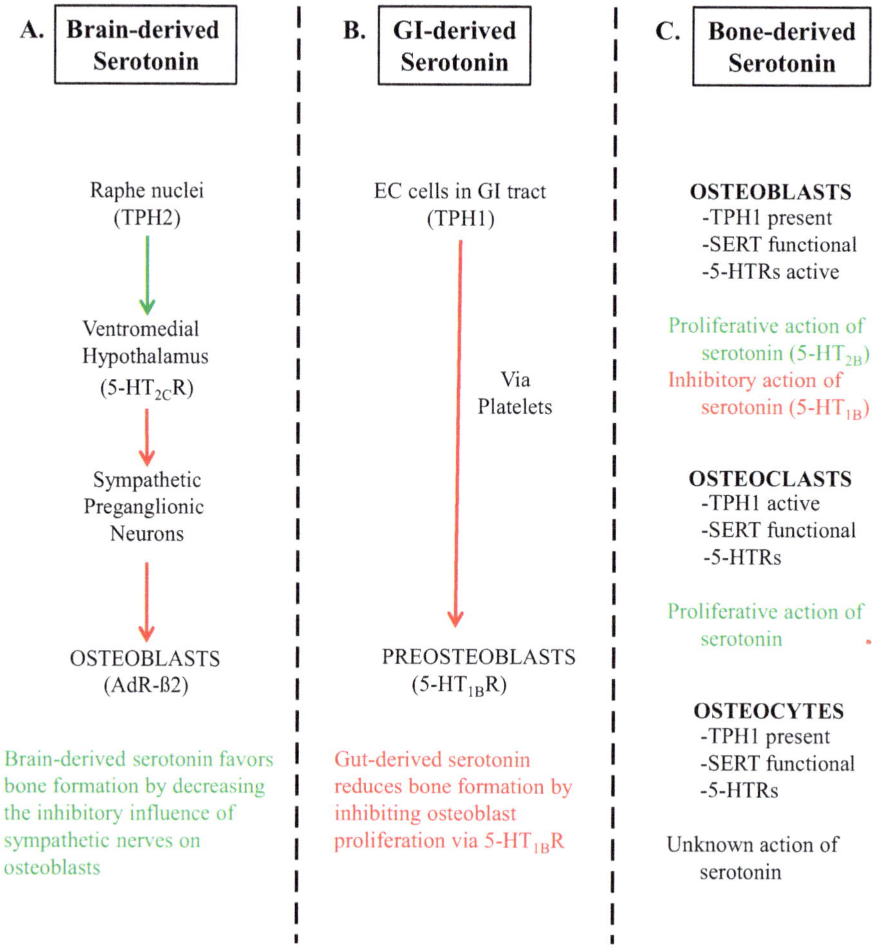

Fig. 1 Summary of the action of serotonin, from different sources, on bone cells. (**a**) Brain-derived serotonin synthesized by tryptophan hydroxylase 2 (TPH2) in raphe nuclei decrease the inhibitory action of the sympathetic neurons by acting on $5\text{-}HT_{2c}$ receptors on ventromedial hypothalamic neurons leading to increased bone formation. (**b**) GI-derived serotonin, synthesized in enterochromaffin (EC) cells by tryptophan hydroxylase 1 (TPH1), enters the blood stream where it accumulates in serotonin-specific reuptake transport (SERT)-expressing platelets. Circulating serotonin inhibits osteoblasts proliferation by its binding on $5\text{-}HT_{1B}$ receptors expressed by pre-osteoblasts. (**c**) Bone-derived serotonin. Bone cells express TPH1, SERT, and 5-HTRs. While serotonin can potentially either promote or inhibit osteoblast proliferation, depending on the receptor that is activated, it has mainly a proliferative action on osteoclasts. Highlights in red are associated with inhibitory pathways, while green highlights represent activation pathways

Therefore, serotonin signaling in the brain can have a positive impact on bone accrual, and this action may be suppressed by the effects of leptin on serotonergic neurons in the raphe nuclei.

Gut-Derived Serotonin

Interest in the role played by gut-derived serotonin in the regulation of bone metabolism developed when a link was identified between a member of the low-density lipoprotein receptor (Lrp) family, Lrp5, and the level of TPH1 expression in the gut [21]. Lineage studies have linked mutations of the Lrp5 gene with changes in bone formation. Loss-of-function Lrp5 mutations are associated with osteoporosis-pseudoglioma syndrome (OPPG), while gain-of-function changes lead to high bone mass (HBM) syndrome [22, 23]. Lrp5 acts as a co-receptor in the canonical Wnt signaling pathway modulating signal transduction in osteoblasts. Using conditional Lrp5 knockout mice, Yadav and Karsenty described bone abnormalities that were different than those seen by inactivation of Wnt signaling in osteoblasts or in osteocytes [21], suggesting that the Lrp5 that influences bone density might be at another site. As in OPPG and HBM patients, Lrp5 null mice present no bone abnormalities at birth, and postnatal changes observed in these mice are not reproduced by blocking the Wnt signaling after birth [3, 22–24]. Transcriptome analysis of Lrp5-/- mice revealed that one of the most dramatic changes was an increase in TPH1 expression, which, as described above, is used by intestinal EC cells to synthesize serotonin [21]. These mice also have increased levels of circulating serotonin and decreased bone density. Mice expressing a gain-of-function mutation of the Lrp5 gene have decreased TPH1 gene expression and lower levels of serotonin in blood. In both Lrp5-/- and TPH1-/- mice, phenotypes were restricted to osteoblast formation, with changes only in the expression of cyclin D1, D2, and E1, all of which are related to cell proliferation as opposed to osteoblast or osteoclast differentiation [21].

The action of serotonin on osteoblast proliferation has been linked to the expression of 5-HT_{1B} receptors by pre-osteoblasts [21]. Binding of serotonin to the 5-HT_{1B} receptor affects the cAMP response element-binding protein (CREB) regulation of osteoblast proliferation through transcription factor FOXO1. The balanced interaction of FOXO1 with transcription factor CREB and activating transcription factor 4 (ATF4) promotes normal proliferation of osteoblast. Elevated circulating serotonin levels are associated with a decrease in the association of FOXO1 with CREB, and this leads to inhibition of bone formation [25].

The concept that gut-derived serotonin can decrease the osteoblast to osteoclast ratio and suppress bone growth has led to the theory that some forms of osteoporosis might be alleviated by downregulating enteric serotonin levels using inhibitors of TPH1. Mouse and rat models of osteoporosis have been studied using LP533401, a TPH1 and TPH2 inhibitor that is not well absorbed and does not readily cross the blood-brain barrier; therefore, its actions are thought to be limited primarily to EC cells [26]. Daily treatment with this antagonist leads to decreased circulating serotonin levels and a significant elevation in bone formation in ovariectomized mice [27].

It is important to note that a study by another group failed to detect evidence for a role for GI-specific Lrp5 action in bone metabolism. Cui and colleagues were unable to detect bone defects when they inactivate Lrp5 gene specifically in the GI tract [28]. They instead proposed that bone-specific Lrp5 is responsible for changes of function in the osteocyte population through the canonical Wnt pathway [28]. In addition, they were unable to reproduce the anabolic effect of TPH1 inhibitors on bone mass.

This discrepancy has led to an unresolved debate in the field about the importance of gut-derived serotonin as it relates to bone density [29–33], but as proposed, serotonin released from EC cells and entering the blood stream has a negative impact on bone formation by inhibiting osteoblast formation (Fig. 1b). Some differences in the studies most likely arose due to differences in mouse models (strain, age, and sex of the mice), as well as methods and assays that were employed by the different research teams. In the view of the recent data, both mechanisms (1-direct regulation of Lrp5 on osteoblast through a canonical Wnt pathway and 2-indirect paracrine regulation via increase circulatory serotonin and action on pre-osteoblast 5-HT_{1B}) appear to be plausible, and potentially complementary, physiological processes. That said, the results of Yadav's studies raise an additional important issue regarding circulated serotonin: for increased circulatory serotonin to affect bone formation, it would need to be released near its targets, most likely in a regulated manner (pre-osteoblasts or other 5-HTR-bearing cells). Because of its effect on smooth muscle tone, serotonin is efficiently stored in platelets, leaving negligible levels of free serotonin in the serum. At this point little is known about how such targeted release occurs, so more studies of the dynamics of platelet storage and release are needed.

Bone-Derived 5-HT

In addition to the osteogenic actions of serotonin originating in the brain and the gut, another source of serotonin was recently added to the equation: serotonin synthesized within the bone itself (Fig. 1c). Serotonin is synthesized by osteoclast precursors in the presence of NF-KB ligand (RANKL), and downregulation of TPH1 expression by pre-osteoclasts leads to decreased bone resorption and, therefore, increased density. Using mice in which TPH1 is constitutively inactivated, Chabbi-Achengli and colleagues noted that osteoclasts locally produced serotonin that promotes bone formation by increasing the activity of osteoblasts while decreasing the production of osteoclasts [15]. Adult TPH1-/- mice have normal bone density, but elevated bone density was reported at earlier ages, indicating that compensatory mechanisms may correct for altered serotonin signaling.

Bone Expression of SERT and 5-HT Receptors

In addition to expressing TPH1, osteoblasts, osteoclasts, and osteocytes all express functional serotonin transporters [34–37]. As described above, SERT is an important player in serotonin signaling, as it is required for the termination of serotonin signaling. The presence of SERT on bone cells can serve to quickly remove serotonin from the interstitial space. This is supported by knowledge of the actions of serotonin-selective reuptake inhibitors (SSRIs) in the brain, which not only increase serotonin availability and activation of receptors at sites of release but can also lead to longer-term changes such as receptor desensitization and altered levels of receptor expression [38, 39].

A number of 5-HT receptors are also expressed on primary bone cells, including 5-HT_{1A}, 5-HT_{2A}, and 5-HT_{2B} receptors [34, 37, 40–42]. The inhibitory action of GI-derived serotonin on 5-HT_{1B} receptor, expressed on pre-osteoblast, has been previously described by Yadav and collaborators [21] (Fig. 1b). It is also known that mice lacking 5-HT_{2B} receptor exhibit osteopenia due to impaired bone formation. The absence of the 5-HT_{2B} receptor causes impaired osteoblast proliferation, recruitment, and matrix mineralization involving nuclear PPAR receptors and prostacyclin [43, 44]. Inhibition of 5-HT_{2A} receptors in mice reduces bone mass by decreasing osteoblasts differentiation [45]. While 5-HT_{1A} receptors are found on osteoblasts and osteoclasts, the physiological role of this receptor in vivo is still unclear.

The 5-HT_6 receptor is also highly expressed in the bone [46]. Expression of the 5-HT_6 receptor is increased during bone remodeling and osteoblast differentiation. In primary cultures of osteoblasts, activation of 5-HT_6 inhibits alkaline phosphatase activity and bone mineralization, while in vivo activation of 5-HT_6 inhibited bone regeneration and delayed bone development. Thus, serotonin receptor activity contributes to major anabolic functions of the osteoblast for bone formation.

Nonconventional Action of Serotonin in Periphery

In addition to acting conventionally through the activation of receptors expressed on cytoplasmic membranes, serotonin has been shown in a number of systems to mediate its actions via a mechanism called serotonylation. Serotonylation is a receptor-independent process in which serotonin activates intracellular processes. Cui and Kaartinen have reported that serotonylation occurs in the bone [47]. They demonstrated that serotonin can be incorporated into plasma fibronectin by transglutaminase-mediated serotonylation and altered its function. Serotonin interferes with the role pFN in extracellular matrix assembly in osteoblasts and might lead to weaker bones.

Serotonin and Bone Remodeling in Disease

One of the first indications that serotonin plays an important role in bone metabolism came from multiple reports of increased incidence of bone fractures and osteoporosis in patients taking SSRIs to treat depression [48, 49]. These compounds increase serotonin availability by inhibiting SERT, and in the case of serotonin secreted from the gut, their use can lead to increased amounts of serotonin entering the circulation. Many studies have also linked depression to osteoporosis, with more severe depression correlating with higher decreased in bone mineral density. Besides the use of SSRIs, there are a number of confounding factors that can lead to the bone loss observed in depressive patients. Behavioral factors (increased cigarette or alcohol use), biological factors (increased cortisol level and inflammation), and increased incidence of Crohn's disease and diabetes, which are observed in this population, can influence serotonin signaling [50].

The need for a better understanding of the effect of serotonin on bone growth is underscored by the increased use of SSRIs to treat depression, including treatment of adolescents with growing bones and of pregnant women. SSRIs used during pregnancy can cross the maternal-fetal blood barrier leading to potential modulation of fetal serotonin. While some studies have reported increased risks of cleft palates, recent studies have shown normal bone growth during pregnancy although newborns had smaller head circumference and height [51]. Decreased growth has also been observed when youths are prescribed SSRIs to treat depression [52].

The observation that both wild-type mice treated with multiple doses of SSRIs and mice lacking the SERT gene have reduced bone growth supported a direct link between serotonin signaling and bone dynamic [53]. Similar results were obtained using perinatal treatment with tranylcypromine (TCP), a monoamine oxidase (MAO) inhibitor in rats, leading to persistent changes in serotonin availability [54]. Perinatal exposure to TCP leads to decreased TPH1 expression in the peripheral compartment and increased bone volume and trabecular number.

These studies as well as the study of a mouse model of depression [55, 56] suggest that peripheral and central serotonin compartments have different mechanisms to react to 5-HT imbalances and suggest a predominant role for gut-derived serotonin in the regulation of bone maintenance under these conditions. While SSRIs increase 5-HT levels and alleviate both depression and unbalanced sympathetic tone, it has deleterious effects on bone metabolism. Of note is the fact that the negative effects of increased serotonin in the periphery appear to outweigh the bone accrual effects that would be expected to result from the enhanced serotonin signaling in the brain.

Table 1 Summary of serotonin signaling-related proteins expressed in bone.

Model	Net effect on bone metabolism	References
TPH1-/-	Increased osteoblast formation	Yadav et al. [21]
TPH1-/-	No effect on bone metabolism	Cui et al. [28]
TPH1-/-	Decreased bone resorption	Chabbi-Achengli et al. [15]
TPH2-/-	Decreased bone formation	Brommage et al. [17]
SERT-/-	Decreased bone formation	Warden et al. [35]
5-HT1A	Expressed on osteoblasts and osteoclasts; action not yet identified	Bliziotes et al. [34][a]
		Bliziotes et al. [42][a]
5-HT1B-/-	Increased osteoblast proliferation	Yadav et al. [21]
5-HT2A inhibition	Decreased osteoblast differentiation	Hirai et al. [41][a]
		Tanaka et al. [45]
5-HT2B-/-	Reduced bone formation	Collet et al. [44]
5-HT2B-/-	Impaired osteoblast proliferation, recruitment, and matrix mineralization	Chabbi-Achengli et al. [43]
5-HT2c-/-	Decreased bone formation and increased bone resorption	Oury et al. [18]
5-HT6 activation	Decreased osteoblast proliferation	Yun et al. [46]

[a]Indicates observations made in cell culture

Concluding Remarks

Serotonin has emerged as another key player in a list of signaling molecules to consider when contemplating the regulation of bone formation and remodeling (Table 1). Within the bone, the biosynthetic enzyme for serotonin (TPH1), receptors for mediating serotonin's actions, and the transporter for terminating its signals are all expressed by the cells that are integral to bone formation and resorption. Furthermore, serotonin arising from sources outside of the bone can impact bone density. Neural pathways in the brain that influence bone density are initiated by serotonergic projections from the brainstem to the hypothalamus, which ultimately promote bone accrual by releasing bone from the inhibitory influence of sympathetic inputs. The predominant source of serotonin in the body is the gut, and there are strong data indicating that gut-derived serotonin can have a negative impact on bone growth by inhibiting osteoblast formation.

The importance of serotonin in bone metabolism is underscored by the fact that, in animal studies, bone density can be dramatically influenced by deletion of the gene for SERT or by its pharmacological inhibition. This is supported by clinical reports of bone fragility in patients treated chronically with SSRIs. The complexity of studying the actions of serotonin on its multiple possible targets is further complicated by the presence of genetic variations of key factors including SERT and $5HT_{1B}$, which can affect the outcome in drug studies. Furthermore, inconsistencies with regard to animal models, strains, sex, ages, diets, sampling techniques, and approaches to bone density analysis have likely contributed to apparently conflicting results in the literature. With this in mind, it is important to gain a more thorough understanding of the interactions between serotonin and the bone as drugs modulating elements of the serotonin signaling pathway (including receptor agonists and antagonists and SSRIs) are increasingly prescribed to treat depression and other disorders in youth and adults. Furthermore, a more complete understanding of the roles of serotonin in bone metabolism could lead to novel therapeutic strategies for alleviating bone pathologies.

Acknowledgments Work conducted in the author's laboratories has been supported by NIH grants DK62267 (to GMM) and R37 DE012528 (to JBL).

References

1. Rosen CJ, Bouillon R, Compston JE, Rosen V, editors. Primer on the metabolic bone diseases and disorders of mineral metabolism. 8th ed. Ames: Wiley-Blackwell; 2013.
2. Dimitri P, Rosen C. The central nervous system and bone metabolism: an evolving story. Calcified tissue international. 2016.
3. Ducy P, Karsenty G. The two faces of serotonin in bone biology. J Cell Biol. 2010;191(1):7–13.
4. Erspamer V. Experimental research on the biological significance of enterochromaffin cells. Arch Fisiol. 1937:156–9.
5. Rapport MM, Green AA, Page IH. Serum vasoconstrictor, serotonin; isolation and characterization. J Biol Chem. 1948;176(3):1243–51.
6. Page IH, Rapport MM, Green AA. The crystallization of serotonin. J Lab Clin Med. 1948;33(12):1606.
7. Erspamer V, Asero B. Identification of enteramine, the specific hormone of the enterochromaffin cell system, as 5-hydroxytryptamine. Nature. 1952;169(4306):800–1.
8. Twarog BM, Page IH. Serotonin content of some mammalian tissues and urine and a method for its determination. Am J Phys. 1953;175(1):157–61.
9. Hensler JG. Serotonin. In: Brady ST, editor. Basic neurochemistry principles of molecular, cellular, and medical neurobiology. 8th ed. Academic; Waltham, MA. 2012. p. 300–22.
10. Mawe GM, Hoffman JM. Serotonin signalling in the gut – functions, dysfunctions and therapeutic targets. Nat Rev Gastroenterol Hepatol. 2013;10(8):473–86.
11. Bertaccini G. Tissue 5-hydroxytryptamine and urinary 5-hydroxyindoleacetic acid after partial or total removal of the gastro-intestinal tract in the rat. J Physiol. 1960;153(2):239–49.
12. Stunes AK, Reseland JE, Hauso O, Kidd M, Tommeras K, Waldum HL, et al. Adipocytes express a functional system for serotonin synthesis, reuptake and receptor activation. Diabetes Obes Metab. 2011;13(6):551–8.
13. Paulmann N, Grohmann M, Voigt JP, Bert B, Vowinckel J, Bader M, et al. Intracellular serotonin modulates insulin secretion from pancreatic beta-cells by protein serotonylation. PLoS Biol. 2009;7(10):e1000229.
14. Kim H, Toyofuku Y, Lynn FC, Chak E, Uchida T, Mizukami H, et al. Serotonin regulates pancreatic beta cell mass during pregnancy. Nat Med. 2010;16(7):804–8.
15. Chabbi-Achengli Y, Coudert AE, Callebert J, Geoffroy V, Cote F, Collet C, et al. Decreased osteoclastogenesis in serotonin-deficient mice. Proc Natl Acad Sci U S A. 2012;109(7):2567–72.
16. Yadav VK, Oury F, Suda N, Liu ZW, Gao XB, Confavreux C, et al. A serotonin-dependent mechanism explains the leptin regulation of bone mass, appetite, and energy expenditure. Cell. 2009;138(5):976–89.

17. Brommage R, Liu J, Doree D, Yu W, Powell DR, Melissa Yang Q. Adult Tph2 knockout mice without brain serotonin have moderately elevated spine trabecular bone but moderately low cortical bone thickness. BoneKEy Rep. 2015;4:718.
18. Oury F, Yadav VK, Wang Y, Zhou B, Liu XS, Guo XE, et al. CREB mediates brain serotonin regulation of bone mass through its expression in ventromedial hypothalamic neurons. Genes Dev. 2010;24(20):2330–42.
19. Motyl KJ, Rosen CJ. Understanding leptin-dependent regulation of skeletal homeostasis. Biochimie. 2012;94(10):2089–96.
20. Lam DD, Leinninger GM, Louis GW, Garfield AS, Marston OJ, Leshan RL, et al. Leptin does not directly affect CNS serotonin neurons to influence appetite. Cell Metab. 2011;13(5):584–91.
21. Yadav VK, Ryu JH, Suda N, Tanaka KF, Gingrich JA, Schutz G, et al. Lrp5 controls bone formation by inhibiting serotonin synthesis in the duodenum. Cell. 2008;135(5):825–37.
22. Gong Y, Slee RB, Fukai N, Rawadi G, Roman-Roman S, Reginato AM, et al. LDL receptor-related protein 5 (LRP5) affects bone accrual and eye development. Cell. 2001;107(4):513–23.
23. Boyden LM, Mao J, Belsky J, Mitzner L, Farhi A, Mitnick MA, et al. High bone density due to a mutation in LDL-receptor-related protein 5. N Engl J Med. 2002;346(20):1513–21.
24. Kato M, Patel MS, Levasseur R, Lobov I, Chang BH, Glass DA 2nd, et al. Cbfa1-independent decrease in osteoblast proliferation, osteopenia, and persistent embryonic eye vascularization in mice deficient in Lrp5, a Wnt coreceptor. J Cell Biol. 2002;157(2):303–14.
25. Kode A, Mosialou I, Silva BC, Rached MT, Zhou B, Wang J, et al. FOXO1 orchestrates the bone-suppressing function of gut-derived serotonin. J Clin Invest. 2012;122(10):3490–503.
26. Yadav VK, Balaji S, Suresh PS, Liu XS, Lu X, Li Z, et al. Pharmacological inhibition of gut-derived serotonin synthesis is a potential bone anabolic treatment for osteoporosis. Nat Med. 2010;16(3):308–12.
27. Inose H, Zhou B, Yadav VK, Guo XE, Karsenty G, Ducy P. Efficacy of serotonin inhibition in mouse models of bone loss. J Bone Miner Res Off J Am Soc Bone Miner Res. 2011;26(9):2002–11.
28. Cui Y, Niziolek PJ, MacDonald BT, Zylstra CR, Alenina N, Robinson DR, et al. Lrp5 functions in bone to regulate bone mass. Nat Med. 2011;17(6):684–91.
29. Cui Y, Niziolek PJ, MacDonald BT, Alenina N, Matthes S, Jacobsen CM, et al. Reply to Lrp5 regulation of bone mass and gut serotonin synthesis. Nat Med. 2014;20(11):1229–30.
30. Kode A, Obri A, Paone R, Kousteni S, Ducy P, Karsenty G. Lrp5 regulation of bone mass and serotonin synthesis in the gut. Nat Med. 2014;20(11):1228–9.
31. Monroe DG, McGee-Lawrence ME, Oursler MJ, Westendorf JJ. Update on Wnt signaling in bone cell biology and bone disease. Gene. 2012;492(1):1–18.
32. Warden SJ, Robling AG, Haney EM, Turner CH, Bliziotes MM. The emerging role of serotonin (5-hydroxytryptamine) in the skeleton and its mediation of the skeletal effects of low-density lipoprotein receptor-related protein 5 (LRP5). Bone. 2010;46(1):4–12.
33. de Vernejoul MC, Collet C, Chabbi-Achengli Y. Serotonin: good or bad for bone. BoneKEy Rep. 2012;1:120.
34. Bliziotes MM, Eshleman AJ, Zhang XW, Wiren KM. Neurotransmitter action in osteoblasts: expression of a functional system for serotonin receptor activation and reuptake. Bone. 2001;29(5):477–86.
35. Warden SJ, Bliziotes MM, Wiren KM, Eshleman AJ, Turner CH. Neural regulation of bone and the skeletal effects of serotonin (5-hydroxytryptamine). Mol Cell Endocrinol. 2005;242(1–2):1–9.
36. Warden SJ, Nelson IR, Fuchs RK, Bliziotes MM, Turner CH. Serotonin (5-hydroxytryptamine) transporter inhibition causes bone loss in adult mice independently of estrogen deficiency. Menopause (New York, NY). 2008;15(6):1176–83.
37. Westbroek I, van der Plas A, de Rooij KE, Klein-Nulend J, Nijweide PJ. Expression of serotonin receptors in bone. J Biol Chem. 2001;276(31):28961–8.
38. Raap DK, Van de Kar LD. Selective serotonin reuptake inhibitors and neuroendocrine function. Life Sci. 1999;65(12):1217–35.

39. Le Poul E, Boni C, Hanoun N, Laporte AM, Laaris N, Chauveau J, et al. Differential adaptation of brain 5-HT1A and 5-HT1B receptors and 5-HT transporter in rats treated chronically with fluoxetine. Neuropharmacology. 2000;39(1):110–22.
40. Bliziotes M, Gunness M, Eshleman A, Wiren K. The role of dopamine and serotonin in regulating bone mass and strength: studies on dopamine and serotonin transporter null mice. J Musculoskelet Neuronal Interact. 2002;2(3):291–5.
41. Hirai T, Tokumo K, Tsuchiya D, Nishio H. Expression of mRNA for 5-HT2 receptors and proteins related to inactivation of 5-HT in mouse osteoblasts. J Pharmacol Sci. 2009;109(2):319–23.
42. Bliziotes M, Eshleman A, Burt-Pichat B, Zhang XW, Hashimoto J, Wiren K, et al. Serotonin transporter and receptor expression in osteocytic MLO-Y4 cells. Bone. 2006;39(6):1313–21.
43. Chabbi-Achengli Y, Launay JM, Maroteaux L, de Vernejoul MC, Collet C. Serotonin 2B receptor (5-HT2B R) signals through prostacyclin and PPAR-ss/delta in osteoblasts. PLoS One. 2013;8(9):e75783.
44. Collet C, Schiltz C, Geoffroy V, Maroteaux L, Launay JM, de Vernejoul MC. The serotonin 5-HT2B receptor controls bone mass via osteoblast recruitment and proliferation. FASEB J Off Publ Fed Am Soc Exp Biol. 2008;22(2):418–27.
45. Tanaka K, Hirai T, Ishibashi Y, Izumo N, Togari A. Modulation of osteoblast differentiation and bone mass by 5-HT2A receptor signaling in mice. Eur J Pharmacol. 2015;762:150–7.
46. Yun HM, Park KR, Hong JT, Kim EC. Peripheral serotonin-mediated system suppresses bone development and regeneration via serotonin 6 G-protein-coupled receptor. Sci Rep. 2016;6:30985.
47. Cui C, Kaartinen MT. Serotonin (5-HT) inhibits factor XIII-A-mediated plasma fibronectin matrix assembly and crosslinking in osteoblast cultures via direct competition with transamidation. Bone. 2015;72:43–52.
48. Warden SJ, Fuchs RK. Do selective serotonin reuptake inhibitors (SSRIs) cause fractures? Curr Osteoporos Rep. 2016;14(5):211–8.
49. Rizzoli R, Cooper C, Reginster JY, Abrahamsen B, Adachi JD, Brandi ML, et al. Antidepressant medications and osteoporosis. Bone. 2012;51(3):606–13.
50. Mezuk B, Eaton WW, Golden SH. Depression and osteoporosis: epidemiology and potential mediating pathways. Osteoporos Int: J Established Results Cooperation Between Eur Found Osteoporos Natl Osteoporos Found USA. 2008;19(1):1–12.
51. Dubnov-Raz G, Hemilä H, Vurembrand Y, Kuint J, Maayan-Metzger A. Maternal use of selective serotonin reuptake inhibitors during pregnancy and neonatal bone density. Early Hum Dev. 2012;88(3):191–4.
52. Weintrob N, Cohen D, Klipper-Aurbach Y, Zadik Z, Dickerman Z. Decreased growth during therapy with selective serotonin reuptake inhibitors. Arch Pediatr Adolesc Med. 2002;156(7):696–701.
53. Warden SJ, Robling AG, Sanders MS, Bliziotes MM, Turner CH. Inhibition of the serotonin (5-hydroxytryptamine) transporter reduces bone accrual during growth. Endocrinology. 2005;146(2):685–93.
54. Blazevic S, Erjavec I, Brizic M, Vukicevic S, Hranilovic D. Molecular background and physiological consequences of altered peripheral serotonin homeostasis in adult rats perinatally treated with tranylcypromine. J Physiol Pharmacol: Off J Pol Physiol Soc. 2015;66(4):529–37.
55. Bab I, Yirmiya R. Depression, selective serotonin reuptake inhibitors, and osteoporosis. Curr Osteoporos Rep. 2010;8(4):185–91.
56. Yirmiya R, Goshen I, Bajayo A, Kreisel T, Feldman S, Tam J, et al. Depression induces bone loss through stimulation of the sympathetic nervous system. Proc Natl Acad Sci U S A. 2006;103(45):16876–81.

Gut Microbiota and Bone Health

Darin Quach and Robert A. Britton

Introduction

Bone is a dynamic organ that undergoes constant remodeling during all stages of life. This remodeling process requires the coordinated actions from two cell types, osteoclasts and osteoblasts. Osteoclasts carry out the action of bone resorption, while osteoblasts take part in the formation of new bone. In early human life, bone formation outweighs bone resorption and contributes to increased bone deposition until a plateau is reached in early adulthood [1]. Then, a shift in favor of bone resorption takes place with bone loss gradually increasing over time. Moreover, this imbalance may become even more pronounced under certain conditions with one example being inflammation.

Osteoclasts originate from monocytic precursors in the bone marrow, and several studies have demonstrated its interaction and regulation by immune cells including B and T cells [2–5]. Different pathologies can accelerate bone loss by impacting the activities of osteoblasts and osteoclasts. For example, bone loss mediated by estrogen deficiency is due in part to increased inflammation (i.e., TNF-α, IL-1, IL-6) and the activation of immune cells promoting the differentiation and activity of osteoclasts [6–11]. Under many different pathological bone conditions, the primary driver of bone loss is caused by increased osteoclastic bone resorption that is often times mediated by immune signaling [12, 13]. Interestingly, the drivers of inflammation in the bone marrow remain poorly understood, and it has been proposed that bacteria derived

D. Quach
Department of Microbiology and Molecular Genetics, Michigan State University, East Lansing, MI, USA

R.A. Britton (✉)
Baylor College of Medicine, Department of Molecular Virology and Microbiology, Alkek Center for Metagenomics and Microbiome Research, Houston, TX, USA
e-mail: Robert.Britton@bcm.edu

L.R. McCabe, N. Parameswaran (eds.), *Understanding the Gut-Bone Signaling Axis*, Advances in Experimental Medicine and Biology 1033,
DOI 10.1007/978-3-319-66653-2_4

signals from the gut driving T cell activation in bone. As a result, the topic of immune regulation and the term "osteoimmunology" have been coined for the focus of interactions taking place between the immune system and bone health [4, 14].

The gastrointestinal (GI) tract is the organ with the largest surface area in the human body. The digestive tract, which starts at the mouth and extends through the intestines to the rectum, has an immense surface area at approximately 30–40 m^2 [15]. The gut is responsible for a number of functions including digestion, absorption of essential nutrients, education and adaptation of the immune system, and prevention of pathogen invasion. Furthermore, the gut harbors its own nervous system that allows for events occurring in the intestinal tract to be disseminated to distal parts of the body. Finally, communication between the intestinal microbiota and host is modulated via the epithelial barrier of the gut. Many of these functions are greatly impacted by the microbial communities that inhabit different sections of the intestinal tract. Indeed, several key functions are dramatically altered in germ-free mice, highlighting the importance of the intestinal microbiota working in concert with the host.

Recent evidence strongly indicates that humans have coevolved with bacteria, and this relationship provides benefits to both the bacteria and the host in most cases. However, due to alterations in our diet, antibiotic usage, and general improvements in our hygiene that have dramatically altered the communities that inhabit humans, it is becoming increasingly clear that dysbiotic microbial communities contribute to the development of disease. For example, this has been studied in the context of several diseases ranging from diabetes, obesity, and atherosclerosis to asthma [16–18]. Furthermore, the importance of the gut microbiota in bone health has become an area of focus following the discovery that the GI microbiota has the potential to regulate bone mass in mice.

Bone metabolism is maintained through a combination of several regulated host mechanisms including mineral absorption, hormonal control, and immunomodulation [14, 19, 20]. While the role of the GI tract in mineral absorption is a well-studied physiological phenomenon through the characterization of vitamin D in calcium and phosphate absorption and its impact on bone, more recently, studies have been investigating the direct impact of bacteria on bone health. In this chapter, possible mechanisms driving the interconnected relationship of the gut-bone axis and bone health in the context of the microbiota will be discussed along with possible future avenues of research that links the brain, gut, and bone.

The Role of Prebiotics and Probiotics in Bone Health

Some of the early indications for the regulation of bone health by bacteria revolved around studies with prebiotics, which are defined as "non-digestible food ingredients that beneficially affect the host by selectively stimulating the growth and/or activity of one or a limited number of bacterial species already established in the colon, and thus in effect improve host health" [21]. Although prebiotics may function by stimulating gut microbes that are beneficial for bone, evidence suggests they may also stimulate calcium absorption. Several studies have demonstrated an increase in calcium absorption following the administration of nondigestible oligosaccharides [22–24]. In an experiment with ovariectomized Sprague Dawley rats, the consumption of inulin and fructooligosaccharides led to an increase in calcium absorption and bone density [24]. In addition, feeding estrogen-sufficient Sprague Dawley rats with galactooligosaccharides was also shown to effectively increase calcium and magnesium absorption along with bone density [22]. Interestingly, there was also an observed increase in the abundance of bifidobacteria. Furthermore, in a human-controlled feeding study with adolescent females where some individuals were randomly assigned to a diet consisting of soluble corn fiber, calcium absorption was once again increased in comparison to the control group [23]. Significant differences in the microbial community diversity were also observed as increases in *Parabacteroides* and *Clostridium* correlated with increasing soluble corn fiber consumption and calcium absorption [23]. Since it is difficult to attribute these changes in calcium absorption to the changes in community structure, more directed studies are required in order to better understand the contribution of microbes to mineral absorption. Readers interested in additional information into the functions of prebiotics in bone health are directed to this excellent review [25].

Probiotics have also been investigated for the ability to ameliorate bone loss in a number of bone disease models. Probiotics are defined as "live microorganisms that when administered in adequate amounts can confer health benefits to the host" [26]. Many of these studies have revolved around the use of lactic acid bacteria. Examples include strains from the genera *Lactococcus* and *Lactobacillus*, which are among the most commonly used probiotics for human consumption. Initial studies with probiotics identified that bacteria could ameliorate bone loss symptoms in different animal models of disease.

One of the earliest studies demonstrating an impact of probiotics on bone health was demonstrated in mice. Using a senescence-accelerated mouse model, oral administration of *Lactococcus lactis* was capable of suppressing bone loss that is associated with aging [27]. Interestingly, splenocytes from mice that received *L. lactis* produced more IL-12 and IFN-γ compared to the control condition suggesting that the immune response was modulated in a pro-inflammatory manner. Since inflammation is primarily associated with bone resorption, this result suggests that the suppression of bone loss following *L. lactis* treatment may be driven by another mechanism not solely dependent on immune regulation or one where the environment in the spleen is independent of the bone marrow compartment.

In another set of studies investigating the impact of *Lactobacillus helveticus* on bone health, rats in the early stages of life as well as ovariectomized rats benefitted from supplementation of the probiotic factors [28, 29]. Groups that received *L. helveticus* fermented milk had increased bone mineral density and bone mineral content in comparison to their respective control conditions. While the levels of calcium do not appear to be impacted in these rats, a human trial investigating the impact of the same strain of *L. helveticus* in postmenopausal women demonstrated that *L. helveticus* fermented milk supplementation increased serum calcium levels and reduced serum parathyroid hormone (PTH) levels [28–30]. The long-term effect of this on bone health in this cohort of postmenopausal women was not measured, but the changes in calcium and PTH suggest a benefit to bone health. Nevertheless, additional longitudinal studies will aid in understanding the impact of *L. helveticus* to bone density in women especially since the mechanisms of action appear to be different in rats in comparison to humans.

More recently, the probiotic bacterium *Lactobacillus reuteri* 6475 was demonstrated to improve bone health in several different murine mouse models [31–33]. First, it was shown that administration of *L. reuteri* was capable of decreasing intestinal inflammation while increasing bone density in healthy male C57BL/6 mice [31]. Interestingly, this phenomenon only held true in male mice as bone density in healthy female mice was not affected by probiotic treatment. In another study, *L. reuteri* treatment was shown to improve bone health in an ovariectomized mouse model [32]. Under estrogen deficiency, mice given *L. reuteri* had increased bone density following 4 weeks of treatment. The population of $CD4^+$ T cells, which was previously shown to promote osteoclastogenesis during estrogen-deficient conditions, decreased in the bone marrow following *L. reuteri* treatment. Additionally, osteoclast differentiation was identified as a key pathway that was targeted in vivo and in vitro since it was observed that osteoclastogenesis was suppressed by *L. reuteri* [32]. Lastly, using a model of mild inflammation induced by dorsal surgical excision, it was demonstrated that *L. reuteri* increased bone density in female mice that were estrogen sufficient but subjected to inflammation due to surgery [33]. However, as opposed to the *L. reuteri* effect observed in healthy male mice, intestinal inflammation was not impacted [31]. Instead, the expression levels of genes (e.g., RANKL, OPG, and IL-10) involved in osteoclastogenesis were modulated by *L. reuteri* [33]. Previously, this specific strain of *L. reuteri* has been demonstrated to exhibit immunomodulatory activity in vitro through the suppression of TNF-α production [34, 35]. Coupled with the fact that the heightened state of inflammation that is induced by estrogen deficiency promotes osteoclastic bone resorption, these studies suggest that the beneficial impact of *L. reuteri* on bone health relies on immunomodulation of key pathways involved in osteoclastogenesis and estrogen signaling.

Microbiome and Bone Health

Recently, more emphasis has been placed on the role of the microbiota in skeletal health [25, 36]. The rationale for this is due to the known impact of the microbiota on multiple facets of the immune system, the importance of the immune system in bone remodeling, and the GI tract that serves as a messenger between these two entities [37–39]. In addition to impacting the local immune response in the gut, the microbiota has also been shown to regulate the immune response and hematopoiesis at distant sites including the bone marrow [40]. Despite these connections linking the gut microbiota, the immune system, and bone health, very few studies assessing the interactions between these entities have been conducted to date (Table 1).

Table 1 Microbiome studies on bone health

Subject	Study design	Outcome	Citation
C57BL/6 mice, 7–9 weeks old	Compared GF and conventionalized mice	Presence of microbiota led to:	Sjögren et al. [41]
Female		(1) Decreased BMD	
		(2) Increased CD4+ T cells in bone marrow	
C57BL/6 mice, 20 weeks old	Compared GF and mice undergoing low-dose penicillin (LDP) treatment	Low-dose penicillin treatment led to:	Cho et al. [42]
Male and female		(1) Increased BMD, BMC in female mice	
		(2) Decreased BMC in male mice	
		(3) Increased central adiposity in male mice	
C57BL/6 mice, 7 weeks old	Compared conventionalized mice with and without antibiotic treatment[a]	Antibiotic treatment led to:	Cox et al. [43]
		(1) Increased BMD at 3 weeks but not at 7 weeks	
		(2) Increased body fat but no change in weight	
C57BL/6 mice, 20 weeks old	Compared GF and conventionalized mice, with/without leuprolide (suppresses estrogen production)	Presence of microbiota led to:	
Female		(1) Decreased BVF *only* during estrogen deficiency	
		(2) Increased CD4+ T cells in bone marrow *only* during estrogen deficiency	
		(3) Increased osteoclastogenic cytokines *only* during estrogen deficiency	
Wistar rats, 22–24 months old	Compared ovariectomized conventionalized rats with and without minocycline treatment	Minocycline treatment led to:	Williams et al. [44]
Female		(1) Increased BMD	
		(2) Increased bone formation rate	
BALB/c mice, 8 weeks old	Compared GF and conventionalized mice	Presence of microbiota led to:	Schwarzer et al. [45]
Male		(1) Increased BMD, BVF, femur length	
		(2) Increased body weight, body length	
		(3) Increased insulin-like growth factor-1	

BMD bone mineral density, *BVF* bone volume fraction

[a]Penicillin, vancomycin, penicillin plus vancomycin, chlortetracycline

To better understand the role that the microbiota plays in bone health, GF mouse models were used in early studies. GF animals are raised in sterile conditions and do not harbor a microbiome on any body surface. In one study, it was demonstrated that female GF mice at 8 weeks of age in the C57BL/6 genetic background possessed 50% higher bone density compared to CONV-R animals. Subsequent conventionalization of GF mice normalized bone to the level similar to CONV-R mice [41]. Flow cytometry and histomorphometric measurements demonstrated that the presence of a gut microbiota increased the amount of osteoclast precursor cells ($CD4^+/GR1^-$) in the bone marrow compartment and osteoclasts on bone surfaces [41]. Additionally, an expansion in $CD4^+$ cells in the bone marrow and inflammatory markers (e.g., TNF-α, IL-6) from bone mRNA also resulted following the conventionalization of mice [41]. Taken together, these results suggest that the gut microbiota modulates the immune system and drives osteoclastic bone resorption.

Further support for the role of intestinal bacteria in regulating bone mass revolves around a set of studies investigating the impact of antibiotics on bone health in mice. Consistent with the results obtained by Sjogren et al. [41], antibiotic treatment in the early stages (3 weeks) of life increased bone mineral density in female C57BL/6 mice. However, by 7 weeks, no changes were observed between the control and antibiotic-treated groups [42]. Conversely, a longer-lasting effect was observed when the administration of low-dose penicillin either at birth or at weaning led to an increase in bone mineral density of female C57BL/6 mice at 20 weeks of age but not in male mice [43]. Thus, both sex and age appear to be contributing factors.

While a few studies support the notion that the intestinal microbiota is detrimental to bone health, more recent studies suggest the interaction between the microbiota is more complex and the general notion that microbes promote reduced bone density is probably context dependent. Postmenopausal osteoporosis is a common bone disease that results from estrogen deficiency when ovarian function decreases in the later stages of life. In a model of estrogen deficiency, GF female C57BL/6 mice were spared from bone loss following continuous leuprolide treatment used to blunt estrogen production [46]. Accordingly, CONV-R mice were shown to lose bone following leuprolide treatment. Since the bone density of GF mice did not differ from CONV-R mice, the intestinal microbiota was shown to contribute to bone loss only during estrogen deficiency. Thus this work showed no difference in bone density between GF mice and mice colonized with an intestinal microbiota. Consistent with the results observed by Britton et al. [32], the administration of the probiotic *Lactobacillus rhamnosus* GG or probiotic cocktail VSL#3 was able to suppress bone loss during leuprolide-induced estrogen deficiency. Moreover, treatment with the antibiotic minocycline was able to prevent bone loss during estrogen deficiency [44]. Therefore, these studies suggest that the impact from the microbiota on bone health is both context dependent and microbe specific.

Currently, the studies investigating the impact of the gut microbiota on bone health are mixed, and much remains to be learned. Several factors that could be contributing to the discrepancies in results include the genetic background, sex, or age of the animals. The disease process (e.g., estrogen deficiency) being investigated also seems to play a crucial role in determining the impact of gut bacteria on bone health. Additionally, the composition of the microbial community likely plays an important role. Since there were no published community analysis results in most of the studies described so far, it is impossible to compare the studies with regard to the composition of the microbial communities present. However, it has been well documented that mice purchased from different animal vendors and housed at different mouse facilities foster microbial community colonization that is location specific [47]. Taken together, the likelihood that dissimilar microbiotas were present in the studies described, coupled with varying phenotypes in bone health, suggests that specific microbial members impact bone metabolism in different ways.

Lastly, another study investigating the impact of the intestinal microbiota on bone health was conducted in a different mouse genetic background. In this study, 8-week-old male GF mice in the BALB/c background were compared to CONV-R mice, and it was reported that CONV-R mice had increased cortical and trabecular bone in the femur [45]. Monocolonization with a strain of *Lactobacillus plantarum* was shown to promote bone growth as well. In addition to the studies performed by Li et al. [46], this serves as another example where the presence of the microbiota was not detrimental to bone health. Mechanistically, the hypothalamic-pituitary axis appeared to be impacted by microbial colonization as serum growth hormone was significantly increased in CONV-R or monocolonized animals when compared to GF conditions. In this case, it appears that neuroendocrine functioning is being affected by the presence of intestinal microbiota to improve bone health.

Exploring a Potential Brain-Gut-Bone Axis

Since the development of bone loss is highly impacted by cross talk between immune cells and monocytes that can differentiate into osteoclasts, early focus has been placed on the dysregulation of osteoclastogenesis by the immune system [14, 48, 49]. Other studies have also investigated the role of gut bacteria on calcium reabsorption since that is one of the main functions of the GI tract, which drives bone metabolism [23, 25, 50, 51]. However, our attempts to understand the determinants of bone health are still ongoing. An area of research that has recently been gaining momentum involves the regulation of bone metabolism by the brain [52].

The enteric nervous system is responsible for communication that exists between the brain and gastrointestinal tract [53, 54]. Bidirectional signaling allows the brain to sense what the GI tract comes into contact with from the outside world and regulate a number of pathways including gastrointestinal motility, appetite, and central metabolism [55, 56]. The hypothalamus is a very important portion of the brain involved with the secretion of endocrine hormones that are involved with many physiological processes including growth, stress responses, sexual development, and blood pressure control [57–59]. The hypothalamus is surrounded by a capillary system that is semipermeable to the blood-brain barrier, allowing for the brain to sense systemic signals. Through the production and secretion of hormones into the systemic circulation at this site, communication between the brain and the rest of the body is made possible. For example, the hypothalamus can respond to changes in the levels of gut-derived hormones such as neuropeptide Y (NPY) or leptin in the bloodstream to regulate energy homeostasis and bone mass [60–62]. Serotonin, an important neurotransmitter produced primarily in the small intestine, has also been shown to regulate bone mass [20, 63, 64]. Taken together, these studies support a relationship between bone metabolism and the neural signaling.

In the last decade, studies have been conducted to investigate the impact of the gut microbiota on various neuropeptides. Examples include how obesity and metabolic disease are linked through the regulation of NPY and leptin by the gut microbiota [65–67]. By comparing GF and CONV-R mice, it was demonstrated that the presence of the gut microbiota had reduced expression levels of NPY as well as sensitivity to leptin [65]. In high-fat-fed Sprague Dawley rats, changes in the microbiota correlated with elevated levels of NPY and were protective against obesity [66]. While cause and effect cannot be extrapolated from these correlations, this paves the way for future studies that are directed at identifying microbes or microbial communities that have the ability to modulate neural signaling.

In addition to the intestinal microbiome, individual probiotics have been linked to the regulation of neurotransmitters and peptides [68, 69]. For example, it has been demonstrated that *L. reuteri* 6475 promoted wound healing and hair growth in a murine mouse model [69]. These observations were linked to an increase in oxytocin and testosterone, which are both hormones regulated by the hypothalamic-pituitary axis in the brain [69–71]. Interestingly, *L. reuteri* was ineffective in vagotomized mice, which demonstrated the necessity of a neural connection between the brain and gut [69]. Vagal innervation was also demonstrated to be crucial for the anxiolytic effects provided by *Lactobacillus rhamnosus* treatment in a murine model [68]. Lastly, the ability of *L. reuteri* 6475 to stimulate oxytocin has also been linked to the ability to relieve autism spectrum disorder behaviors in mice [72]. Despite the fact that bone health was not the primary focus of these studies, the established links that have been made between microbes, gut hormones, the brain, and bone support the possibility of a novel gut-brain-bone regulatory axis. Indeed, oxytocin effects on bone health are now well documented in animal models, and high levels of oxytocin in women correlate with high bone mineral density [73, 74]. What has been described alludes to the complexities involved in bone metabolism and serves as the impetus to continue exploring novel regulatory pathways such as one between the enteric nervous system, gut microbiota, and bone in what can be referred to as a brain-gut-bone axis.

Conclusions

With the discovery of the importance of the microbiota in regulating health and disease, more emphasis has been placed on how bone health is regulated by gut bacteria. While the pioneering study in this field demonstrated a deleterious effect on bone health by the gut microbiota, recent studies have supported the notion that this is context and microbiota dependent. The presence of gut bacteria does not always result in bone loss as evidenced by the use of probiotics and the protective effect on bone health. While there are similarities in disease processes underlying bone disease in that osteoclastic bone resorption is promoted, the pathological basis for disease may differ. Thus, what will become increasingly important is a better understanding of disease processes that contribute toward creating an imbalance in bone remodeling and the identification of microbes regulating those changes. The advancement of sequencing technology (e.g., metagenomics, metatranscriptomics, metabolomics) in the past decade will aid in better characterizing the disease process and identification of the microbial community members that correlate with those changes. Ultimately, this will allow for the identification of next-generation probiotics and defined microbial community therapeutics that can be utilized in a controlled and disease-specific context.

References

1. Riggs BL, Melton Iii LJ, Robb RA, Camp JJ, Atkinson EJ, Peterson JM, et al. Population-based study of age and sex differences in bone volumetric density, size, geometry, and structure at different skeletal sites. J Bone Miner Res. 2004;19(12):1945–54.
2. Martin GHKQR. Activated T lymphocytes support osteoclast formation in vitro. Biochem Biophys Res Commun. 1999;265:144–50.
3. Pappalardo A, Thompson K. Activated γδ T cells inhibit osteoclast differentiation and resorptive activity in vitro. Clin Exp Immunol. 2013;174(2):281–91.
4. Nakashima T, Takayanagi H. Osteoimmunology: crosstalk between the immune and bone systems. J Clin Immunol. 2009;29(5):555–67.
5. Asagiri M, Takayanagi H. The molecular understanding of osteoclast differentiation. Bone. 2007;40(2):251–64.
6. Kimble RB, Srivastava S, Ross FP, Matayoshi A, Pacifici R. Estrogen deficiency increases the ability of stromal cells to support murine osteoclastogenesis via an interleukin-1and tumor necrosis factor- mediated stimulation of macrophage colony-stimulating factor production. J Biol Chem. 1996;271(46):28890–7.
7. Feng X. Regulatory roles and molecular signaling of TNF family members in osteoclasts. Gene. 2005;350(1):1–13.
8. Srivastava S, Weitzmann MN, Cenci S, Ross FP, Adler S, Pacifici R. Estrogen decreases TNF gene expression by blocking JNK activity and the resulting production of c-Jun and JunD. J Clin Invest. 1999;104(4):503–13.
9. Weitzmann MN. The role of inflammatory cytokines, the RANKL/OPG axis, and the immunoskeletal interface in physiological bone turnover and osteoporosis. Scientifica (Cairo). 2013;2013:125705.
10. Most W, Van der Wee-Pals L, Ederveen A, Papapoulos S, Löwik C. Ovariectomy and orchidectomy induce a transient increase in the osteoclastogenic potential of bone marrow cells in the mouse. Bone. 1997;20(1):27–30.
11. Bellido T, Jilka RL, Boyce BF, Girasole G, Broxmeyer H, Dalrymple SA, et al. Regulation of interleukin-6, osteoclastogenesis, and bone mass by androgens: the role of the androgen receptor. J Clin Invest. 1995;95(6):2886–95.
12. Karieb S, Fox SW. Suppression of T cell-induced osteoclast formation. Biochem Biophys Res Commun. Elsevier Inc. 2013;436(4):619–24.
13. Salamanna F, Maglio M, Borsari V, Giavaresi G, Aldini NN, Fini M. Peripheral blood mononuclear cells spontaneous osteoclastogenesis: mechanisms driving the process and clinical relevance in skeletal disease. J Cell Physiol. 2016;231(3):521–30.
14. D'Amelio P, Sassi F. Osteoimmunology: from mice to humans. Bonekey Rep. 2016;5:802.
15. Helander HF, Fändriks L. Surface area of the digestive tract – revisited. Scand J Gastroenterol. 2014;49(6):681–9.

16. Cani PD, Possemiers S, Van de Wiele T, Guiot Y, Everard A, Rottier O, et al. Changes in gut microbiota control inflammation in obese mice through a mechanism involving GLP-2-driven improvement of gut permeability. Gut. 2009;58(8):1091–103.
17. Koeth RA, Wang Z, Levison BS, Buffa JA, Org E, Sheehy BT, et al. Intestinal microbiota metabolism of l-carnitine, a nutrient in red meat, promotes atherosclerosis. Nat Med. Nature Publishing Group. 2013;19:576–85.
18. Korpela K, Salonen A, Virta LJ, Kekkonen RA, Forslund K, Bork P, et al. Intestinal microbiome is related to lifetime antibiotic use in Finnish pre-school children. Nat Commun. 2016;7:10410.
19. Carmeliet G, Dermauw V, Bouillon R. Vitamin D signaling in calcium and bone homeostasis: a delicate balance. Best Pract Res Clin Endocrinol Metab. Elsevier Ltd. 2015;29(4):621–31.
20. Sharan K, Yadav VK. Hypothalamic control of bone metabolism. Best Pract Res Clin Endocrinol Metab. Elsevier Ltd. 2014;28(5):713–23.
21. Gibson GR, Scott KP, Rastall RA, Tuohy KM, Hotchkiss A, Dubert-Ferrandon A, et al. Dietary prebiotics: current status and new definition. Food Sci Technol Bull Funct Foods. 2010;7(1):1–19.
22. Weaver CM, Martin BR, Nakatsu CH, Armstrong AP, Clavijo A, Mccabe LD, et al. Galactooligosaccharides improve mineral absorption and bone properties in growing rats through gut fermentation. J Agric Food Chem (ACS Publ). 2011;59:6501–10.
23. Whisner CM, Martin BR, Nakatsu CH, Story JA, MacDonald-Clarke CJ, McCabe LD, et al. Soluble corn fiber increases calcium absorption associated with shifts in the gut microbiome: a randomized dose-response trial in free-living pubertal females. J Nutr. 2016;146(7):1298–306.
24. Zafar TA, Weaver CM, Zhao YD, Martin BR, Wastney ME. Nondigestible oligosaccharides increase calcium absorption and suppress bone resorption in ovariectomized rats. J Nutr. 2004;134(2):399–402.
25. McCabe L, Britton RA, Parameswaran N. Prebiotic and probiotic regulation of bone health: role of the intestine and its microbiome. Curr Osteoporos Rep. 2015;13(6):363–71.
26. Morelli L, Capurso L. FAO/WHO guidelines on probiotics. J Clin Gastroenterol. 2012;46:S1–2.
27. Kimoto-Nira H, Suzuki C, Kobayashi M, Sasaki K, Kurisaki J-I, Mizumachi K. Antiageing effect of a lactococcal strain: analysis using senescence-accelerated mice. Br J Nutr. 2007;98(6):1178–86.
28. Narva M, Collin M, Lamberg-Allardt C, Kärkkäinen M, Poussa T, Vapaatalo H, et al. Effects of long-term intervention with lactobacillus helveticus-fermented milk on bone mineral density and bone mineral content in growing rats. Ann Nutr Metab. 2004;48(4):228–34.
29. Narva M, Rissanen J, Halleen J, Vapaatalo H, Väänänen K, Korpela R. Effects of bioactive peptide, Valyl-Prolyl-Proline (VPP), and *Lactobacillus helveticus* fermented milk containing VPP on bone loss in ovariectomized rats. Ann Nutr Metab. 2007;51(1):65–74.
30. Narva M, Nevala R, Poussa T, Korpela R. The effect of Lactobacillus helveticus fermented milk on acute changes in calcium metabolism in postmenopausal women. Eur J Nutr. 2004;43(2):61–8.
31. McCabe LR, Irwin R, Schaefer L, Britton RA. Probiotic use decreases intestinal inflammation and increases bone density in healthy male but not female mice. J Cell Physiol. 2013;228(8):1793–8.
32. Britton RA, Irwin R, Quach D, Schaefer L, Zhang J, Lee T, et al. Probiotic L. reuteri treatment prevents bone loss in a menopausal ovariectomized mouse model. J Cell Physiol. Wiley-Liss Inc. 2014;229(11):1822–30.
33. Collins FL, Irwin R, Bierhalter H, Schepper J, Britton RA, Parameswaran N, et al. Lactobacillus reuteri 6475 increases bone density in intact females only under an inflammatory setting. PLoS One. 2016;11(4):e0153180.
34. Jones SE, Whitehead K, Saulnier D, Thomas CM, Versalovic J, Britton RA. Cyclopropane fatty acid synthase mutants of probiotic human-derived lactobacillus reuteri are defective in TNF inhibition. Gut Microbes. 2011;2(2):69–79.

35. Thomas CM, Hong T, van Pijkeren JP, Hemarajata P, Trinh DV, Hu W, et al. Histamine derived from probiotic lactobacillus reuteri suppresses tnf via modulation of pka and erk signaling. PLoS One. 2012;7(2):e31951.
36. Ohlsson C, Sjögren K. Effects of the gut microbiota on bone mass. Trends Endocrinol Metab. 2015;26(2):69–74.
37. Pacifici R. Estrogen deficiency, T cells and bone loss. Cell Immunol. 2008;252(1–2):68–80.
38. Pacifici R. Role of T cells in ovariectomy induced bone loss-revisited. J Bone Miner Res. 2012;27(2):231–9.
39. Manolagas SC. From estrogen-centric to aging and oxidative stress: a revised perspective of the pathogenesis of osteoporosis. Endocr Rev. 2010;31(3):266–300.
40. Khosravi A, Yáñez A, Price JG, Chow A, Merad M, Goodridge HS, et al. Gut microbiota promote hematopoiesis to control bacterial infection. Cell Host Microbe [Internet]. 2014;15(3):374–81. Available from: http://www.ncbi.nlm.nih.gov/pubmed/24629343
41. Sjögren K, Engdahl C, Henning P, Lerner UH, Tremaroli V, Lagerquist MK, et al. The gut microbiota regulates bone mass in mice. J Bone Miner Res. 2012;27(6):1357–67.
42. Cho I, Yamanishi S, Cox L, Methé BA, Zavadil J, Li K, et al. Antibiotics in early life alter the murine colonic microbiome and adiposity. Nature. 2012;488(7413):621–6.
43. Cox LM, Yamanishi S, Sohn J, Alekseyenko AV, Leung JM, Cho I, et al. Altering the intestinal microbiota during a critical developmental window has lasting metabolic consequences. Cell. 2014;158(4):705–21.
44. Williams S, Wakisaka A, Zeng QQ, Barnes J, Martin G, Wechter WJ, et al. Minocycline prevents the decrease in bone mineral density and trabecular bone in ovariectomized aged rats. Bone. 1996;19(6):637–44.
45. Schwarzer M, Makki K, Storelli G, Machuca-gayet I, Hudcovic T, Heddi A, et al. Lactobacillus plantarum strain maintains growth of infant mice during chronic undernutrition. Science. 2016;351(6275):854–7.
46. Li J-Y, Chassaing B, Tyagi AM, Vaccaro C, Luo T, Adams J, et al. Sex steroid deficiency-associated bone loss is microbiota dependent and prevented by probiotics. J Clin Invest. 2016;126(6):2049–63.
47. Ericsson AC, Davis JW, Spollen W, Bivens N, Givan S, Hagan CE, et al. Effects of vendor and genetic background on the composition of the fecal microbiota of inbred mice. PLoS One. 2015;10(2):1–19.
48. Weitzmann MN, Pacifici R. Estrogen regulation of immune cell bone interactions. Ann N Y Acad Sci. 2006;1068(1):256–74.
49. Hirayama T, Danks L, Sabokbar A, Athanasou NA. Osteoclast formation and activity in the pathogenesis of osteoporosis in rheumatoid arthritis. Rheumatology (Oxford). 2002;41(11):1232–9.
50. Dimitrov V, White JH. Vitamin D signaling in intestinal innate immunity and homeostasis. Mol Cell Endocrinol. Elsevier Ireland Ltd. 2017;3:68–78.
51. Adachi T, Kakuta S, Aihara Y, Kamiya T, Watanabe Y, Osakabe N, et al. Visualization of probiotic-mediated Ca(2+) signaling in intestinal epithelial cells in vivo. Front Immunol. 2016;7:601.
52. Zofková I, Matucha P. New insights into the physiology of bone regulation: the role of neurohormones. Physiol Res. 2014;63(4):421–7.
53. Forsythe P, Kunze WA. Voices from within: gut microbes and the CNS. Cell Mol Life Sci. 2013;70(1):55–69.
54. Hyland NP, Cryan JF. Microbe-host interactions: influence of the gut microbiota on the enteric nervous system. Dev Biol. 2016;417(2):182–7.
55. Das UN. Obesity: genes, brain, gut, and environment. Nutrition. Elsevier Ltd. 2010;26(5):459–73.
56. Cluny NL, Reimer RA, Sharkey KA. Cannabinoid signalling regulates inflammation and energy balance: the importance of the brain-gut axis. Brain Behav Immun. Elsevier Inc. 2012;26(5):691–8.

57. Han C, Rice MW, Cai D. Neuroinflammatory and autonomic mechanisms in diabetes and hypertension. Am J Physiol Endocrinol Metab. 2016;311(1):E32–41.
58. Mooney-Leber SM, Brummelte S. Neonatal pain and reduced maternal care: early-life stressors interacting to impact brain and behavioral development. Neurosci IBRO. 2017;342:21–36.
59. Cavadas C, Aveleira CA, Souza GFP, Velloso LA. The pathophysiology of defective proteostasis in the hypothalamus – from obesity to ageing. Nat Rev Endocrinol. Nature Publishing Group. 2016;12(12):723–33.
60. Baldock PA, Lee NJ, Driessler F, Lin S, Allison S, Stehrer B, et al. Neuropeptide Y knockout mice reveal a central role of NPY in the coordination of bone mass to body weight. PLoS One. 2009;4(12):e8415.
61. Baldock PA, Allison S, McDonald MM, Sainsbury A, Enriquez RF, Little DG, et al. Hypothalamic regulation of cortical bone mass: opposing activity of Y2 receptor and leptin pathways. J Bone Miner Res. 2006;21(10):1600–7.
62. Steppan CM, Crawford DT, Chidsey-Frink KL, Ke H, Swick AG. Leptin is a potent stimulator of bone growth in ob/ob mice. Regul Pept. 2000;92(1–3):73–8.
63. Hage MP, El-Hajj Fuleihan G. Bone and mineral metabolism in patients undergoing Roux-en-Y gastric bypass. Osteoporos Int. 2014;25(2):423–39.
64. Quiros-Gonzalez I, Yadav VK. Central genes, pathways and modules that regulate bone mass. Arch Biochem Biophys. Elsevier Inc. 2014;561:130–9.
65. Schéle E, Grahnemo L, Anesten F, Halleń A, Bäckhed F, Jansson JO. The gut microbiota reduces leptin sensitivity and the expression of the obesity-suppressing neuropeptides proglucagon (Gcg) and brain-derived neurotrophic factor (Bdnf) in the central nervous system. Endocrinology. 2013;154(10):3643–51.
66. Sun H, Nie X, Wang N, Cang Z, Zhu C, Zhao L, et al. Modulation of microbiota-gut-brain axis by berberine resulting in improved metabolic status in high-fat diet-fed rats. Obes Facts. 2016;9:365–78.
67. Federico A, Dallio M, Tolone S, Gravina AG, Patrone V, Romano M, et al. Gastrointestinal hormones, intestinal microbiota and metabolic homeostasis in obese patients: effect of bariatric surgery. In Vivo. 2016;30(3):321–30.
68. Bravo JA, Forsythe P, Chew MV, Escaravage E, Savignac HM, Dinan TG, et al. Ingestion of Lactobacillus strain regulates emotional behavior and central GABA receptor expression in a mouse via the vagus nerve. Proc Natl Acad Sci U S A. 2011;108(38):16050–5.
69. Poutahidis T, Kearney SM, Levkovich T, Qi P, Varian BJ, Lakritz JR, et al. Microbial symbionts accelerate wound healing via the neuropeptide hormone oxytocin. PLoS One. 2013;8(10):e78898.
70. Poutahidis T, Springer A, Levkovich T, Qi P, Varian BJ, Lakritz JR, et al. Probiotic microbes sustain youthful serum testosterone levels and testicular size in aging mice. PLoS One. 2014;9(1):e84877.
71. Ferrini M, Wang C, Swerdloff RS, Sinha Hikim AP, Rajfer J, Gonzalez-Cadavid NF. Aging-related increased expression of inducible nitric oxide synthase and cytotoxicity markers in rat hypothalamic regions associated with male reproductive function. Neuroendocrinology. 2001;74(1):1–11.
72. Buffington SA, Di Prisco GV, Auchtung TA, Ajami NJ, Petrosino JF, Costa-Mattioli M. Microbial reconstitution reverses maternal diet-induced social and synaptic deficits in offspring. Cell. Elsevier Inc. 2016;165(7):1762–75.
73. Beranger GE, Pisani DF, Castel J, Djedaini M, Battaglia S, Amiaud J, et al. Oxytocin reverses ovariectomy-induced osteopenia and body fat gain. Endocrinology. 2014;155(4):1340–52.
74. Breuil V, Panaia-Ferrari P, Fontas E, Roux C, Kolta S, Eastell R, et al. Oxytocin, a new determinant of bone mineral density in post-menopausal women: analysis of the OPUS cohort. J Clin Endocrinol Metab. 2014;99(4):634–41.

Immunology of Gut-Bone Signaling

Fraser L. Collins, Jonathan D. Schepper, Naiomy Deliz Rios-Arce, Michael D. Steury, Ho Jun Kang, Heather Mallin, Daniel Schoenherr, Glen Camfield, Saima Chishti, Laura R. McCabe, and Narayanan Parameswaran

The Digestive Tract

The gastrointestinal tract forms a continuous tube from the mouth to the anus with the intestine consisting of two major segments: the small and large intestines. Each intestinal segment consists of specialized regions that mediate nutrient digestion and absorption (small intestine), control water and electrolyte absorption (small and large intestine), absorb vitamin B12 and bile absorption (ileum), regulate the

Laura R McCabe and Narayanan Parameswaran contributed equally to this work and are co-senior authors.

F.L. Collins • J.D. Schepper • M.D. Steury • H.J. Kang • H. Mallin • D. Schoenherr
G. Camfield • S. Chishti
Department of Physiology, Michigan State University, East Lansing, MI, USA
e-mail: narap@msu.edu

N.D. Rios-Arce • N. Parameswaran (✉)
Comparative Medicine and Integrative Biology Program, East Lansing, MI, USA

Department of Physiology, Michigan State University, East Lansing, MI, USA

L.R. McCabe (✉)
Department of Physiology and Department of Radiology, Biomedical Imaging Research Centre, Michigan State University, East Lansing, MI, USA
e-mail: mccabel@msu.edu

L.R. McCabe, N. Parameswaran (eds.), *Understanding the Gut-Bone Signaling Axis*, Advances in Experimental Medicine and Biology 1033,
DOI 10.1007/978-3-319-66653-2_5

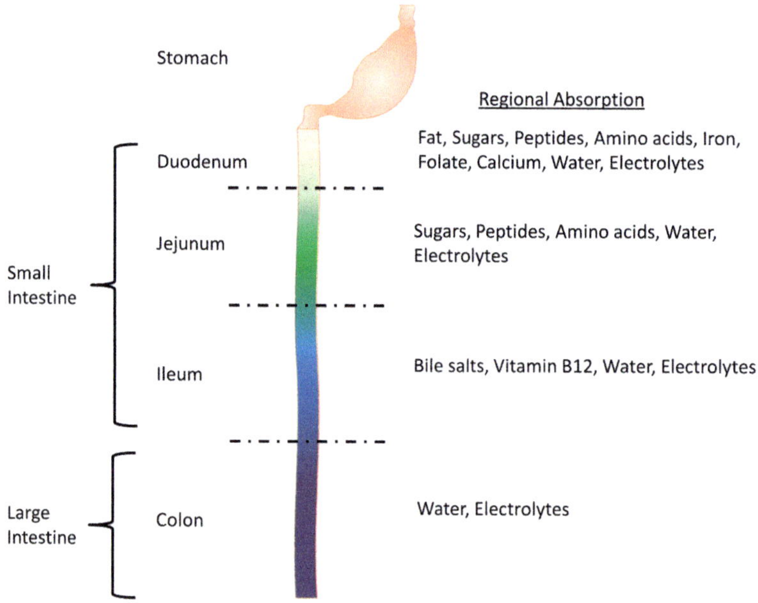

Fig. 1 Regions of the intestine. The intestine is composed of two main segments: the small intestine and the large intestine. The small intestine is further divided into the duodenum, jejunum, and ileum, while the large intestine consists of the cecum, colon, and anus. These regions are specialized with regard to the digestion and absorption of nutrients required to maintain homeostasis

movement of digested products/chyme through the tract, as well as interact with and influence the intestinal microbiome (Fig. 1). Partly due to this constant interaction with intestinal flora, the digestive tract has one of the largest immune compartments in the body [1]. By providing the body with the ability to sense the environment (dietary composition, pathogens, etc.) and to absorb water and nutrients needed to maintain homeostasis, the gut is a crucial organ for survival.

Descending from the stomach, the human small intestine is approximately 5 m in length and has three sections (based on histology and function): the duodenum, jejunum, and ileum. The small intestine is characterized by plicae circulares (folds in the duodenum), villi which are fingerlike projections that extend out into the lumen throughout the small intestine, and microvilli at the apical surface of epithelial cells; together these structures vastly increase the surface area for nutrient absorption; while once thought to be greater than the area of a tennis court, more recent studies indicate the surface area is more on the order of half a badminton court (32 m) [2]. The large intestine consists of the cecum; ascending, transverse, and descending colon; and rectum and ends in the anus. While shorter than the small intestine, approximately 1.5 m, the large intestine is wider in diameter (4.8 cm vs 2.5 cm diameter in the small intestine). In contrast to the small intestine, the large intestine does not contain villi [1].

Though different in gross morphology due to their divergent functions, all sections of the intestinal tract are comprised of four layers visualized cross-sectionally: the mucosa, submucosa, muscularis externa, and the serosa. The mucosa is vital for the regulation of many immunological processes aiding in the regulation and appropriate tolerance of the microbiota. It is made up of three sub-layers: (1) the epithelial layer which contains strong cell/cell connections (tight junctions) and forms a strong barrier that has low permeability; (2) the underlying lamina propria where connective tissues, blood vessels, and immune cells are located; and (3) the muscularis mucosae, a thin muscular layer under the lamina propria (Fig. 2). The mucosa plays a key role in the intestinal immunological defenses ensuring that luminal pathogens do not cross the intestinal barrier [1]. In this chapter we will focus on gut and bone immunology separately and then discuss how the immune system can serve as a link between the two organs.

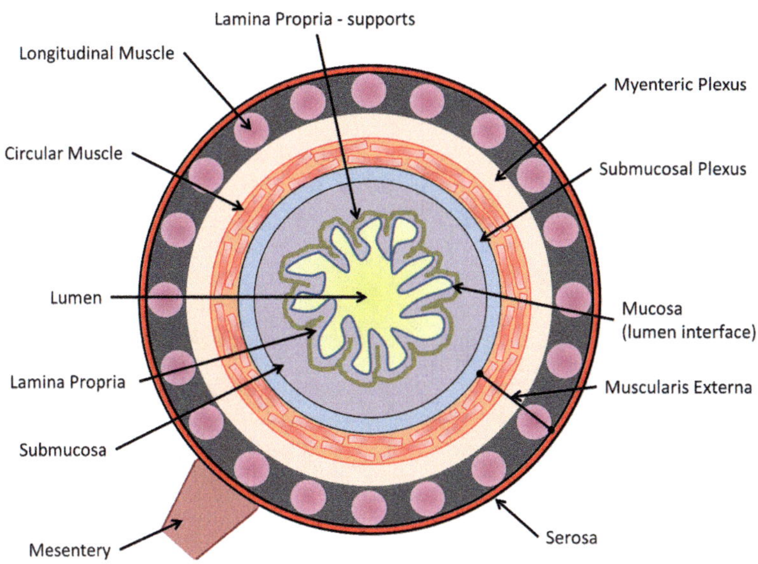

Fig. 2 Layers of the intestinal tract. A cross section of the intestine displaying the different layers. The serosa, the outermost layer, covers the intestine. The muscular layer (muscularis externa) consists of two types of muscle, the inner circular muscle and the outer longitudinal muscle; coordinated contraction of these muscles, known as peristalsis, moves chyme/feces through the intestinal tract. The submucosa consists of dense and irregular connective tissue containing blood vessels, lymphatics, and nerves that branch into the mucosa. The mucosa is the innermost layer of the intestine and consists of the epithelium (in contact with the lumen), the lamina propria, and the muscularis mucosae

The Intestinal Immune System

The adult human intestine is residence to 10^{13}–10^{14} microorganisms [3, 4]. Gut microbiota, dietary proteins, and pathogens are recognized by the immune system as nonself and therefore can elicit an immune response. The gut must both protect the host from invading pathogens while maintaining immune tolerance against beneficial dietary protein and harmless commensal microbes [5–8]. Given that immune surveillance is critical, it is not surprising that the intestinal tract contains the largest number of immune cells of any organ in the body, known as the gut-associated lymphoid tissue (GALT). In addition, the intestine has specialized epithelial cells (Paneth and goblet cells) that secrete antibacterial factors and has intraepithelial lymphocytes to sample luminal antigens (Fig. 3). The intestine also contains organized lymphoid structures comprised of Peyer's patches, solitary isolated lymphoid tissues, lamia propria, and draining lymph nodes. In the following section, we will discuss each of these barriers and surveillance components and how they influence gut defense.

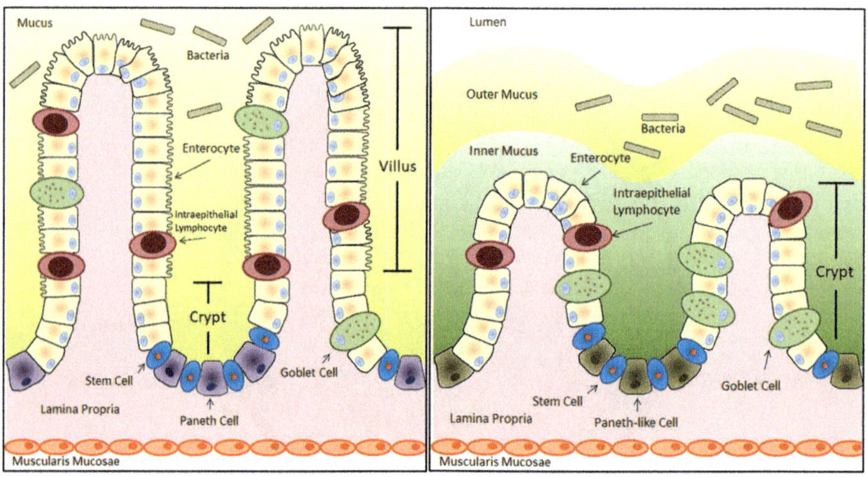

Fig 3 Simplified schematic representation of the intestinal layers and epithelial barrier of the small (*left*) and large (*right*) intestine. The small intestine consists of projections into the lumen known as villi. These vastly increase the absorptive surface area. In contrast, the colon is relatively smooth and contains crypts. The epithelial barrier is made up of numerous cell types. Enterocytes, found in the small intestine, are the absorptive cell and have microvilli on their apical surface further increasing the surface area involved in digestion. Interspersed between the enterocytes are goblet cells and intraepithelial lymphocytes (IELs). Goblet cells produce mucus that forms a protective layer over the epithelial barrier. In the small intestine, this mucus layer is loosely adherent, while in the large intestine, the mucus layer is split into two sections: a loose outer layer and an adherent inner layer. IELs have numerous functions including but not limited to forming a frontline of defense against pathogens in the lumen, suppressing excessive inflammation, and ensuring the integrity of the epithelium. Within the intestinal crypt, Paneth cells produce antimicrobial peptides, defensins, and lysozymes. These Paneth cells are interspersed between stem cells which renew the intestinal epithelium (Adapted from Mowat et al. [1])

Paneth Cells

Paneth cells are specialized, long-lived cells that migrate to the base of crypts after differentiating from stem cells. They produce antimicrobial peptides including regenerating islet-derived protein IIIγ (REGIIIγ), defensins, and lysozymes, which are secreted into crypt/lumen area to reduce bacterial survival near the epithelial barrier. Paneth cells are only found in the small intestine with the highest concentration observed in the ileum. They play a key role in small intestinal homeostasis. Dysregulation in their function results in an increased risk of developing adverse conditions such as Crohn's disease [9].

Goblet Cells

Goblet cells are responsible for the production and secretion of mucus. The frequency of goblet cells increases along the gastrointestinal tract, comprising ~10% of epithelial cells in the upper small intestine and ~25% of epithelial cells in the distal colon. The mucus layer produced by goblet cells covers the mucosa/epithelial cells and forms a barrier to luminal toxins and bacteria. In the small intestine, the secreted mucus exists as a single layer and is loosely adherent. In contrast, the mucus layer in the colon is divided into two sections: a loose outer layer that contains bacteria and a dense bacteria-free inner layer that is firmly attached to the epithelium [10–12]. In addition to being a physical barrier, mucus contains glycol proteins, IgA, antimicrobial peptides, and other substances that are toxic to many bacteria [1, 13].

Intraepithelial Lymphocytes

The majority of intraepithelial lymphocytes (IELs) are antigen-experienced T lymphocytes belonging to both the T cell receptor-γδ (TCRγδ) and TCRαβ lineages and form a frontline of immune defense against invading pathogens [14]. IELs are heterogeneous with alternate distribution patterns in the small and large intestine based upon the physiological conditions. TCRγδ T cells make up approximately 60% of IELs in the small intestine. The majority of IELs can express effector cytokines including interleukin (IL)-2, IL-4, IL-17, and interferon gamma (IFNγ) [14, 15]. Under normal physiological conditions, IELs display regulatory functions and suppress excessive inflammation that can damage the delicate single-cell layer epithelial barrier. Thus their primary role is to ensure the integrity of the intestinal epithelium and maintain local immune quiescence [16, 17]. However, in some pathological conditions, excessive or uncontrolled IEL cytotoxicity can promote inflammation which may initiate or exacerbate inflammatory bowel diseases or celiac disease [14, 18].

Peyer's Patches

Peyer's patches are the best characterized lymphoid structures in the GALT (Fig. 4). Peyer's patches are macroscopic structures that are visible on the antimesenteric side of the small intestine. The size and number of Peyer's patches increase from the jejunum to the ileum with the highest concentration found in the distal ileum; Peyer's patches are rarely observed in the duodenum [19]. Peyer's patches are characterized by an overlying follicle-associated epithelium. The follicle-associated epithelium contains microfold cells (M cells) that are specialized for the uptake and transport of antigens from the lumen into Peyer's patches, where they are taken up by dendritic cells and presented to the lymphocytes for adaptive immune response [20]. Peyer's patches are composed of germinal centers with B cell lymphoid

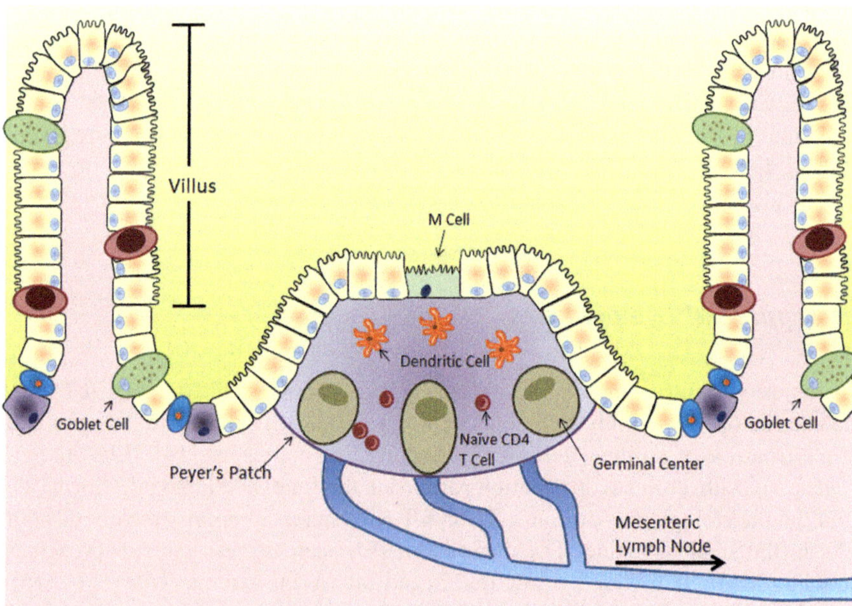

Fig 4 Simplified schematic diagram of the Peyer's patch. Peyer's patches are lymphoid structures visible on the antimesenteric side of the small intestine and are composed of B cell germinal centers surrounded by T cells. Peyer's patches are characterized by the presence of the follicle-associated epithelium which contains M cells. M cells transport the antigens present in the lumen into the Peyer's patch where they are up taken by DCs. DCs present these antigens to the lymphocytes which initiate the adaptive immune response (Adapted from Mowat et al. [23])

follicles that are flanked by smaller T cell areas [21]. While the large intestine does not have Peyer's patches per se, equivalent macroscopic M cell-containing structures can be found; around the ileocecal valve, these structures are known as cecal patches, while throughout the colon and rectum, these structures are known as colonic patches [22].

Solitary Isolated Lymphoid Tissues

Solitary isolated lymphoid tissues (SILTs) can only be detected microscopically and range in size from small cryopatches to larger mature isolated lymphoid follicles [24]. In contrast to Peyer's patches, the isolated lymphoid follicles consist of B cells with no clear T cell zone but do contain germinal centers [25]. SILTs increase in number down the human small intestine, with tenfold more in the distal ileum compared to the proximal jejunum. In the colon numbers of SILTs triple from the ascending colon to the rectosigmoid colon [24]. The role of SILTs is relatively unknown when compared to the larger lymphoid tissues; however, studies have suggested that they may have a similar role to Peyer's patches in the induction of an immune response [26, 27].

Lamina Propria

The lamina propria is comprised of loosely packed connective tissue forming the supportive structure for the villi and also contains the blood supply, lymph drainage, and nervous system for the mucosa [1]. In addition, the lamina propria contains many cells of the innate and adaptive immune system: dendritic cells (DCs), macrophages, eosinophils, mast cells, B cells, and T cells (Fig. 5). $CD4^+$ and $CD8^+$ T cells are found in a ratio of approximately 2:1 with the majority displaying an effector memory phenotype [28]. The $CD4^+$ population of T cells in the lamina propria is very diverse with T helper (Th) 1, T_h2, T_h17, forkhead box P3-positive (Foxp3) regulatory T cells (T_{reg}), and Foxp3-negative (T_R1) subsets present. T_h17 cells decrease in number from the duodenum to the colon, whereas T_{reg} numbers are highest in the colon. Numbers of T_h1 and T_h2 cells have been reported to be consistent along the length of the intestine [29].

The lamina propria also contains large numbers of B cells and plasma cells, with the highest density at the proximal and distal ends of the intestinal tract [30]. IgA plasma cells are the most abundant throughout the intestine; with the frequency increasing from approximately 75% of plasma cells in the duodenum to 90% in the colon. The majority of the remaining plasma cells secrete IgM [1].

The most abundant leukocyte in the healthy lamina propria is the macrophage which is present in the small and large intestine. Intestinal macrophages are involved in maintaining intestinal homeostasis, phagocytosing, and degrading microorgan-

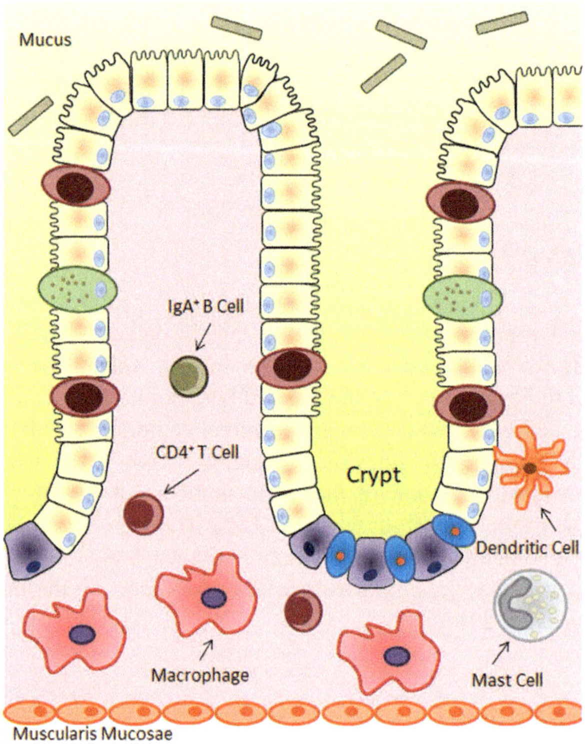

Fig 5 Immune cells of the lamina propria. The lamina propria is comprised of loosely packed connective tissue that forms a supportive structure for the villi and contains the mucosa blood supply, lymph drainage, and nervous system. The lamina propria cell population is variable containing cells of the innate immune system (DCs, macrophages, eosinophils, and mast cells) and adaptive immune system (T lymphocytes ($CD4^+$ T_h1, T_h2, T_h17, T_{reg}, and T_R1; $CD8^+$) and B lymphocytes (B cells and IgA- and IgM-producing plasma cells)) (Adapted from Mowat et al. [1])

isms and dead cells while releasing factors that stimulate renewal of the epithelium. Lamina propria macrophages produce large quantities of IL-10, an anti-inflammatory cytokine that promotes the survival and function of $Foxp3^+$ T_{reg} cells [31, 32].

Intestinal DCs are characterized by the expression of the markers $CD11c^+$ and MHC II^+ and the absence of CD64 and F4/80. Four main subsets of DCs have been described in the mouse lamina propria and are classified on the basis of their expression of CD103 and CD11b [33]. These subsets have been suggested to have distinct roles in intestinal immune homeostasis. DCs continuously migrate from the gut to the draining lymph nodes, a process required for tolerance induction to "self" or food antigens [34]. Following antigen stimulation DCs can activate T cells to induce the appropriate inflammatory response, promote T cell differentiation into effector or regulatory subsets, and induce B cell isotype switching as well as mucosal homing properties to T and B cells [34].

Mast cells are present throughout the healthy intestine and account for 2–3% of lamina propria cells under normal conditions [35]. They produce factors that mediate epithelial barrier integrity, peristalsis, vascular tone, and permeability [1]. Furthermore, mast cells express toll-like receptors suggesting that these cells can also detect microorganisms and/or the associated structures [35]. Eosinophils are believed to have a role in tissue repair in both physiological and pathological conditions in both the small and large intestines. In addition, studies have suggested that eosinophils may have a role in maintaining IgA^+ plasma cell numbers, DCs, and $Foxp3^+$ T_{reg} cells in the lamina propria [36].

Draining Lymph Nodes

The intestinal lymph nodes are the largest in the body. In rodents, different segments of the intestine drain to specific nodes; the duodenum primarily drains to a small node located in the pancreatic tissue; the jejunum drains to the middle mesenteric lymph node (MLN); the distal ileum, cecum, and ascending colon drain to the distal MLN; the traverse colon drains to two small nodes buried in the pancreas; and the descending colon and rectum drain to the caudal lymph node. Similar region-specific differences in lymph drainage are also observed in humans [1]. The intestinal lymphatic system plays a key role in acquired immunity and tolerance; recirculating lymphocytes pass from the blood to tissue and then to the lymph nodes before returning to the circulation [37]. Studies have revealed that the MLNs are key sites for the induction of tolerance to food and commensal bacteria antigens. Antigen-laden DCs migrate from the intestinal epithelial barrier to the MLNs where they can "educate" MLN-resident T cells. In addition, the MLNs act as a "fire wall" stopping commensal bacteria-laden DCs from entering the thoracic duct lymph system and reaching the systemic circulation. Bacteria that may be released by DCs upon cell death are eliminated by MLN macrophages [38].

Inflammatory Intestinal Pathologies

Dysregulation of the intestinal immune system is involved in a number of serious pathological conditions including celiac disease, Crohn's disease, and ulcerative colitis. While celiac disease is localized to the duodenum, Crohn's disease is most common in the distal small intestine and ulcerative colitis in the colon [39, 40]. The specific cytokines dysregulated in the two major forms of inflammatory bowel disease, Crohn's disease and ulcerative colitis, are also different. Crohn's disease is driven by a T_h1/T_h17 response in which IL-12, IL-23, IL-17, and IFNγ play key roles, while ulcerative colitis is driven by a T_h2-like response in which IL-5 and IL-13 play a key role [41, 42]. Interestingly, these intestinal pathologies are also associated with bone pathology with a significant decrease in trabecular bone leading to the development of osteoporosis [43, 44].

Osteoimmunology: The Bone Immune System

Bone remodeling is a coordinated process that maintains the rate of osteoblast (OB) bone formation and osteoclast (OC) bone resorption, ensuring the maintenance of skeletal integrity. Under normal physiological conditions, the osteoblast and osteoclast engage in cross talk, with each cell able to regulate the differentiation and action of the other. The process of canonical osteoclast differentiation is regulated by the action of three critical cytokines released by the osteoblast: macrophage colony-stimulating factor (MCSF), receptor activator of nuclear factor kappa-B ligand (RANKL), and the soluble RANKL receptor osteoprotegerin (OPG) [45]. Conversely, the osteoclast can modulate osteoblast differentiation through the secretion of coupling factors including BMP-6 and Wnt10b [46, 47] and membrane-bound components including ephrin/Eph family members [47] (Fig. 6). In addition

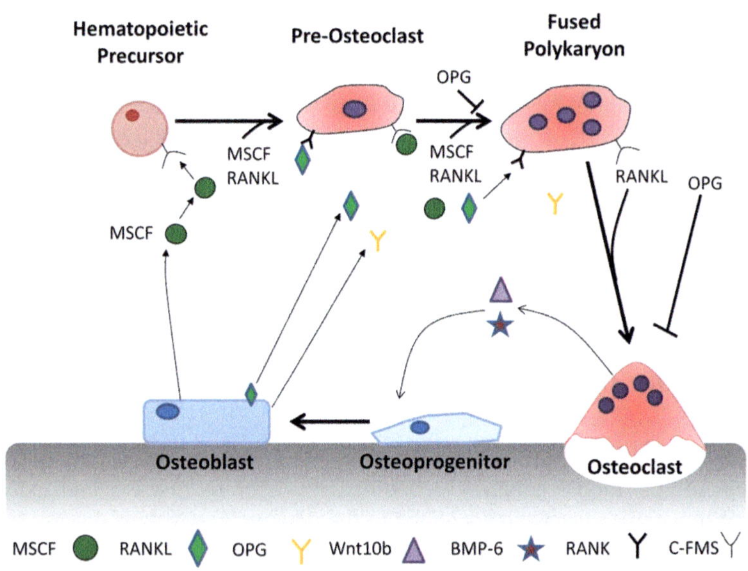

Fig. 6 Osteoblast and osteoclast cross talk. Osteoblasts regulate osteoclast differentiation and activity through the expression of three critical cytokines: macrophage colony-stimulating factor (MCSF), receptor activator of nuclear factor kappa-B ligand (RANKL), and osteoprotegerin (OPG), a soluble decoy receptor for RANKL. Conversely, osteoclasts can modulate osteoblast differentiation through expression of the osteogenic factors Wnt10b and bone morphogenetic protein (BMP)-6

to the cross talk between bone cells, the immune system has been demonstrated to play a significant role in regulating bone remodeling. This section will discuss the role of the immune system in the regulation of bone cell activity.

Immune Cells and Bone

T Lymphocyte Regulation of Bone

As discussed before, T cells are classed into two main categories based on the subunits of the T cell receptor (TCR) they express: $\alpha\beta$ or $\gamma\delta$. The majority of T cells express the $\alpha\beta$ subunits and can be further divided based on the expression of the CD4 and CD8 markers [48]. The $CD4^+$ T cells can additionally be subdivided based on their cytokine profiles: T_h1, T_h2, T_{reg}, and T_h17. While T lymphocytes are critical mediators of the adaptive immune response, they are also capable of modulating bone remodeling.

Under physiological conditions T cells are not major contributors to the levels of RANKL because T cell-deficient nude mice do not exhibit diminished RANKL mRNA in their BM [49]. However, resting T cells are demonstrated to have a bone-protective role, stimulating B-lymphocyte-mediated OPG production through CD40-CD40L interaction [49]. Consistent with this, T cell-deficient nude mice have baseline OC numbers and reduced bone density compared to controls [49, 50]. In addition, CD4/CD8 T cell-depleted mice exhibit significantly elevated in vitro osteoclastogenesis [51]. In contrast, activation of T cells by infection and inflammation leads to an increase in T cell-derived osteoclastogenic factors such as RANKL and tumor necrosis factor alpha (TNFα), subsequently resulting in increased bone resorption as seen in conditions such as osteoporosis, inflammatory arthritides, and periodontitis [52]. It should be noted, nonetheless, that distinct T cell subsets have differing functions when it comes to regulating bone turnover (Fig. 7).

$CD4^+$ T Cell Subsets and Their Effects on Bone

Following antigen stimulation naïve CD4 T cells proliferate and undergo differentiation into distinct effector subsets that are classified according to their cytokine profiles. T helper 1 (T_h1)cells are characterized by the expression of TNFα, interferon gamma (IFNγ), and interleukin (IL)-2; T_h2 cells produce IL-4, IL-5, IL-6, IL-9, IL-10, and IL-13; Th17 cells produce IL-17A, IL-17F, IL-21, and IL-22; and T_{regs} express CD25 and Foxp3 and produce IL-10 and TGFβ [53, 54]. Of these subsets T_h1, T_h2, and T_{reg} cells suppress osteoclast formation, while T_h17 cells stimulate osteoclastogenesis. Inhibition by Th1 and Th2 is in part dependent on the number of T cells as well as on the expression of IFNγ and IL-4 [55].

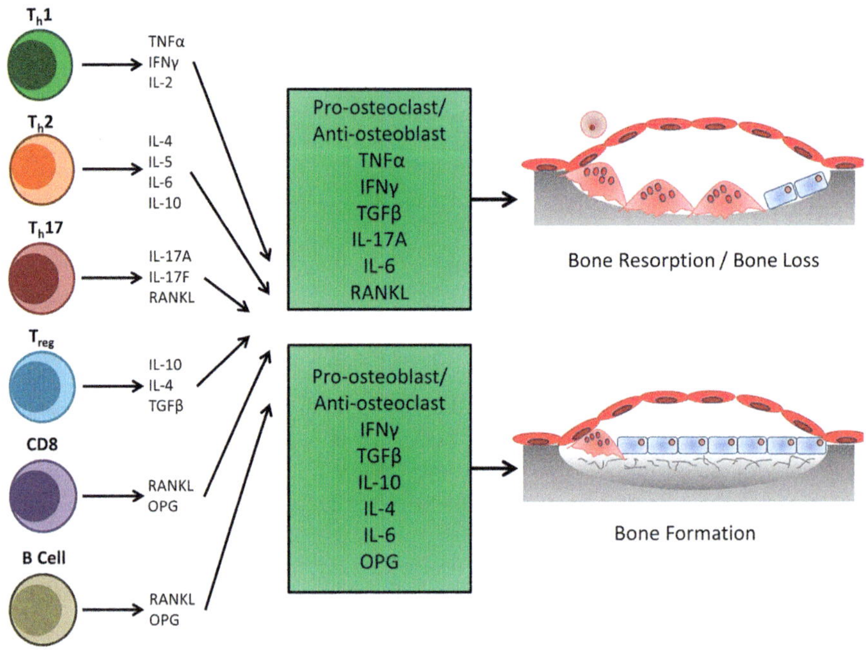

Fig. 7 Lymphocyte modulation of bone remodeling. Bone marrow lymphocytes can modulate the differentiation and function of osteoclasts and osteoblasts through the expression of pro- and anti-inflammatory cytokines as well as RANKL and OPG

T_{regs} can modulate osteoclast formation via direct cell-to-cell contact as well as through cytokine production [56]. Mechanism of direct cell contact is through cytotoxic T-lymphocyte antigen 4 (CTLA-4) expressed on Tregs [56]. Furthermore, addition of CTLA-4 to osteoclast precursor cells dose dependently suppresses osteoclast formation, a process reversed by anti-CTLA-4 antibodies [56]. These data suggest that CTLA-4 may directly suppress osteoclastogenesis by binding to CD80/CD86 on mononuclear OC precursors [57]. In addition to the direct suppressive effect of T_{regs} on osteoclast differentiation, expression of TGFβ, IL-4, and IL-10 was observed to contribute to the inhibitory effect on osteoclastogenesis [56]. The role of these cytokines in the mechanism by which T_{regs} inhibit osteoclast formation is supported in studies by Luo et al. [58] and Kim et al. [59] who demonstrated that CD4 T_{regs} suppressed osteoclast formation from peripheral blood mononuclear cells (PBMCs) via the expression of the cytokines TGFβ, IL-4, and IL-10.

Compared to the above T helper subsets, T_h17 cells primarily support osteoclast differentiation through the production of IL-17A and are believed to play a crucial role in inflammation and the development of autoimmune diseases such as rheumatoid arthritis. Furthermore, T_h17 cells express RANKL and TNFα which can directly act on osteoclast precursors to induce osteoclastogenesis [55, 60, 61].

$CD8^+$ T Lymphocyte Regulation of Bone

Activated $CD8^+$ T cells express high levels of RANKL and OPG. When cultured with osteoclast precursors, $CD8^+$ T cells, however, inhibit soluble RANKL-induced osteoclast formation. Interestingly, blocking OPG with an antibody does not reverse the inhibitory effect of the $CD8^+$ T cells suggesting the involvement of other factors [62]. The bone-protective role of $CD8^+$ T cells has also been reported by John et al. [63] who revealed that depletion of $CD8^+$ T cells from murine bone marrow increases osteoclast differentiation.

$CD8^+$ T cells also play a key role in the anabolic effect of parathyroid hormone (PTH). Treatment of mice with intermittent PTH (iPTH) significantly increases the production of Wnt10b by bone marrow $CD8^+$ T cells subsequently activating canonical Wnt signaling in osteoblasts [64]. $CD8^+$ T cell-dependent anabolic effect of PTH was demonstrated in T cell-null mice which displayed diminished Wnt signaling in pre-osteoblasts and blunted osteoblastic commitment, proliferation, differentiation, and life span [64]. In addition, conditional silencing of PTH receptor 1 in T cells blunted the ability of iPTH to induce T cell Wnt10b production, activate osteoblastic Wnt signaling, and increase bone turnover [65].

$CD8^+$ T Regulatory Cells in Bone Health

In contrast to $CD4^+$ T_{regs}, $CD8^+$ T_{regs} have not been extensively studied partly due to their low abundance in lymphoid tissues [57], though like their CD4 counterparts they express both CD25 and the transcription factor Foxp3 [66]. In vitro, the osteoclast-induced $Foxp3^+$ $CD8^+$ T_{reg} cells potently and directly suppressed osteoclast differentiation and resorptive activity, but did not affect mature osteoclast survival. This suppressive effect of $Foxp3^+$ $CD8^+$ T_{regs} was relieved following antibody blockade against IL-6, IL-10, and IFNγ [67]. Using adoptive transfer experiments, $Foxp3^+$ $CD8^+$ T_{regs} were shown to limit the bone loss induced by RANKL administration as well as protect against ovariectomy-induced bone loss [68].

B Lymphocytes and Their Regulation of Bone Cells

B cells have a close and multifaceted relationship with bone cells [69]. B cells are differentiated from hematopoietic stem cells (HSCs) in niches found on the endosteal bone surface [57, 69]. These niches are supported by osteoblasts which sustain the HSCs and B cell differentiation [69]. The role of B cells in normal bone remodeling is not fully clear; however, cells of the B cell lineage have been demonstrated to be responsible for 64% of total BM OPG production, with 45% derived from mature B cells [49]. In support of a role for B cells in physiological bone remodeling, B cell knockout (KO) mice have an osteoporotic phenotype, due to elevated osteoclast bone resorption and decreased BM OPG expression, a phenotype rescued by B cell reconstitution [49]. In addition to a role in physiological bone remodeling, activated B cells have been implicated in many inflammatory diseases associated with osteoporosis, such as estrogen deficiency, multiple myeloma, and periodontal disease [62, 70].

In conditions of estrogen depletion, B lymphogenesis is increased. Consistent with this, estrogen treatment downregulates B lymphogenesis [71, 72]. Furthermore, B cells isolated from the bone marrow of postmenopausal women have also been shown to express RANKL [73] suggesting a role for B cells in menopausal bone loss. This is further supported in a study by Onal et al. [74] using B cell-specific RANKL KO mice. These KO mice were partially protected from ovariectomy-induced bone loss. This protection occurred in the trabecular bone, but not the cortical bone, and was associated with a failure to increase osteoclast numbers. Conversely however, in an in vitro model of osteoclastogenesis, B cells isolated from the peripheral blood inhibited OC formation via the secretion of TGFβ, a cytokine that induces OC apoptosis [75], suggesting that the role of B cells in estrogen deficiency-induced bone loss may not be clear-cut.

In multiple myeloma B cells and B cell-derived plasma cells have been reported to support osteoclastogenesis directly through expression of RANKL and indirectly through expression of IL-7 [76, 77]. In addition, malignant B cell-derived plasma cells have been shown to secrete cytokines such as sclerostin and DKK1 which inhibit osteoblastogenesis [77, 78], affecting the balance of remodeling on two fronts. In human chronic periodontitis, B lineage cells (B lymphocytes and plasma cells) have been revealed to dominate the inflammatory infiltrate suggesting that these cells play a key role in the associated bone loss [79]. This is supported by murine studies in the ligature-induced periodontitis model and the *P. gingivalis* oral infection model, where reduced bone loss was observed in B cell-deficient mice (μMT) compared to wild-type controls [80, 81]. Furthermore, antibody neutralization of a proliferation-inducing ligand (APRIL) or B-lymphocyte stimulator (BLyS), two cytokines important for the maturation and survival of B cells, diminished the number of B cells in the gingival tissue and inhibited bone loss in wild-type but not in B cell-deficient mice [80].

Regulation of Bone by Neutrophils

Neutrophils are implicated with bone loss associated with murine models of inflammatory arthritis [82–84] and human periodontitis [85]. Human and murine neutrophils strongly upregulate their expression of RANKL following LPS treatment, stimulating osteoclastic bone resorption through direct cell-cell contact with osteoclasts [86] (Fig. 8).

Role of Dendritic Cells in Bone Health

Dendritic cells (DCs) are broadly composed of two main subsets: conventional dendritic cell (cDC) and plasmacytoid dendritic cell (pDC). The main function of cDCs is to process and present antigens to antigen-specific naïve T cells, whereas the function of pDCs is the secretion of IFNα/IFNβ in response to viral infection [87]. While DCs do not appear to play a role in physiological bone remodeling in chronic inflammatory conditions, such as RA, immature cDCs have the potential to serve as osteoclast precursors [88, 89]. These DC-derived osteoclasts are functionally active and able to resorb bone [90, 91] (Fig. 8).

Immune Cytokines and Bone

TNFα

TNFα can affect bone remodeling in multiple ways, acting on the osteoclast and the osteoblast [92]. TNFα increases osteoclast formation by upregulating osteoblast expression of RANKL and MCSF [93, 94]. Furthermore, TNFα can act on

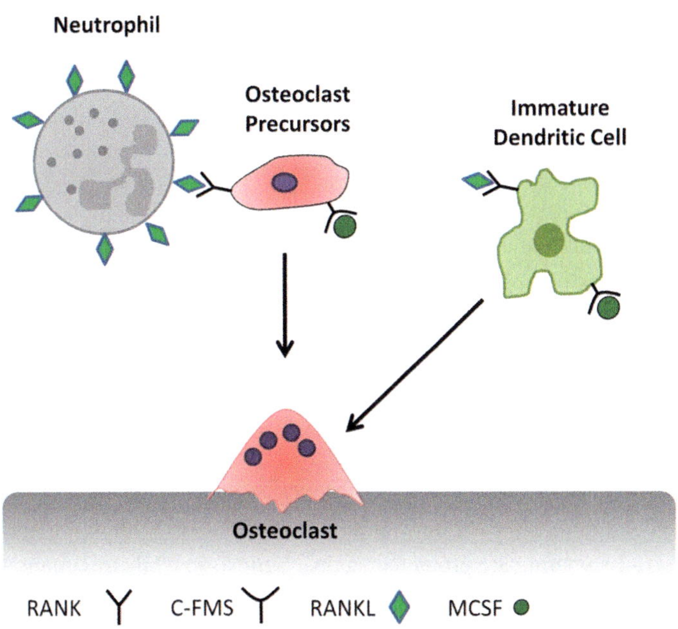

Fig. 8 Role of myeloid cells in osteoclast differentiation. Under inflammatory conditions neutrophils upregulate expression of RANKL which can stimulate osteoclast differentiation. Furthermore, following stimulation with RANKL and MCSF, immature dendritic cells can transdifferentiate into functional osteoclasts

osteoclast precursors in permissive levels of RANKL inducing their differentiation by enhancing its effects [95–97]. Blockade of TNF activity in bone pathological conditions such as inflammatory arthritis and estrogen deficiency has been shown to protect against the associated bone loss [98, 99]. In addition to stimulating osteoclast formation and bone resorption, TNFα also inhibits osteoblast proliferation, differentiation, and matrix deposition [100, 101]. Moreover, TNFα is proapoptotic for osteoblasts, potentially through Fas-FasL signaling [102, 103], though the significance of this pathway in osteoblast apoptosis has been disputed in recent studies [104]. Interestingly however, low levels of TNF have been revealed to induce osteogenic differentiation via upregulation of the transcription factors Runx2 and Osx [92, 105] demonstrating a complex role for TNFα in the regulation of bone remodeling.

Other TNF Superfamily Members

TNF-like protein 1A (TL1A) has been demonstrated to enhance RANKL-induced osteoclastogenesis in murine and human models [106, 107]; furthermore, TL1A has been linked to the adverse bone pathogenesis observed in inflammatory arthritis [106–108]. Studies have also suggested that TL1A, signaling through death receptor 3 (DR3), may affect osteoblast differentiation and mineralization [109].

LIGHT (homologous to lymphotoxins exhibiting inducible expression and competing with herpes simplex virus glycoprotein D for herpesvirus entry mediator [HVEM], a receptor expressed by T lymphocytes) is elevated in conditions such as RA and multiple myeloma [110, 111]. In vitro, LIGHT has been reported to induce osteoclastogenesis in RANKL-dependent and RANKL-independent manners and indirectly inhibit osteoblastogenesis [110, 111].

TNF-related apoptosis-inducing ligand (TRAIL) is another member of the TNFSF that has biphasic effects on bone [112]. TRAIL has been shown to induce apoptosis of osteoclasts but not osteoblasts [113, 114]. In addition, in vivo injection of TRAIL for 8 days in 4-week.-old mice induces an increase in bone mass [115].

Interleukin-1

IL-1 plays an important role in pathological bone loss as seen in conditions such as RA and estrogen deficiency. In murine models of menopause (ovariectomy), mice deficient in IL-1 signaling are protected against bone loss, while treatment with IL-1 receptor antagonists reduces osteoclast formation and activity [116, 117]. In the collagen-induced arthritis model, anti-IL-1β and anti-IL-1R treatment reduced arthritis severity and protected against focal bone erosions [118]. IL-1 acts on the bone cells in a number of ways [119]; IL-1 enhances RANKL expression from osteoblasts/stromal cells [93], indirectly stimulates osteoclast differentiation by stimulating prostaglandin E_2 synthesis in osteoblasts [120], induces osteoclast precursor fusion into the multinucleated osteoclast [121], enhances osteoclast resorptive activity [121], and directly affects the survival of osteoclasts [122, 123].

Interleukin-6

The role of IL-6 in osteoclast bone resorption is currently debated with studies revealing that it can either stimulate bone resorption, inhibit osteoclast differentiation, or has no effect depending on the model used [124–126]. However, in menopausal women, levels of IL-6 are elevated, while in the murine OVX model, IL-6 has been linked to increased osteoclast development [127]. The role of IL-6 in estrogen deficiency-induced bone loss is further supported in the IL-6 knockout mouse which is protected against development of osteoporosis [128]. The role for IL-6 in bone loss has also been reported in diseases such as Paget's disease and rheumatoid arthritis [129, 130].

Interleukin-7

IL-7 is an osteoclastogenic cytokine, produced in the bone marrow primarily by stromal cells and osteoblasts, which is able to stimulate B and T cells [131]. IL-7 promotes osteoclastogenesis by upregulating T and B cell-derived RANKL and TNFα [50, 132]. In the OVX model, increased production of IL-7 has been postulated to be a key factor in the uncoupling of bone formation from bone resorption [131].

Interleukin-10

IL-10, an anti-inflammatory cytokine, is mainly expressed by activated T and B cells. IL-10 has potent inhibitory effects on osteoclastogenesis both direct and indirect. IL-10 directly suppresses osteoclast formation by reducing nuclear factor of activated T cells, cytoplasmic 1 (NFATc1) expression, and its translocation to the nucleus [133] and by suppressing c-Fos and c-Jun [134]. Indirectly IL-10 has been shown to modulate osteoclast formation by increasing gene expression of OPG and decreasing RANKL gene expression in dental follicle cells [135].

Interleukin-17A

IL-17A acts on osteoclasts via both direct and indirect mechanisms. In an in vitro system, IL-17A has been shown to directly act on human osteoclast precursors, upregulating expression of RANK and rendering the cells more sensitive to the actions of RANKL [136]. In addition, IL-17A elevates stromal cell expression of RANKL and synovial macrophage expression of the osteoclastogenic factors TNFα and IL-1β, indirectly effecting osteoclastogenesis [137]. Conversely, IL-17A has been demonstrated to exhibit strong osteogenic effects when exposed directly to mesenchymal stem cells (MSCs) acting synergistically with bone morphogenetic protein 2 (BMP-2) [138].

Interferon-γ

The effect of IFNγ on osteoclastogenesis is controversial. In vitro, IFNγ has been reported to inhibit osteoclastogenesis [139, 140] and downregulate expression of cathepsin K [141]. Conversely, IFNγ has also been reported to indirectly stimulate osteoclast resorptive activity by enhancing RANKL and TNFα production by T cells [140]. In vivo, treatment of T cell-deficient nude mice with recombinant IFNγ had no effect on bone density; however, adoptive transfer of wild-type T cells into these mice resulted in a significant decrease in vertebral bone mineral density [140].

TGFβ

TGFβ has a complex role in bone remodeling and is thought to have a key contribution in the coupling of the osteoblasts and osteoclasts. TGFβ is expressed in a latent form by numerous cell types, including osteoblasts, stromal cells, and bone marrow cells that must be activated to exert its effect [57]. TGFβ has been reported to stimulate osteoclastogenesis [142–144] and induce osteoclast apoptosis [145, 146]. Likewise in osteoblasts, TGFβ has been reported to suppress differentiation [147], inhibit mineralization [148], induce osteoblast precursor migration and proliferation [149], and block apoptosis [150].

Immune Involvement and Gut-Bone Signaling in Pathophysiology

Celiac Disease: Immune System and Bone Responses

Celiac disease is an enteric autoimmune disorder that is triggered by ingestion of gluten, a protein present in wheat, rye, and barley [44]. Some of the typical gastrointestinal (GI) symptoms of the disease include diarrhea, abdominal pain, distention, weight loss, and failure to thrive in children. Other symptoms may also be present with or without GI symptoms, including anemia, low bone mineral density, elevated liver enzymes, unexpected weight loss, prolonged fatigue, and infertility [44].

Celiac disease can severely affect bone health in patients, causing metabolic bone diseases such as low bone mass, osteoporosis, secondary hyperparathyroidism, and osteomalacia [44]. Multiple mechanisms can be attributed to the development of these bone disorders. Mucosal atrophy in the intestinal lumen due to celiac disease impairs calcium absorption. This causes a significant increase in parathyroid hormone levels that stimulates bone degradation by osteoclasts in an effort to avoid hypocalcemia. Calcium loss from the bones can lead to osteopenia and osteoporosis, increasing the fracture risk. Celiac disease can also cause hypogonadism, which can subsequently have an adverse effect on the bone [44].

Inflammation from celiac disease is characterized by villus atrophy, crypt cell hyperplasia, and infiltration of T lymphocytes in epithelial and lamina propria compartments [151]. The recruited active T cells produce high levels of pro-inflammatory cytokines that can eventually damage intestinal tissue [151]. Studies have revealed that celiac disease is associated with exacerbated production of Th1 cytokines (e.g., IFNγ and IL-21). In addition, expression of IL-17A, IL-1β, IL-6, IL-8, and IL-15 has been shown to be upregulated in patients with active celiac disease. TNFα and IL-6 are elevated, however, in refractory celiac disease (not responsive to a gluten-free diet) [151]. This elevation in serum pro-inflammatory cytokines observed in celiac disease could directly influence bone remodeling by stimulating osteoclastogenesis or indirectly by increasing the RANKL/OPG ratio [44, 152, 153]. The area most affected by bone remodeling in celiac disease is the trabecular bone, which is the most metabolically active bone section [44]. Increased bone resorption in patients with celiac disease leads to thinner and fewer trabeculae, thus resulting in increased fracture risk [44].

Inflammatory Bowel Disease: Intestinal Immune Responses

Inflammatory bowel disease (IBD) is characterized by chronic recurrent inflammation of all or part of the gastrointestinal (GI) tract. The most common types of IBD are Crohn's disease (CD) and ulcerative colitis (UC). Crohn's disease can affect the GI tract anywhere from the mouth to the anus, with the most common area affected being the ileum. Ulcerative colitis, however, is specific to the colon [154, 155]. In 2014 IBD was reported to affect 1.6 million Americans, and as many as 70,000 new cases are reported per year (CCFA). The immune system plays a critical role in the pathogenesis of IBD, with the adaptive immune system thought to be one of the main contributors to IBD pathogenesis [156]. Pro-inflammatory cytokine-producing T-helper cells are thought to be overactive, whereas anti-inflammatory T_{regs} are ineffective at resolving the inflammation [156].

At the onset of IBD, weakening of the epithelial barrier promotes increased permeability resulting in increased contact between the intestinal microbiota and the mucosal immune system. This excessive contact causes massive influx of neutrophils as well as loss of tolerance by activating dendritic cells [157]. In addition to neutrophils, IBD pathology is also associated with an influx of lymphocytes into the mucosa [156]. These cells produce vast quantities of pro-inflammatory cytokines [41, 42, 158]. Increased activation especially of the IL-1 pathway is more active during the onset of inflammation rather than during the sustained phase, especially in the colon [42]. Members of the IL-12 family of heterodimeric cytokines including IL-12, IL-23, IL-27, and IL-35 have been shown to be upregulated in antigen-presenting cells [42, 158], while TNFα has been shown to be enhanced in mononuclear cells. The effects of TNF are suggested to be twofold: promoting inflammation by binding to its receptors as well as causing cell death through activation of receptor-interacting protein kinase 1 and caspase 3 proteins [42, 158].

Among the many immune cell types involved in the pathogenesis of IBD, T lymphocytes are thought to be critical [42]. Depending on the type of IBD, involvement of the specific cytokine also differs. In Crohn's disease, Th1-mediated increases in IFNγ and IL-2 are thought to be important. In contrast, in ulcerative colitis Th2-mediated IL-5 and IL-13 have been shown to be critical in the pathogenesis. However, in UC nonclassical natural killer T cells have also been identified to have an atypical cytokine response [42]. This response has been linked to fibrosis as well as altered tight junction function and apoptosis of intestinal epithelial cells (IECs), driving intestinal epithelial cell ulceration. This response from the NKT cells has been found to be mainly mediated through IL-13 [42].

Regulatory T cells (T_{regs}) ($CD4^{+}CD25^{+}FoxP3^{+}$) have been shown to have a protective role in colitis, through expression of the anti-inflammatory cytokines IL-10 and transforming growth factor beta (TGFβ) [42]. In CD and UC patients, T_{reg} numbers are increased in the inflamed lamina propria compared to uninflamed mucosa and healthy controls. Why these cells fail to control the inflammation is currently unknown; however, it has been suggested that effector T cells are resistant to the effects of inhibitory cytokines such as TGF-β in the context of inflammation [40].

Inflammatory Bowel Disease: Bone Immune Responses

IBD is associated with significant and divergent responses in lymphoid and myeloid hematopoiesis in the bone marrow. In a study by Trottier et al. [159], the dextran sodium sulfate (DSS)-induced model of colitis resulted in a 60% and 80% increase in monocytic and granulocytic lineages, respectively, consistent with the massive influx of monocytes and neutrophils into the gut. In contrast, the early lineages of B and T cells declined in the marrow and thymus. Furthermore, a significant 40% decrease was also observed in cells of erythrocytic lineage. Because bone marrow is present within the bone structure, it is conceivable that intestinal inflammation such as what is seen in IBD can have profound influences not only on bone marrow composition but also on bone cells themselves, and this is aided by changes in and influence of inflammatory cytokines. Previous studies in animal models of IBD demonstrate that inflammatory cytokines such as TNFα, IL-1α, IL-1β, IL-6, IL-11, and IL-17A are linked to decreased bone volume, bone formation rates, and osteoid surface [154, 160–164]. In some studies, osteoclast surface is increased which further exacerbates bone loss [160]. Patients with IBD have elevated levels of pro-inflammatory osteoclastogenic cytokines including TNFα, IL-1, IL-6, IL-11, IL-17, TGFα, epidermal growth factor, and prostaglandin E2 [154, 161]. These osteoclast activators presumably act on the osteoclasts and their precursors promoting osteoclast formation and osteoclast bone resorption. In addition, some of these cytokines can inhibit osteoblast differentiation and mineralization. The dual effects of inflammation lead to a substantial decrease in bone density subsequently resulting in osteoporosis. For a detailed review of how IBD affects bone health, please see elsewhere in this book (chapter 7 – Francisco Sylvester).

Enteritis

Pathogenic enteric infections may lead to drastic nutrient loss due to sustained diarrhea. These nutrient losses have been associated with a decrease in linear growth observed in children, where such infections are common [165]. However, intestinal pathogens can also potentially affect bone growth through pathways other than reduced nutrient absorption. The immune response initiated by an intestinal pathogen may affect the process of long bone growth in children [165].

To study the response of the intestinal immune system to enteric pathogens, Noel et al. treated primary human macrophage-enteroid coculture with enterotoxigenic and enteropathogenic *Escherichia coli* strains (ETEC and EPEC) [166]. ETEC is the most common pathogen that causes traveler's diarrhea, while EPEC is associated with cases of fatal diarrhea in children [167, 168]. EPEC infection resulted in increased macrophage adherence to and projection across a filter into the enteroid monolayer, simulating an immune response where macrophages extend between enterocytes to interact with the pathogen. Macrophages present in ETEC infection also interacted with the bacteria through the filter and even reduced viable ETEC number, suggesting successful phagocytic action. In the absence of macrophages, overnight infection resulted in a loss of barrier function. However, their presence appeared to lessen this effect, further elucidating their protective role in such bacterial infections. While changes in intestinal cytokine expression levels of IFNγ, IL-8, and IL-6 have previously been implicated in affecting bone health [169], expression of these cytokines did not change in ETEC infection [166]. This is further supported by the finding that ETEC infection is not associated with impaired linear growth [170]. In contrast, enteroaggregative *E. coli* (EAEC) has been demonstrated to impair growth in infected children [171]. Infected children showed a significant decrease in height when compared to the population. Interestingly, the decline in growth occurred regardless of diarrhea, discounting any effect of major nutrient loss. To examine a possible inflammatory mechanism, IL-8 levels were measured from stool samples and from EAEC-treated Caco 2 cells. IL-8 was elevated in children with persistent diarrhea and was shown to be increased upon exposure to two EAEC strains in in vitro studies [171]. Considering previous studies on the role of IL-8 and bone [169], it is possible that an increase in this pro-inflammatory cytokine induced by EAEC infection results in reduced bone growth.

Furthermore, sera TNFα and IL-10 are elevated during enteritis from several bacterial species [172]. In rotavirus and norovirus infections, IL-6 and IL-8 have been shown to increase according to the severity of disease [173]. Although these studies do not reveal a bone phenotype in relation to the infection, these cytokines have been related to bone formation in other studies. For example, the combination of TNFα and IL-6 has been shown to induce bone resorption via differentiation of osteoclast-like cells [174]. The role of TNFα in promoting osteoclast formation is well established [95]. In a model of childhood inflammatory bowel disease, a decrease in the proliferative zone of the tibia was observed and correlated with an increase in serum IL-6 concentration [175]. It is possible that the reduction in bone growth seen in affected children is due to the activity of these cytokines. While pathogenic intestinal infection invokes immune responses such as increased macrophage projection and release of inflammatory cytokines, the associated decline in bone growth observed in some pathogenic infections in children is not clearly understood. Studies that measure bone parameters in addition to measuring inflammatory markers and other immune system responses are needed to accurately detail any effect of infection, if any, on bone health.

Dysbiosis and Gut-Bone Immune Link

Dysbiosis is characterized as "a compositional and functional alteration in the microbiota that is driven by a set of environmental and host-related factors that perturb the microbial ecosystem to an extent that exceeds its resistance and resilience capabilities" [176]. This imbalance of "good" versus "bad" microbiota can be caused by various environmental factors: genetics of the host, diet, obesity, disease, or medical intervention [177]. The use of antibiotics, for example, disrupts the composition of the gut microorganisms by reducing diversity and deleting or expanding bacterial communities [178]. Specifically, streptomycin, one of the most commonly used antibiotics, reduced *Firmicutes* and depleted *Lactobacillus* spp., group D *Streptococcus* spp., and *Enterococcus* spp. but expanded *Enterobacteriaceae* [179]; this demonstrates that the use of antibiotics can lead to dysbiosis. In addition to drug intervention, studies have shown that dysbiosis is observed in patients with a wide variety of diseases including inflammatory bowel disease (Crohn's disease and ulcerative colitis) [180], asthma [181], and type II diabetes [182]. Diet also plays a role in altering the composition of the gut microbes. According to Zhang et al. [183], mice fed with a high-fat diet had a reduction in *Bifidobacteria* spp., while a high-fat and high-sugar diet increased the amount of *Clostridium innocuum, Catenibacterium mitsuokai, and Enterococcus* spp. in the gut [184]. In addition, obesity can also change the composition of gut microbiota. Obese humans tend to have a decreased abundance of *Bacteroidetes* compared to healthy humans [185]. Weight loss in obese humans can restore the *Bacteroidetes* population [186] indicating that obesity has a negative impact on the composition of the human gut microbiota, which is reversible. Taken together, these findings indicate that a variety of environmental factors can cause dysbiosis which can have serious consequences through altering the intestinal immune system and promoting inflammation [187]. Furthermore, these changes can have far-reaching effects as seen in bone health and the development of osteoporosis [188].

Dysbiosis also has negative influences on development and homeostasis of host tissue, including the bone [189]. Specifically, dysbiosis can lead to the promotion of osteoclast differentiation and resorptive activity through the activation of pro-inflammatory T helper 17 (T_H17) cells, resulting in reduced bone mass [190]. Moreover, Yan et al. [191] found that colonization of conventional specific pathogen-free gut microbiota in germ-free mice increased bone formation and resorption, suggesting that the gut microbiota can have a critical effect on bone remodeling. This link between dysbiosis and bone health can have important implications in infant and adult humans.

Pediatric Disease and Gut-Bone Immune Link

The importance of the gastrointestinal tract and the associated microbiome in the maturation and development of the immune system has been demonstrated in studies utilizing germ-free (GF) mice. In the absence of the microbiome, germ-free mice have (a) significantly fewer Peyer's patches that are hypoplastic and contain few germinal centers, (b) smaller mesenteric lymph nodes, and (c) reduced lamina propria $CD4^+$ T cells [192–194]. Moreover, GF mice have less than one tenth of the number of IgA-producing B cells compared to a conventional animal [193]. Interestingly, exposure of the GF animals to a conventional microbiome results in the development of a "normal" immune system, further supporting the key role of the intestinal bacteria in immune maturation [192, 193]. The gut bacteria have been demonstrated to influence the development of T_h17 cells, T_{regs} cells, and memory T cells [195, 196]. Studies have shown that segmented filamentous bacteria are required for T_h17 development [197] and that members of the *Clostridium* spp. induce colonic T_{reg} differentiation [198].

Emerging evidence supports that early life exposure to microbes may have an effect on later-life susceptibility to disease [199]. Dysbiosis of the microbiome in young children can lead to the development of pediatric Crohn's disease (CD) and ulcerative colitis (UC) [200, 201]. Patients who develop these conditions as children or young adults are at an increased risk of bone fracture. Impairment of linear growth and failure to gain weight are common in children with CD and UC. As a consequence of this, these patients have a reduced peak bone mass with significantly lower bone mineral density and bone mineral content, significantly increasing the risk of developing osteoporosis in later life [202].

Aging and the Intestinal/Bone Immune Systems

In the elderly both the bone and gut immune systems undergo age-related alterations. This age-related change in the immune system is known as immunosenescence, which contributes to the increased susceptibility of the elderly to infectious disease, cancer, and autoimmunity [203]. Immunosenescence affects various cell types located in the bone marrow and thymus, lymphocytes located in secondary lymphoid organs and the peripheral blood, as well as aspects of the innate immune system [203–205]. In bone this change can lead to bone loss resulting in deteriorated structure and function and eventually osteoporosis [206]. While in the intestine, this can result in a low-grade inflammatory status, potentially developing into a more severe pathological condition such as IBD. This can subsequently have adverse effects on bone health [207].

During aging, the hematopoietic compartment of the bone marrow decreases in size and is replaced by fatty deposits [208]. Unexpectedly the number of hematopoietic stem cells (HSCs) in the bone marrow increases with age; however, these cells exhibit several functional defects including diminished regenerative potential [209]. Aging HSCs also exhibit skewed differentiation bias with decreased output of lymphoid and erythrocyte lineage cells. In contrast, output of myeloid cells has been reported to decrease, remain the same, or increase [204, 209, 210]. However, aging has been shown to impair the ability of bone marrow macrophages to secrete cytokines such as TNFα, IL-1β, and IL-6 [210]. The number of T cell and B cell progenitors present in the bone marrow and thymus is also significantly reduced with age. The progenitors are also of an inferior quality when compared to their young counterparts as they exhibit reduced proliferative ability and increased apoptosis [211–213]. Aging has also been shown to suppress expression of IL-7 by bone marrow stromal cells, a cytokine that is essential for the survival of developing lymphocytes [205, 214]. What direct effects these changes in bone marrow immune function have on bone health is not well understood.

In addition to changes in the bone immune system, the gastrointestinal lymphoid tissue (GALT) also undergoes significant age-induced modifications which are likely a major factor in the increase in the incidence and severity of infections in the elderly [215]. Studies have observed that aging is linked to decreased Peyer's patch size; this is potentially due to impaired movement of immune cells in and out of the Peyer's patches [216]. Studies by Ogino et al. [217] and Schmucker et al. [218] report reduced immune cell migration with age. Further evidence for age-induced changes in the gastrointestinal immune systems is seen in the reduced number of intraepithelial regulatory-type T cell subsets and reduced expression of TGFβ and IL-10 [219]. Whether these effects of aging on the gut immune system directly affect bone marrow immune function and if this impacts bone health negatively is not well known.

The Effect of Antibiotics on Gut-Bone Immune Signaling

The use of antibiotics in the past 80 years has saved millions of human lives from bacterial infection. However, oral antibiotic treatment kills both pathogenic and commensal microbes. Commensal microbes have been shown to perform an array of important functions in the human body including regulation of bone health [220, 221]. Thus it is conceivable that oral antibiotics can impair gut immune function that can negatively influence bone health. While there is evidence for this, the mechanisms of how antibiotic treatment alters the interaction between the microbiome and the immune system are not well understood. There have been relatively few studies that have investigated the effect of antibiotic depletion of the gut microbiome on bone health. In a study by Cho et al. [222], weaned female C57BL/6 mice were treated with antibiotics (penicillin, vancomycin, chlortetracycline) for 3 weeks and displayed an increase in total body bone mineral density (BMD) compared to the untreated mice. Interestingly, after 7 weeks of treatment, BMD was comparable between the treated and untreated groups [222]. Cox et al. [223] treated C57BL/6 male and female mice with low-dose penicillin for 20 weeks (either before birth or from weaning) and found a small but significant increase in bone mineral content

(BMC) and BMD in female mice. In contrast, males, given antibiotics before birth, exhibited a significant decrease in BMC [223]. These effects were thought to be linked to the microbiome affecting the differentiation of stem cells into osteoblasts, adipocytes, or myocytes. Taken together, these studies emphasize the complexity of bone responses to antibiotic treatment and highlight the importance of variables such as length of treatment, mouse strain/sex and age, and potentially mouse microbiome.

Conclusion

The intestine is residence to the largest number of immune cells in the body. These immune cells are intricately associated with the intestinal microbiota, and subsequent changes in the microbiota can have both local and systemic beneficial or adverse effects. Recent studies implicate a gut-bone axis where changes to the intestine can affect distal bone health. Indeed, dysbiosis of the intestinal microbiota, either through disease, aging, or antibiotics, can alter the intestinal immune system leading to increased inflammation and subsequently resulting in bone loss and the development of osteoporosis.

References

1. Mowat AM, Agace WW. Regional specialization within the intestinal immune system. Nat Rev Immunol. 2014;14(10):667–85.
2. Helander HF, Fändriks L. Surface area of the digestive tract – revisited. Scand J Gastroenterol. 2014;49(6):681–9.
3. Savage DC. Microbial ecology of the gastrointestinal tract. Annu Rev Microbiol. 1977;31:107–33.
4. Oozeer R, Rescigno M, Ross RP, Knol J, Blaut M, Khlebnikov A, et al. Gut health: predictive biomarkers for preventive medicine and development of functional foods. Br J Nutr. 2010;103(10):1539–44.
5. Tenorio MD, Espinosa-Martos I, Préstamo G, Rupérez P. Soybean whey enhance mineral balance and caecal fermentation in rats. Eur J Nutr. 2010;49(3):155–63.
6. Sjögren K, Engdahl C, Henning P, Lerner UH, Tremaroli V, Lagerquist MK, et al. The gut microbiota regulates bone mass in mice. J Bone Miner Res. 2012;27(6):1357–67.
7. Kelsall BL, Leon F. Involvement of intestinal dendritic cells in oral tolerance, immunity to pathogens, and inflammatory bowel disease. Immunol Rev, 2005. 206:132–48.
8. Alpan O. Oral tolerance and gut-oriented immune response to dietary proteins. Curr Allergy Asthma Rep. 2001;1(6):572–7.
9. Ouellette AJ. Paneth cells and innate mucosal immunity. Curr Opin Gastroenterol. 2010;26(6):547–53.
10. Cornick S, Tawiah A, Chadee K. Roles and regulation of the mucus barrier in the gut. Tissue Barriers. 2015;3(1–2):e982426.
11. Johansson MEV, Hansson GC. Immunological aspects of intestinal mucus and mucins. Nat Rev Immunol. 2016;16(10):639–49.

12. Hansson GC, Johansson MEV. The inner of the two Muc2 mucin-dependent mucus layers in colon is devoid of bacteria. Gut Microbes. 2010;1(1):51–4.
13. Johansson MEV, Ambort D, Pelaseyed T, Schütte A, Gustafsson JK, Ermund A, et al. Composition and functional role of the mucus layers in the intestine. Cell Mol Life Sci. 2011;68(22):3635–41.
14. Cheroutre H, Lambolez F, Mucida D. The light and dark sides of intestinal intraepithelial lymphocytes. Nat Rev Immunol. 2011;11(7):445–56.
15. Shires J, Theodoridis E, Hayday AC. Biological insights into TCRgammadelta+ and TCRalphabeta+ intraepithelial lymphocytes provided by serial analysis of gene expression (SAGE). Immunity. 2001;15(3):419–34.
16. Kühl AA, Pawlowski NN, Grollich K, Loddenkemper C, Zeitz M, Hoffmann JC. Aggravation of intestinal inflammation by depletion/deficiency of gammadelta T cells in different types of IBD animal models. J Leukoc Biol. 2007;81(1):168–75.
17. Komano H, Fujiura Y, Kawaguchi M, Matsumoto S, Hashimoto Y, Obana S, et al. Homeostatic regulation of intestinal epithelia by intraepithelial gamma delta T cells. Proc Natl Acad Sci U S A. 1995;92(13):6147–51.
18. Kawaguchi-Miyashita M, Shimada S, Kurosu H, Kato-Nagaoka N, Matsuoka Y, Ohwaki M, et al. An accessory role of TCRgammadelta (+) cells in the exacerbation of inflammatory bowel disease in TCRalpha mutant mice. Eur J Immunol. 2001;31(4):980–8.
19. Cornes JS. Number, size, and distribution of Peyer's patches in the human small intestine: part I the development of Peyer's patches. Gut. 1965;6(3):225–9.
20. Mabbott NA, Donaldson DS, Ohno H, Williams IR, Mahajan A. Microfold (M) cells: important immunosurveillance posts in the intestinal epithelium. Mucosal Immunol. 2013;6(4):666–77.
21. Jung C, Hugot J-P, Barreau F. Peyer's patches: the immune sensors of the intestine. Int J Inflam. 2010;2010:823710.
22. Owen RL, Piazza AJ, Ermak TH. Ultrastructural and cytoarchitectural features of lymphoreticular organs in the colon and rectum of adult BALB/c mice. Am J Anat. 1991;190(1):10–8.
23. Mowat AM. Anatomical basis of tolerance and immunity to intestinal antigens. Nat Rev Immunol. 2003;3(4):331–41.
24. Herbrand H, Bernhardt G, Förster R, Pabst O. Dynamics and function of solitary intestinal lymphoid tissue. Crit Rev Immunol. 2008;28(1):1–13.
25. Tsuji M, Suzuki K, Kitamura H, Maruya M, Kinoshita K, Ivanov II, et al. Requirement for lymphoid tissue-inducer cells in isolated follicle formation and T cell-independent immunoglobulin A generation in the gut. Immunity. 2008;29(2):261–71.
26. Pabst O, Herbrand H, Worbs T, Friedrichsen M, Yan S, Hoffmann MW, et al. Cryptopatches and isolated lymphoid follicles: dynamic lymphoid tissues dispensable for the generation of intraepithelial lymphocytes. Eur J Immunol. 2005;35(1):98–107.
27. Pabst O, Herbrand H, Friedrichsen M, Velaga S, Dorsch M, Berhardt G, et al. Adaptation of solitary intestinal lymphoid tissue in response to microbiota and chemokine receptor CCR7 signaling. J Immunol. 2006;177(10):6824–32.
28. Sathaliyawala T, Kubota M, Yudanin N, Turner D, Camp P, Thome JJC, et al. Distribution and compartmentalization of human circulating and tissue-resident memory T cell subsets. Immunity. 2013;38(1):187–97.
29. Wolff MJ, Leung JM, Davenport M, Poles MA, Cho I, Loke P. TH17, TH22 and Treg cells are enriched in the healthy human cecum. PLoS One. 2012;7(7):e41373.
30. Brandtzaeg P. Function of mucosa-associated lymphoid tissue in antibody formation. Immunol Investig. 2010;39(4–5):303–55.
31. Cerovic V, Bain CC, Mowat AM, Milling SWF. Intestinal macrophages and dendritic cells: what's the difference? Trends Immunol. 2014;35(6):270–7.
32. Ueda Y, Kayama H, Jeon SG, Kusu T, Isaka Y, Rakugi H, et al. Commensal microbiota induce LPS hyporesponsiveness in colonic macrophages via the production of IL-10. Int Immunol. 2010;22(12):953–62.

33. Persson EK, Jaensson E, Agace WW. The diverse ontogeny and function of murine small intestinal dendritic cell/macrophage subsets. Immunobiology. 2010;215(9–10):692–7.
34. Rescigno M. Intestinal dendritic cells. Adv Immunol. 1st ed. 2010;107(C):109–38.
35. Bischoff SC. Physiological and pathophysiological functions of intestinal mast cells. Semin Immunopathol. 2009;31(2):185–205.
36. Chu VT, Beller A, Rausch S, Strandmark J, Zänker M, Arbach O, et al. Eosinophils promote generation and maintenance of immunoglobulin-A-expressing plasma cells and contribute to gut immune homeostasis. Immunity. 2014;40(4):582–93.
37. Alexander JS, Ganta VC, Jordan PA, Witte MH. Gastrointestinal lymphatics in health and disease. Pathophysiol Off J Int Soc Pathophysiol. 2010;17(4):315–35.
38. Macpherson AJ, Smith K. Mesenteric lymph nodes at the center of immune anatomy. J Exp Med. 2006;203(3):497–500.
39. Green PHR, Cellier C. Celiac disease. N Engl J Med. 2007;357(17):1731–43.
40. Kaser A, Zeissig S, Blumberg RS. Inflammatory bowel disease. Annu Rev Immunol. 2010;28(5):573–621.
41. Strober W, Fuss IJ. Proinflammatory cytokines in the pathogenesis of inflammatory bowel diseases. Gastroenterology. 2011;140(6):1756–67.
42. Neurath MF. Cytokines in inflammatory bowel disease. Nat Rev Immunol. 2014;14(5):329–42.
43. Bjarnason I, Macpherson A, Mackintosh C, Buxton-Thomas M, Forgacs I, Moniz C. Reduced bone density in patients with inflammatory bowel disease. Gut. 1997;40(2):228–33.
44. Zanchetta MB, Longobardi V, Bai JC. Bone and celiac disease. Curr Osteoporos Rep. 2016;14(2):43–8.
45. Teitelbaum SL. Osteoclasts: what do they do and how do they do it? Am J Pathol. 2007;170(2):427–35.
46. Pederson L, Ruan M, Westendorf JJ, Khosla S, Oursler MJ. Regulation of bone formation by osteoclasts involves Wnt/BMP signaling and the chemokine sphingosine-1-phosphate. Proc Natl Acad Sci U S A. 2008;105(52):20764–9.
47. Sims NA, Martin TJ. Coupling the activities of bone formation and resorption: a multitude of signals within the basic multicellular unit. Bonekey Rep. 2014;3(August 2013):481.
48. Broere F, Apasov SG, Sitkovsky MV, Van EW. T cell subsets and T cell-mediated immunity. In: Nijkamp F, Parnham MJ, editors. Principles of immunopharmacology. 3rd revise ed. Basel: Birkhäuser; 2011. p. 15–28.
49. Li Y, Toraldo G, Li A, Yang X, Zhang H, Qian WP, et al. B cells and T cells are critical for the preservation of bone homeostasis and attainment of peak bone mass in vivo. Blood. 2007;109(9):3839–48.
50. Toraldo G, Roggia C, Qian W-P, Pacifici R, Weitzmann MN. IL-7 induces bone loss in vivo by induction of receptor activator of nuclear factor kappa B ligand and tumor necrosis factor alpha from T cells. Proc Natl Acad Sci U S A. 2003;100(1):125–30.
51. Grcević D, Lee SK, Marusić A, Lorenzo JA. Depletion of CD4 and CD8 T lymphocytes in mice in vivo enhances 1,25-dihydroxyvitamin D3-stimulated osteoclast-like cell formation in vitro by a mechanism that is dependent on prostaglandin synthesis. J Immunol. 2000;165(8):4231–8.
52. Boyce BF, Li P, Yao Z, Zhang Q, Badell IR, Schwarz EM, et al. TNF-alpha and pathologic bone resorption. Keio J Med. 2005;54(3):127–31.
53. Romagnani S. T-cell subsets (Th1 versus Th2). Ann Allergy Asthma Immunol. 2000;85(1):9–18. 21
54. Luckheeram RV, Zhou R, Verma AD, Xia B. CD4$^+$T cells: differentiation and functions. Clin Dev Immunol. 2012;2012:925135.
55. Sato K, Suematsu A, Okamoto K, Yamaguchi A, Morishita Y, Kadono Y, et al. Th17 functions as an osteoclastogenic helper T cell subset that links T cell activation and bone destruction. J Exp Med. 2006;203(12):2673–82.

56. Zaiss MM, Axmann R, Zwerina J, Polzer K, Gückel E, Skapenko A, et al. Treg cells suppress osteoclast formation: a new link between the immune system and bone. Arthritis Rheum. 2007;56(12):4104–12.
57. Mori G, D'Amelio P, Faccio R, Brunetti G. The interplay between the bone and the immune system. Clin Dev Immunol. 2013;2013:720504.
58. Luo CY, Wang L, Sun C, Li DJ. Estrogen enhances the functions of CD4(+)CD25(+)Foxp3(+) regulatory T cells that suppress osteoclast differentiation and bone resorption in vitro. Cell Mol Immunol. 2011;8(1):50–8.
59. Kim YG, Lee C-K, Nah S-S, Mun SH, Yoo B, Moon H-B. Human CD4+CD25+ regulatory T cells inhibit the differentiation of osteoclasts from peripheral blood mononuclear cells. Biochem Biophys Res Commun. 2007;357(4):1046–52.
60. Kikuta J, Wada Y, Kowada T, Wang Z, Sun-Wada G-HH, Nishiyama I, et al. Dynamic visualization of RANKL and Th17-mediated osteoclast function. J Clin Invest. 2013;123(2):866–73.
61. Pène J, Chevalier S, Preisser L, Vénéreau E, Guilleux M-HM-H, Ghannam S, et al. Chronically inflamed human tissues are infiltrated by highly differentiated Th17 lymphocytes. J Immunol. 2008;180(11):7423–30.
62. Choi Y, Woo KM, Ko SH, Lee YJ, Park SJ, Kim HM, et al. Osteoclastogenesis is enhanced by activated B cells but suppressed by activated CD8+ T cells. Eur J Immunol. 2001;31(7):2179–88.
63. John V, Hock JM, Short LL, Glasebrook AL, Galvin RJ. A role for CD8+ T lymphocytes in osteoclast differentiation in vitro. Endocrinology. 1996;137(6):2457–63.
64. Terauchi M, Li JY, Bedi B, Baek KH, Tawfeek H, Galley S, et al. T lymphocytes amplify the anabolic activity of parathyroid hormone through Wnt10b signaling. Cell Metab. 2009;10(3):229–40.
65. Bedi B, Li J-Y, Tawfeek H, Baek K-H, Adams J, Vangara SS, et al. Silencing of parathyroid hormone (PTH) receptor 1 in T cells blunts the bone anabolic activity of PTH. Proc Natl Acad Sci U S A. 2012;109(12):E725–33.
66. Kiesel JR, Buchwald ZS, Aurora R. Cross-presentation by osteoclasts induces FoxP3 in CD8+ T cells. J Immunol. 2009;182(9):5477–87.
67. Buchwald ZS, Kiesel JR, DiPaolo R, Pagadala MS, Aurora R. Osteoclast activated FoxP3 + CD8 + T-cells suppress bone resorption in vitro. PLoS ONE. 2012;7(6).
68. Buchwald ZS, Kiesel JR, Yang C, DiPaolo R, Novack DV, Aurora R. Osteoclast-induced Foxp3+ CD8 T-cells limit bone loss in mice. Bone. 2013;56(1):163–73.
69. Horowitz MC, Fretz JA, Lorenzo JA. How B cells influence bone biology in health and disease. Bone. 2010;47(3):472–9.
70. Manilay JO, Zouali M. Tight relationships between B lymphocytes and the skeletal system. Trends Mol Med. 2014;20(7):405–12.
71. Masuzawa T, Miyaura C, Onoe Y, Kusano K, Ohta H, Nozawa S, et al. Estrogen deficiency stimulates B lymphopoiesis in mouse bone marrow. J Clin Invest. 1994;94(3):1090–7.
72. Erlandsson MC, Jonsson CA, Islander U, Ohlsson C, Carlsten H. Oestrogen receptor specificity in oestradiol-mediated effects on B lymphopoiesis and immunoglobulin production in male mice. Immunology. 2003;108(3):346–51.
73. Eghbali-Fatourechi G, Khosla S, Sanyal A, Boyle WJ, Lacey DL, Riggs BL. Role of RANK ligand in mediating increased bone resorption in early postmenopausal women. J Clin Invest. 2003;111(8):1221–30.
74. Onal M, Xiong J, Chen X, Thostenson JD, Almeida M, Manolagas SC, et al. Receptor activator of nuclear factor kappa B ligand (RANKL) protein expression by B lymphocytes contributes to ovariectomy-induced bone loss. J Biol Chem. 2012;287(35):29851–60.
75. Weitzmann MN, Cenci S, Haug J, Brown C, DiPersio J, Pacifici R. B lymphocytes inhibit human osteoclastogenesis by secretion of TGFβ. J Cell Biochem. 2000;78(2):318–24.
76. Heider U, Langelotz C, Jakob C, Zavrski I, Fleissner C, Eucker J, et al. Expression of receptor activator of nuclear factor kappaB ligand on bone marrow plasma cells correlates with osteolytic bone disease in patients with multiple myeloma. Clin Cancer Res. 2003;9(4):1436–40.

77. Oranger A, Carbone C, Izzo M, Grano M. Cellular mechanisms of multiple myeloma bone disease. Clin Dev Immunol. 2013;2013(8):289458.
78. Colucci S, Brunetti G, Oranger A, Mori G, Sardone F, Specchia G, et al. Myeloma cells suppress osteoblasts through sclerostin secretion. Blood Cancer J. 2011;1(6):e27.
79. Thorbert-Mros S, Larsson L, Berglundh T. Cellular composition of long-standing gingivitis and periodontitis lesions. J Periodontal Res. 2014;4:535–43.
80. Abe T, AlSarhan M, Benakanakere MR, Maekawa T, Kinane DF, Cancro MP, et al. The B cell-stimulatory cytokines BLyS and APRIL are elevated in human periodontitis and are required for B cell-dependent bone loss in experimental murine periodontitis. J Immunol. 2015;195(4):1427–35.
81. Oliver-Bell J, Butcher JP, Malcolm J, Macleod MKL, Adrados Planell A, Campbell L, et al. Periodontitis in the absence of B cells and specific anti-bacterial antibody. Mol Oral Microbiol. 2015;30(2):160–9.
82. Wang ECY, Newton Z, Hayward OA, Clark SR, Collins F, Perks WV, et al. Regulation of early cartilage destruction in inflammatory arthritis by death receptor 3. Arthritis Rheumatol (Hoboken, NJ). 2014;66(10):2762–72.
83. Tanaka D, Kagari T, Doi H, Shimozato T. Essential role of neutrophils in anti-type II collagen antibody and lipopolysaccharide-induced arthritis. Immunology. 2006;119(2):195–202.
84. Wipke BT, Allen PM. Essential role of neutrophils in the initiation and progression of a murine model of rheumatoid arthritis. J Immunol. 2001;167(3):1601–8.
85. Kantarci A, Oyaizu K, Van Dyke TE. Neutrophil-mediated tissue injury in periodontal disease pathogenesis: findings from localized aggressive periodontitis. J Periodontol. 2003;74(1):66–75.
86. Chakravarti A, Raquil MA, Tessier P, Poubelle PE. Surface RANKL of toll-like receptor 4-stimulated human neutrophils activates osteoclastic bone resorption. Blood. 2009;114(8):1633–44.
87. Wythe SE, Nicolaidou V, Horwood NJ. Cells of the immune system orchestrate changes in bone cell function. Calcif Tissue Int. 2014;94(1):98–111.
88. Rivollier A, Mazzorana M, Tebib J, Piperno M, Aitsiselmi T, Rabourdin-Combe C, et al. Immature dendritic cell transdifferentiation into osteoclasts: a novel pathway sustained by the rheumatoid arthritis microenvironment. Blood. 2004;104(13):4029–37.
89. Speziani C, Rivollier A, Gallois A, Coury F, Mazzorana M, Azocar O, et al. Murine dendritic cell transdifferentiation into osteoclasts is differentially regulated by innate and adaptive cytokines. Eur J Immunol. 2007;37(3):747–57.
90. Alnaeeli M, Penninger JM, Teng Y-TA. Immune interactions with CD4+ T cells promote the development of functional osteoclasts from murine CD11c+ dendritic cells. J Immunol. 2006;177(5):3314–26.
91. Maitra R, Follenzi A, Yaghoobian A, Montagna C, Merlin S, Cannizzo ES, et al. Dendritic cell-mediated in vivo bone resorption. J Immunol. 2010;185(3):1485–91.
92. Osta B, Benedetti G, Miossec P. Classical and paradoxical effects of TNF-alpha on bone homeostasis. Front Immunol. 2014;5(FEB):1–9.
93. Hofbauer LC, Lacey DL, Dunstan CR, Spelsberg TC, Riggs BL, Khosla S. Interleukin-1B and tumor necrosis factor-a, but not interleukin-6, stimulate osteoprotegerin ligand gene expression in human osteoblastic cells. Bone. 1999;25(3):255–9.
94. Wei S, Kitaura H, Zhou P, Ross FP, Teitelbaum SL. IL-1 mediates TNF-induced osteoclastogenesis. J Clin Invest. 2005;115(2):282–90.
95. Azuma Y, Kaji K, Katogi R, Takeshita S, Kudo A. Tumor necrosis factor-alpha induces differentiation of and bone resorption by osteoclasts. J Biol Chem. 2000;275(7):4858–64.
96. Lam J, Takeshita S, Barker JE, Kanagawa O, Ross FP, Teitelbaum SL. TNF-alpha induces osteoclastogenesis by direct stimulation of macrophages exposed to permissive levels of RANK ligand. J Clin Invest. 2000;106(12):1481–8.
97. Kitaura H, Kimura K, Ishida M, Kohara H, Yoshimatsu M, Takano-Yamamoto T. Immunological reaction in TNF alpha-mediated osteoclast formation and bone resorption in vitro and in vivo. Clin Dev Immunol. 2013;2013
98. Feldmann M. Development of anti-TNF therapy for rheumatoid arthritis. Nat Rev Immunol. 2002;2(5):364–71.

99. Kimble RB, Bain S, Pacifici R. The functional block of TNF but not of IL-6 prevents bone loss in ovariectomized mice. J Bone Miner Res. 1997;12(6):935–41.
100. Frost A, Jonsson KB, Nilsson O, Ljunggren O. Inflammatory cytokines regulate proliferation of cultured human osteoblasts. Acta Orthop Scand. 1997;68(2):91–6.
101. Gilbert LC, Chen H, Lu X, Nanes MS. Chronic low dose tumor necrosis factor-alpha (TNF) suppresses early bone accrual in young mice by inhibiting osteoblasts without affecting osteoclasts. Bone. 2013;56(1):174–83.
102. Jilka RL, Weinstein RS, Bellido T, Parfitt AM, Manolagas SC. Osteoblast programmed cell death (apoptosis): modulation by growth factors and cytokines. J Bone Miner Res. 1998;13(5):793–802.
103. Kawakami A, Eguchi K, Matsuoka N, Tsuboi M, Koji T, Urayama S, et al. Fas and Fas ligand interaction is necessary for human osteoblast apoptosis. J Bone Min Res. 1997;12(10):1637–46.
104. Kovacić N, Lukić IK, Grcević D, Katavić V, Croucher P, Marusić A, et al. The Fas/Fas ligand system inhibits differentiation of murine osteoblasts but has a limited role in osteoblast and osteoclast apoptosis. J Immunol. 2007;178(6):3379–89.
105. Huang H, Zhao N, Xu X, Xu Y, Li S, Zhang J, et al. Dose-specific effects of tumor necrosis factor alpha on osteogenic differentiation of mesenchymal stem cells. Cell Prolif. 2011;44(5):420–7.
106. Bull MJ, Williams AS, Mecklenburgh Z, Calder CJ, Twohig JP, Elford C, et al. The death receptor 3-TNF-like protein 1A pathway drives adverse bone pathology in inflammatory arthritis. J Exp Med. 2008;205(11):2457–64.
107. Collins FL, Williams JO, Bloom AC, Singh RK, Jordan L, Stone MD, et al. CCL3 and MMP-9 are induced by TL1A during death receptor 3 (TNFRSF25)-dependent osteoclast function and systemic bone loss. Bone. 2017;97:94–104.
108. Bamias G, Siakavellas SI, Stamatelopoulos KS, Chryssochoou E, Papamichael C, Sfikakis PP. Circulating levels of TNF-like cytokine 1A (TL1A) and its decoy receptor 3 (DcR3) in rheumatoid arthritis. Clin Immunol. 2008;129(2):249–55.
109. Collins FL, Williams JO, Bloom AC, Stone MD, Choy E, Wang ECY, et al. Death receptor 3 (TNFRSF25) increases mineral apposition by osteoblasts and region specific new bone formation in the axial skeleton of male DBA/1 mice. J Immunol Res. 2015;2015:1–9.
110. Edwards JR, Sun SG, Locklin R, Shipman CM, Adamopoulos IE, Athanasou NA, et al. LIGHT (TNFSF14), a novel mediator of bone resorption, is elevated in rheumatoid arthritis. Arthritis Rheum. 2006;54(5):1451–62.
111. Brunetti G, Rizzi R, Oranger A, Gigante I, Mori G, Taurino G, et al. LIGHT/TNFSF14 increases osteoclastogenesis and decreases osteoblastogenesis in multiple myeloma-bone disease. Oncotarget. 2014;5(24):12950–67.
112. Lorenzo J, Horowitz M, Choi Y. Osteoimmunology: interactions of the bone and immune system. Endocr Rev. 2008;29(4):403–40.
113. Colucci S, Brunetti G, Cantatore FP, Oranger A, Mori G, Pignataro P, et al. The death receptor DR5 is involved in TRAIL-mediated human osteoclast apoptosis. Apoptosis. 2007;12(9):1623–32.
114. Atkins G, Bouralexis S, Evdokiou A, Hay S, Labrinidis A, Zannettino AC, et al. Human osteoblasts are resistant to Apo2L/TRAIL-mediated apoptosis. Bone. 2002;31(4):448–56.
115. Zauli G, Rimondi E, Stea S, Baruffaldi F, Stebel M, Zerbinati C, et al. TRAIL inhibits osteoclastic differentiation by counteracting RANKL-dependent p27Kip1 accumulation in pre-osteoclast precursors. J Cell Physiol. 2008;214(1):117–25.
116. Lorenzo JA, Naprta A, Rao Y, Alander C, Glaccum M, Widmer M, et al. Mice lacking the type I interleukin-1 receptor do not lose bone mass after ovariectomy. Endocrinology. 1998;139(6):3022–5.
117. Kimble RB, Matayoshi AB, Vannice JL, Kung VT, Williams C, Pacifici R. Simultaneous block of interleukin-1 and tumor necrosis factor is required to completely prevent bone loss in the early postovariectomy period. Endocrinology. 1995;136(7):3054–61.
118. Williams RO, Marinova-Mutafchieva L, Feldmann M, Maini RN. Evaluation of TNF-alpha and IL-1 blockade in collagen-induced arthritis and comparison with combined anti-TNF-alpha/anti-CD4 therapy. J Immunol. 2000;165:7240–5.

119. Lee Y-M, Fujikado N, Manaka H, Yasuda H, Iwakura Y. IL-1 plays an important role in the bone metabolism under physiological conditions. Int Immunol. 2010;22(10):805–16.
120. Akatsu T, Takahashi N, Udagawa N, Imamura K, Yamaguchi A, Sato K, et al. Role of prostaglandins in interleukin-1-induced bone resorption in mice in vitro. J Bone Miner Res. 1991;6(2):183–9.
121. Jimi E, Nakamura I, Duong LT, Ikebe T, Takahashi N, Rodan GA, et al. Interleukin 1 induces multinucleation and bone-resorbing activity of osteoclasts in the absence of osteoblasts/stromal cells. Exp Cell Res. 1999;247(1):84–93.
122. Jimi E, Shuto T, Koga T. Macrophage colony-stimulating factor and interleukin-1 alpha maintain the survival of osteoclast-like cells. Endocrinology. 1995;136(2):808–11.
123. Jimi E, Nakamura I, Ikebe T, Akiyama S, Takahashi N, Suda T. Activation of NF-kappaB is involved in the survival of osteoclasts promoted by interleukin-1. J Biol Chem. 1998;273(15):8799–805.
124. al-Humidan A, Ralston SH, Hughes DE, Chapman K, Aarden L, Russell RG, et al. Interleukin-6 does not stimulate bone resorption in neonatal mouse calvariae. J Bone Miner Res. 1991;6(1):3–8.
125. Ishimi Y, Miyaura C, Jin CH, Akatsu T, Abe E, Nakamura Y, et al. IL-6 is produced by osteoblasts and induces bone resorption. J Immunol. 1990;145(10):3297–303.
126. Yoshitake F, Itoh S, Narita H, Ishihara K, Ebisu S. Interleukin-6 directly inhibits osteoclast differentiation by suppressing receptor activator of NF-kappaB signaling pathways. J Biol Chem. 2008;283(17):11535–40.
127. Jilka RL, Hangoc G, Girasole G, Passeri G, Williams DC, Abrams JS, et al. Increased osteoclast development after estrogen loss: mediation by interleukin-6. Science. 1992;257(5066):88–91.
128. Poli V, Balena R, Fattori E, Markatos A, Yamamoto M, Tanaka H, et al. Interleukin-6 deficient mice are protected from bone loss caused by estrogen depletion. EMBO J. 1994;13(5):1189–96.
129. Mossetti G, Rendina D, De Filippo G, Viceconti R, Di Domenico G, Cioffi M, et al. Interleukin-6 and osteoprotegerin systems in Paget's disease of bone: relationship to risedronate treatment. Bone. 2005;36(3):549–54.
130. Maini RN, Taylor PC, Szechinski J, Pavelka K, Broll J, Balint G, et al. Double-blind randomized controlled clinical trial of the interleukin-6 receptor antagonist, tocilizumab, in European patients with rheumatoid arthritis who had an incomplete response to methotrexate. Arthritis Rheum. 2006;54(9):2817–29.
131. Weitzmann M, Roggia C. Increased production of IL-7 uncouples bone formation from bone resorption during estrogen deficiency. J Clin .... 2002;110(11).
132. D'Amelio P, Grimaldi A, Bernabei P, Pescarmona GP, Isaia G. Immune system and bone metabolism: does thymectomy influence postmenopausal bone loss in humans? Bone. 2006;39(3):658–65.
133. Evans KE, Fox SW. Interleukin-10 inhibits osteoclastogenesis by reducing NFATc1 expression and preventing its translocation to the nucleus. BMC Cell Biol. 2007;8:4.
134. Mohamed SGK, Sugiyama E, Shinoda K, Taki H, Hounoki H, Abdel-Aziz HO, et al. Interleukin-10 inhibits RANKL-mediated expression of NFATc1 in part via suppression of c-Fos and c-Jun in RAW264.7 cells and mouse bone marrow cells. Bone. 2007;41(4):592–602.
135. Liu D, Yao S, Wise GE. Effect of interleukin-10 on gene expression of osteoclastogenic regulatory molecules in the rat dental follicle. Eur J Oral Sci. 2006;114(1):42–9.
136. Adamopoulos IE, Chao C-C, Geissler R, Laface D, Blumenschein W, Iwakura Y, et al. Interleukin-17A upregulates receptor activator of NF-kappaB on osteoclast precursors. Arthritis Res Ther. 2010;12(1):R29.
137. Adamopoulos IE, Bowman EP. Immune regulation of bone loss by Th17 cells. Arthritis Res Ther. 2008;10(5):225.
138. Croes M, Öner FC, van Neerven D, Sabir E, Kruyt MC, Blokhuis TJ, et al. Proinflammatory T cells and IL-17 stimulate osteoblast differentiation. Bone. 2016;84:262–70.

139. Takayanagi H, Ogasawara K, Hida S, Chiba T, Murata S, Sato K, et al. T-cell-mediated regulation of osteoclastogenesis by signalling cross-talk between RANKL and IFN-gamma. Nature. 2000;408(6812):600–5.
140. Gao Y, Grassi F, Ryan MR, Terauchi M, Page K, Yang X, et al. IFN-gamma stimulates osteoclast formation and bone loss in vivo via antigen-driven T cell activation. J Clin Invest. 2007;117(1):122–32.
141. Kamolmatyakul S, Chen W, Li YP. Interferon-gamma down-regulates gene expression of cathepsin K in osteoclasts and inhibits osteoclast formation. J Dent Res. 2001;80(1):351–5.
142. Itonaga I, Sabokbar A, Sun S, Kudo O, Danks L, Ferguson D, et al. Transforming growth factor-β induces osteoclast formation in the absence of RANKL. Bone. 2004;34(1):57–64.
143. Yasui T, Kadono Y, Nakamura M, Oshima Y, Matsumoto T, Masuda H, et al. Regulation of RANKL-induced osteoclastogenesis by TGF-beta through molecular interaction between Smad3 and Traf6. J Bone Miner Res. 2011;26(7):1447–56.
144. Fuller K, Lean JM, Bayley KE, Wani MR, Chambers TJ. A role for TGFbeta(1) in osteoclast differentiation and survival. J Cell Sci. 2000;113(Pt 1):2445–53.
145. Houde N, Chamoux E, Bisson M, Roux S. Transforming growth factor-β1 (TGF-β1) induces human osteoclast apoptosis by up-regulating bim. J Biol Chem. 2009;284(35):23397–404.
146. Murakami T, Yamamoto M, Ono K, Nishikawa M, Nagata N, Motoyoshi K, et al. Transforming growth factor-beta1 increases mRNA levels of osteoclastogenesis inhibitory factor in osteoblastic/stromal cells and inhibits the survival of murine osteoclast-like cells. Biochem Biophys Res Commun. 1998;252(3):747–52.
147. Lian N, Lin T, Liu W, Wang W, Li N, Sun S, et al. Transforming growth factor β suppresses osteoblast differentiation via the vimentin activating transcription factor 4 (ATF4) axis. J Biol Chem. 2012;287(43):35975–84.
148. Matsunobu T, Torigoe K, Ishikawa M, de Vega S, Kulkarni AB, Iwamoto Y, et al. Critical roles of the TGF-beta type I receptor ALK5 in perichondrial formation and function, cartilage integrity, and osteoblast differentiation during growth plate development. Dev Biol. 2009;332(2):325–38.
149. Tang Y, Wu X, Lei W, Pang L, Wan C, Shi Z, et al. TGF-beta1-induced migration of bone mesenchymal stem cells couples bone resorption with formation. Nat Med. 2009;15(7):757–65.
150. Karsdal MA, Larsen L, Engsig MT, Lou H, Ferreras M, Lochter A, et al. Matrix metalloproteinase-dependent activation of latent transforming growth factor-beta controls the conversion of osteoblasts into osteocytes by blocking osteoblast apoptosis. J Biol Chem. 2002;277(46):44061–7.
151. Caruso R, Marafini I, Sedda S, Del Vecchio BG, Giuffrida P, MacDonald TT, et al. Analysis of the cytokine profile in the duodenal mucosa of refractory coeliac disease patients. Clin Sci (Lond). 2014;126(6):451–8.
152. Manavalan JS, Hernandez L, Shah JG, Konikkara J, Naiyer AJ, Lee AR, et al. Serum cytokine elevations in celiac disease: association with disease presentation. Hum Immunol. 2010;71(1):50–7.
153. Fiore CE, Pennisi P, Ferro G, Ximenes B, Privitelli L, Mangiafico RA, et al. Altered osteoprotegerin/RANKL ratio and low bone mineral density in celiac patients on long-term treatment with gluten-free diet. Horm Metab Res. 2006;38(6):417–22.
154. Ciucci T, Ibáñez L, Boucoiran A, Birgy-Barelli E, Pène J, Abou-Ezzi G, et al. Bone marrow Th17 TNFα cells induce osteoclast differentiation, and link bone destruction to IBD. Gut. 2015;64(7):1072–81.
155. Corridoni D, Arseneau KO, Cominelli F. Inflammatory bowel disease. Immunol Lett. 2014;161(2):231–5.
156. Wallace KL, Zheng L-B, Kanazawa Y, Shih DQ. Immunopathology of inflammatory bowel disease. World J Gastroenterol. 2014;20(1):6–21.
157. Fournier BM, Parkos CA. The role of neutrophils during intestinal inflammation. Mucosal Immunol. 2012;5(4):354–66.
158. Grainger JR, Konkel JE, Zangerle-Murray T, Shaw TN. Macrophages in gastrointestinal homeostasis and inflammation. Pflugers Arch. 2017;469(3–4):527–39.

159. Trottier MD, Irwin R, Li Y, McCabe LR, Fraker PJ. Enhanced production of early lineages of monocytic and granulocytic cells in mice with colitis. Proc Natl Acad Sci. 2012;109(41):16594–9.
160. Metzger CE, Narayanan A, Zawieja DC, Bloomfield SA. Inflammatory bowel disease in a rodent model alters osteocyte protein levels controlling bone turnover. J Bone Miner Res. 2017;32(4):802–13.
161. Ali T, Lam D, Bronze MS, Humphrey MB. Osteoporosis in inflammatory bowel disease. Am J Med. 2009;122(7):599–604.
162. Harris L, Senagore P, Young VB, McCabe LR. Inflammatory bowel disease causes reversible suppression of osteoblast and chondrocyte function in mice. Am J Physiol Gastrointest Liver Physiol. 2009;296(5):G1020–9.
163. Irwin R, Lee T, Young VB, Parameswaran N, McCabe LR. Colitis-induced bone loss is gender dependent and associated with increased inflammation. Inflamm Bowel Dis. 2013;19(8):1586–97.
164. Irwin R, Raehtz S, Parameswaran N, McCabe LR. Intestinal inflammation without weight loss decreases bone density and growth. Am J Physiol Regul Integr Comp Physiol. 2016;311(6):R1149–57.
165. Stephensen CB. Burden of infection on growth failure. J Nutr. 1999;129(2S Suppl):534S–8S.
166. Noel G, Baetz NW, Staab JF, Donowitz M, Kovbasnjuk O, Pasetti MF, et al. A primary human macrophage-enteroid co-culture model to investigate mucosal gut physiology and host-pathogen interactions. Sci Rep. 2017;7:45270.
167. Chen HD, Frankel G. Enteropathogenic Escherichia coli: unravelling pathogenesis. FEMS Microbiol Rev. 2005;29(1):83–98.
168. Fleckenstein JM, Hardwidge PR, Munson GP, Rasko DA, Sommerfelt H, Steinsland H. Molecular mechanisms of enterotoxigenic Escherichia coli infection. Microbes Infect. 2010;12(2):89–98.
169. Souza PPC, Lerner UH. The role of cytokines in inflammatory bone loss. Immunol Investig. 2013;42(7):555–622.
170. Black RE, Brown KH, Becker S. Effects of diarrhea associated with specific enteropathogens on the growth of children in rural Bangladesh. Pediatrics. 1984;73(6):799–805.
171. Steiner TS, Lima AA, Nataro JP, Guerrant RL. Enteroaggregative Escherichia coli produce intestinal inflammation and growth impairment and cause interleukin-8 release from intestinal epithelial cells. J Infect Dis. 1998;177(1):88–96.
172. Hsu T-R, Chen S-J, Wu T-C, Chung R-L, Tang R-B. Tumor necrosis factor-alpha and interleukin-10 in viral and bacterial gastroenteritis in children. J Chin Med Assoc. 2005;68(6):250–3.
173. Chen S-M, Lin C-P, Tsai J-D, Chao Y-H, Sheu J-N. The significance of serum and fecal levels of interleukin-6 and interleukin-8 in hospitalized children with acute rotavirus and norovirus gastroenteritis. Pediatr Neonatol. 2014;55(2):120–6.
174. Yokota K, Sato K, Miyazaki T, Kitaura H, Kayama H, Miyoshi F, et al. Combination of tumor necrosis factor α and interleukin-6 induces mouse osteoclast-like cells with bone resorption activity both in vitro and in vivo. Arthritis Rheumatol (Hoboken, NJ). 2014;66(1):121–9.
175. Koniaris SG, Fisher SE, Rubin CT, Chawla A. Experimental colitis impairs linear bone growth independent of nutritional factors. J Pediatr Gastroenterol Nutr. 1997;25(2):137–41.
176. Levy M, Kolodziejczyk AA, Thaiss CA, Elinav E. Dysbiosis and the immune system. Nat Rev Immunol. 2017;17(4):219–32.
177. Petersen C, Round JL. Defining dysbiosis and its influence on host immunity and disease. Cell Microbiol. 2014;16(7):1024–33.
178. Ubeda C, Pamer EG. Antibiotics, microbiota, and immune defense. Trends Immunol. 2012;33(9):459–66.
179. Sekirov I, Tam NM, Jogova M, Robertson ML, Li Y, Lupp C, et al. Antibiotic-induced perturbations of the intestinal microbiota alter host susceptibility to enteric infection. Infect Immun. 2008;76(10):4726–36.

180. Frank DN, St Amand AL, Feldman RA, Boedeker EC, Harpaz N, Pace NR. Molecular-phylogenetic characterization of microbial community imbalances in human inflammatory bowel diseases. Proc Natl Acad Sci U S A. 2007;104(34):13780–5.
181. Abrahamsson TR, Jakobsson HE, Andersson AF, Björkstén B, Engstrand L, Jenmalm MC. Low gut microbiota diversity in early infancy precedes asthma at school age. Clin Exp Allergy. 2014;44(6):842–50.
182. Karlsson FH, Tremaroli V, Nookaew I, Bergström G, Behre CJ, Fagerberg B, et al. Gut metagenome in European women with normal, impaired and diabetic glucose control. Nature. 2013;498(7452):99–103.
183. Zhang C, Zhang M, Wang S, Han R, Cao Y, Hua W, et al. Interactions between gut microbiota, host genetics and diet relevant to development of metabolic syndromes in mice. ISME J. 2010;4(2):232–41.
184. Turnbaugh PJ, Ridaura VK, Faith JJ, Rey FE, Knight R, Gordon JI. The effect of diet on the human gut microbiome: a metagenomic analysis in humanized gnotobiotic mice. Sci Transl Med. 2009;1(6):6ra14.
185. Turnbaugh PJ, Ley RE, Mahowald MA, Magrini V, Mardis ER, Gordon JI. An obesity-associated gut microbiome with increased capacity for energy harvest. Nature. 2006;444(7122):1027–31.
186. Bäckhed F, Ding H, Wang T, Hooper LV, Koh GY, Nagy A, et al. The gut microbiota as an environmental factor that regulates fat storage. Proc Natl Acad Sci U S A. 2004;101(44):15718–23.
187. Brown K, DeCoffe D, Molcan E, Gibson DL. Diet-induced dysbiosis of the intestinal microbiota and the effects on immunity and disease. Forum Nutr. 2012;4(8):1095–119.
188. McCabe LR, Britton RA, Parameswaran N. Prebiotic and probiotic regulation of bone health: role of the intestine and its microbiome. Osteoporos Int. 2015;In press.
189. Ohlsson C, Sjögren K. Effects of the gut microbiota on bone mass. Trends Endocrinol Metab. 2015;26(2):69–74.
190. Sommer F, Bäckhed F. The gut microbiota – masters of host development and physiology. Nat Rev Microbiol. 2013;11(4):227–38.
191. Yan J, Herzog JW, Tsang K, Brennan CA, Bower MA, Garrett WS, et al. Gut microbiota induce IGF-1 and promote bone formation and growth. Proc Natl Acad Sci. 2016;113(47):E7554–63.
192. Round JL, Mazmanian SK. The gut microbiota shapes intestinal immune responses during health and disease. Nat Rev Immunol. 2009;9(5):313–23.
193. Ouwehand A, Isolauri E, Salminen S. The role of the intestinal microflora for the development of the immune system in early childhood. Eur J Nutr. 2002;41(Suppl 1):I32–7.
194. Macpherson AJ, Harris NL. Interactions between commensal intestinal bacteria and the immune system. Nat Rev Immunol. 2004;4(6):478–85.
195. Yang Y, Torchinsky MB, Gobert M, Xiong H, Xu M, Linehan JL, et al. Focused specificity of intestinal TH17 cells towards commensal bacterial antigens. Nature. 2014;510(7503):152–6.
196. Round JL, O'Connell RM, Mazmanian SK. Coordination of tolerogenic immune responses by the commensal microbiota. J Autoimmun. 2010;34(3):J220–5.
197. Ivanov II, Atarashi K, Manel N, Brodie EL, Shima T, Karaoz U, et al. Induction of intestinal Th17 cells by segmented filamentous bacteria. Cell. 2009;139(3):485–98.
198. Atarashi K, Tanoue T, Shima T, Imaoka A, Kuwahara T, Momose Y, et al. Induction of colonic regulatory T cells by indigenous clostridium species. Science. 2011;331(6015):337–41.
199. Gensollen T, Iyer SS, Kasper DL, Blumberg RS. How colonization by microbiota in early life shapes the immune system. Science. 2016;352(6285):539–44.
200. Gevers D, Kugathasan S, Denson LA, Vázquez-Baeza Y, Van Treuren W, Ren B, et al. The treatment-naive microbiome in new-onset Crohn's disease. Cell Host Microbe. 2014;15(3):382–92.
201. Michail S, Durbin M, Turner D, Griffiths AM, Mack DR, Hyams J, et al. Alterations in the gut microbiome of children with severe ulcerative colitis. Inflamm Bowel Dis. 2012;18(10):1799–808.

202. Hill RJ, Brookes DSK, Davies PSW. Bones in pediatric Crohn's disease: a review of fracture risk in children and adults. Inflamm Bowel Dis. 2011;17(5):1223–8.
203. Weiskopf D, Weinberger B, Grubeck-Loebenstein B. The aging of the immune system. Transpl Int. 2009;22(11):1041–50.
204. Montecino-Rodriguez E, Berent-Maoz B, Dorshkind K. Causes, consequences, and reversal of immune system aging. J Clin Invest. 2013;123(3):958–65.
205. Gruver AL, Hudson LL, Sempowski GD. Immunosenescence of ageing. J Pathol. 2007;211(2):144–56.
206. Khosla S, Melton LJ, Riggs BL. The unitary model for estrogen deficiency and the pathogenesis of osteoporosis: is a revision needed? J Bone Miner Res. 2011;26(3):441–51.
207. Magrone T, Jirillo E. The interaction between gut microbiota and age-related changes in immune function and inflammation. Immun Ageing. 2013;10(1):31.
208. Rosen CJ, Bouxsein ML. Mechanisms of disease: is osteoporosis the obesity of bone? Nat Clin Pract Rheumatol. 2006;2(1):35–43.
209. Geiger H, de Haan G, Florian MC. The ageing haematopoietic stem cell compartment. Nat Rev Immunol. 2013;13(5):376–89.
210. Linehan E, Fitzgerald DC. Ageing and the immune system: focus on macrophages. Eur J Microbiol Immunol (Bp). 2015;5(1):14–24.
211. Miller JP, Allman D. The decline in B lymphopoiesis in aged mice reflects loss of very early B-lineage precursors. J Immunol. 2003;171(5):2326–30.
212. Min H, Montecino-Rodriguez E, Dorshkind K. Effects of aging on the common lymphoid progenitor to pro-B cell transition. J Immunol. 2006;176(2):1007–12.
213. Min H, Montecino-Rodriguez E, Dorshkind K. Reduction in the developmental potential of intrathymic T cell progenitors with age. J Immunol. 2004;173(1):245–50.
214. Tsuboi I, Morimoto K, Hirabayashi Y, Li G-X, Aizawa S, Mori KJ, et al. Senescent B lymphopoiesis is balanced in suppressive homeostasis: decrease in interleukin-7 and transforming growth factor-beta levels in stromal cells of senescence-accelerated mice. Exp Biol Med (Maywood). 2004;229(6):494–502.
215. Saffrey MJ. Aging of the mammalian gastrointestinal tract: a complex organ system. Age (Omaha). 2014;36(3):1019–32.
216. Fujihashi K, McGhee JR. Mucosal immunity and tolerance in the elderly. Mech Ageing Dev. 2004;125(12. SPEC.ISS.):889–98.
217. Ogino T, Miura S, Komoto S, Hara Y, Hokari R, Tsuzuki Y, et al. Senescence-associated decline of lymphocyte migration in gut-associated lymphoid tissues of rat small intestine. Mech Ageing Dev. 2004;125(3):191–9.
218. Schmucker DL, Owen RL, Outenreath R, Thoreux K. Basis for the age-related decline in intestinal mucosal immunity. Clin Dev Immunol. 2003;10(2–4):167–72.
219. Santiago AF, Alves AC, Oliveira RP, Fernandes RM, Paula-Silva J, Assis FA, et al. Aging correlates with reduction in regulatory-type cytokines and T cells in the gut mucosa. Immunobiology. 2011;216(10):1085–93.
220. McCabe LR, Irwin R, Schaefer L, Britton RA. Probiotic use decreases intestinal inflammation and increases bone density in healthy male but not female mice. J Cell Physiol. 2013;228(8):1793–8.
221. Collins FL, Irwin R, Bierhalter H, Schepper J, Britton RA, Parameswaran N, et al. Lactobacillus reuteri 6475 increases bone density in intact females only under an inflammatory setting. PLoS One. 2016;11(4):e0153180.
222. Cho I, Yamanishi S, Cox L, Methé BA, Zavadil J, Li K, et al. Antibiotics in early life alter the murine colonic microbiome and adiposity. Nature. 2012;488(7413):621–6.
223. Cox LM, Yamanishi S, Sohn J, Alekseyenko AV, Leung JM, Cho I, et al. Altering the intestinal microbiota during a critical developmental window has lasting metabolic consequences. Cell. 2014;158(4):705–21.

Part II
Gut Pathologies and Bone Health

Stomach and Bone

Alice M. Kitay and John P. Geibel

Introduction

Calcium is a pivotal ion in the human body and responsible for the maintenance of physiological processes such as intracellular messaging, membrane integrity, neuronal transmission, and muscle contraction. The free calcium levels in the human body are precisely sustained to an average calcium concentration of ~1.1 mM in the extracellular milieu and more than a 10.000-fold lower concentration in the intracellular milieu. Small disturbances in this balance as the result of an imbalance of calcium homeostasis can have fatal effects on health leading to cardiac arrhythmias, cognitive dysfunction, and neuronal deficits. The three main players of calcium homeostasis are the bone, kidney, and intestine, which are under control of the endocrine regulators: parathyroid hormone (PTH), calcitriol, and calcitonin [1]. The balance is kept between absorption in the intestine, excretion via the kidney, and mobilization from bone, the dynamic calcium reservoir.

This chapter will review how a reduction of gastric acid secretion could adversely affect calcium absorption and consequently increase the risk for bone fractures. In addition to the maintenance of calcium homeostasis and associated regulative processes, we have highlighted the neuroendocrine control of gastric acid secretion, which is essential for calcium ionization. Furthermore, we discuss diseases leading to a systemic calcium-sensing deficiency and therapeutic measures on gastric acid-related diseases with implications on bone health.

A.M. Kitay • J.P. Geibel (✉)
Department of Surgery, Yale School of Medicine, New Haven, CT, USA
e-mail: john.geibel@yale.edu

L.R. McCabe, N. Parameswaran (eds.), *Understanding the Gut-Bone Signaling Axis*, Advances in Experimental Medicine and Biology 1033,
DOI 10.1007/978-3-319-66653-2_6

Overview of Calcium Homeostasis

The skeleton is in a constant remodeling state between calcium release and calcium uptake. When bones are in need of calcium, the first source comes from the destruction of existing bone by osteoclasts and then uptake of this ionized calcium by osteoblasts to remodel the existing bone. Bone calcium release occurs by a proton-induced mechanism in osteoclasts [2], with the release of calcium into the systemic circulation, thereby elevating calcium concentration which then activates the calcium-sensing receptor (CaSR). The CaSR is responsible for sensing extracellular calcium concentration [3]. The CaSR can be found on various tissues, e.g., parathyroid glands, stomach, intestine, and kidney, and also in the skin [1].

In the stomach, the activation of basolateral CaSR results in an increase of gastric acid production due to an activation of the H^+,K^+-ATPase [4]. Low pH in the gastric lumen is not only responsible for sterilization and digestion of foodstuff but ionization of ingested calcium. This seems essential because only calcium in its ionized form can be absorbed by the human body [1]. Calcium absorption takes place in the small intestine, where calcium ions activate the CaSR located on both basolateral and apical membranes [5]. Activation of the apical receptor leads to an increased NaCl, H_2O, and calcium absorption. Simultaneously, elevated blood calcium levels lead to activation of the intestinal CaSR on basolateral side, creating a negative feedback loop and causing an inhibition of the absorption [5]. Absorbed calcium ions reach the bone through blood circulation and precipitate in the presence of phosphate forming mineralized bone. The hormones 1,25$(OH)_2$ vitamin D (calcitriol) and PTH have great impact on bone generation and have the ability to influence the cycle of calcium homeostasis. Serum calcium concentration is preserved and regulated by the calcitriol and PTH and to a lesser extent by calcitonin. These hormones have the capability of influencing the intestine, kidney, and bone. PTH is released by thyroid glands as a result of CaSR activation. Disturbances in regulators of calcium homeostasis can have severe consequences. Genetic mutations of the CaSR cause imbalances in calcium levels with disturbed calciuria, parathyroid dysfunctions can lead to excessive secretion of PTH with increased calcium blood levels and phosphaturia, and long-term use of proton pump inhibitors (PPIs) with elevated gastric pH may lead to a decreased calcium ionization. Consequences and symptoms of these diseases and conditions will be explained in a later section of this chapter (Fig. 1).

The Calcium-Sensing Receptor

Calcium is primarily sensed by the calcium-sensing receptor, abbreviated as CaSR. It is the most important mediator in calcium homeostasis and caretaker of serum calcium levels. CaSR is regulating PTH secretion from the parathyroid gland, subsequently the synthesis of vitamin D and calcium handling in all vitamin

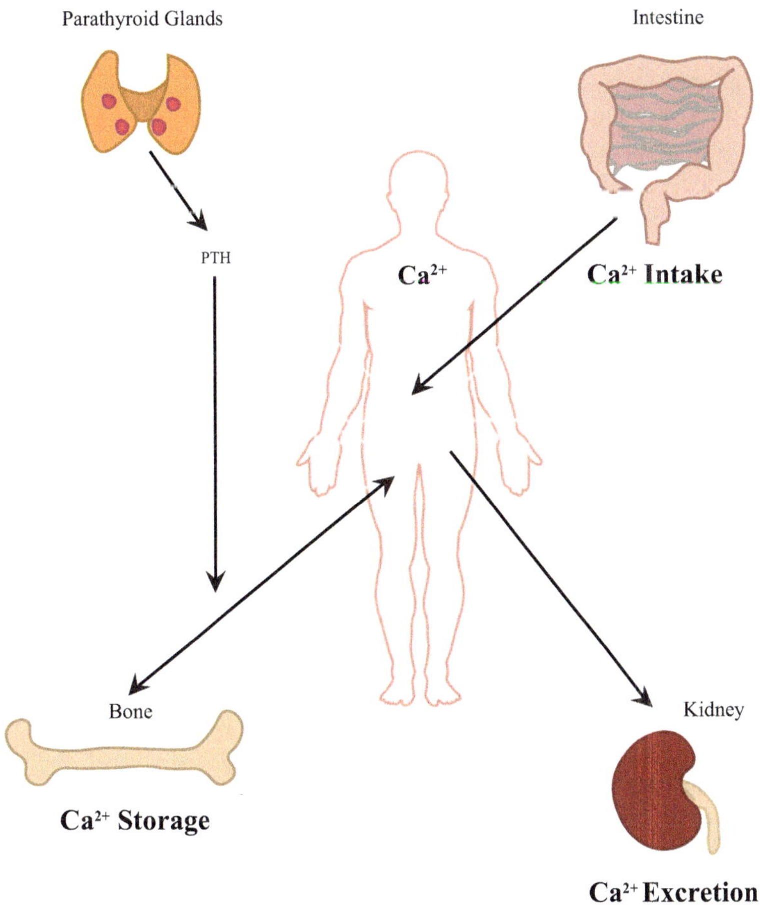

Fig. 1 Role of the CaSR in calcium metabolism: Schematic overview of how organs regulate and maintain a constant extracellular calcium concentration. *Double arrow* between bone and body indicates a permanent state of the bone between mineralization and demineralization dependency of interaction with parathyroid hormone PTH levels. The intestine is primarily responsible for absorption of calcium, the kidney for excretion. All of the organs are equipped with the CaSR

D-sensitive organs, such as the intestine [6], bone [7] and kidney [8]. Thus, the receptor is not only expressed on the parathyroid epithelium but can be found in all vitamin D target tissues. The characteristic localization of the receptor supposes that calcium handling can occur on a local organ level independently of PTH and vitamin D conditions. Compendious and on-site feedback loops assist in a rapid respond to changes in extracellular calcium conditions.

Brown et al. were the first to clone CaSR from bovine parathyroid using a *Xenopus* oocyte expression cloning system in 1993. The protein is a member of the seven-transmembrane domain G protein-coupled receptor family and is arranged as a homodimer on cell surface. The assembly of the receptor seems to occur in the

ER, starting with a dimerization in the form of the formation of disulfide bonds between cysteine and leucine residues in the extracellular part of the receptor [9]. After this process, N-linked glycosylation takes place in the Golgi apparatus, which is playing a very important role for cell surface expression [10]. The trafficking between these organelles is aided and regulated by the protein Rab1. A mutation in this GTP-binding protein leads to a decline of expression of the CaSR on cell surface [11]. The internalization of CaSR is suggested to be regulated by the E3 ligase dorfin, but further investigations are necessary [12].

Unfortunately, to date a complete crystal structure of the CaSR does not exist, but scientists are following the hypothesis that the large extracellular domain of the receptor has two functional lobes that are oscillating in response to calcium and environmental cues; this has led researchers to equate this to the form of a venus flytrap. Thus, the key residue is located in a cavity surrounded by two lobular domains LB1 and LB2. Stimulators of the receptor are not only calcium, but tri- and polyvalent ions such as Mg^{2+}, Pb^{2+}, and Gd^{3+} and larger molecules like spermine and neomycin. Furthermore, calcimimetics can modify the receptor and extend its sensitivity to ligands [1]. This is used for clinical treatment of, e.g., secondary hyperparathyroidism with the small molecule calcimimetic cinacalcet/Sensipar [13]. Amino acids can affect the CaSR, leading to the suggestion that the CaSR functions as not only a calcium sensor but also a nutrient sensor. The CaSR can also be modified by extracellular pH leading to activation or inhibition of the receptor. An acidic milieu leads to a decrease of the affinity of calcium to CaSR, whereas an alkaline milieu increases the affinity. The intracellular domain of the receptor consists of five PKC phosphorylation sides, which are activated during receptor activation. Once activated, an intracellular cascade starting with G-proteins ($G\alpha_{q/11}$, $G\alpha_i$, $G\alpha_{12/13}$) results in accordance with the signal of G-protein in suppression of cAMP and activation of MAPK or increased intracellular calcium concentration via PLC and IP_3.

In the parathyroid the CaSR is responsible for three different types of regulation processes: the release of PTH from secretory granules, de novo synthesis of PTH, and parathyroid cell growth. High serum calcium levels lead to an activation of CaSR following inhibition of PTH release and decrease of serum calcium concentration. This response is mediated by the creation of arachidonic acid through $G\alpha_q$ and PLA_2 pathways. An interesting study by Bourdeau et al. using cultured porcine parathyroid cells verified the increase of arachidonic acids during CaSR stimulation and inhibition of PTH release. Furthermore, this study illustrated the inhibition of PTH release under application of exogenous arachidonic acid [14]. The mechanism of CaSR regulating the de novo synthesis of PTH is linked to the mediation of PTH gene transcription by $1{,}25(OH)_2$ vitamin D. Since $1{,}25(OH)_2$ vitamin D is having an inhibitory effect on PTH gene transcription because binding of $1{,}25(OH)_2$ vitamin D to VDR leads to a decrease in mRNA levels of PTH and CaSR is augmenting the expression of VDR, CaSR potentiates this negative feedback loop [15]. Scientists have shown that activating mutations of CaSR entail decrease of parathyroid size, and conversely loss-of-function mutations result typically in hypertrophic glands [16]. The reason for the dependency of parathyroid cell modulation by CaSR is not known yet, but studies suggest $G\alpha_q$ being responsible for the regulation of cell growth in the parathyroid [17].

The CaSR in the kidney is responsible for water and ion exchange. In the proximal tubule, CaSR is located apically at the base of the brush border. In this part of the nephron, phosphate transport takes place, which is mainly regulated by PTH. A high serum PTH level can lead to an inhibition of phosphate reabsorption from the lumen. This effect can be partially reversed by the activation of CaSR. However, high phosphate levels and PTH can restrict CaSR expression in the proximal tubule [18]. Monovalent and polyvalent ion absorption occurs in the thick ascending loop of Henle and is modified by the CaSR, which can be found on the basolateral membrane. The apical renal outer medullary potassium channel ROMK (Kir1.1) is inhibited by the activation of CaSR. Having a direct effect on the potassium amount in the lumen, CaSR is indirectly affecting the Na^+-K^+-$2Cl^-$ (NKCC2), which is in need of potassium ion efflux, following inhibition of chloride and sodium absorption through the NKCC2 [8]. The effects of CaSR on the NKCC2 can have impact on calcium absorption due to decrease of lumen-positive potential and an effect on the countercurrent multiplication, leading to problems in concentrate the urine properly. This affects calcium absorption, which is thought to occur in the medullary part of the thick ascending limb passively through the paracellular route [19].

The distal convoluted tubule and the connecting tubule are in charge of the precise adjustment of calcium reabsorption, achieved by absorbing calcium via the transcellular pathway against its own electrochemical gradient [20]. The transporter TRPV5, which is necessary for transcellular calcium absorption next to the transporters calbindin-D28k, NCX1, and PMCA1b, can be regulated by $1,25(OH)_2$ vitamin D and PTH and is colocalized by CaSR. Activation of CaSR leads to an increase of calcium absorption through TRPV5, thereby promptly and locally adapting absorption to calcium urine concentrations [21].

In the collecting duct, CaSR is regulating proton and water exchange. Apical V-ATPases in the intercalated cells of the collecting duct acidify the urine and can be stimulated by luminal calcium and neomycin activating the CaSR. This can be considered as autoprotection against calcium kidney stones or nephrolithiasis in general, since the formation of these stones is pH dependent [22]. Besides, stimulation of the CaSR in the principal cells of the collecting duct this also causes diuresis by the inhibition of AQP2-mediated water absorption [23].

The CaSR can be found in nearly all organs of the gastrointestinal tract. In the stomach, the receptor can be found on the basolateral membrane of parietal cells and on both the basolateral and apical surface on G-cells. Elevated serum calcium levels lead to an increase of gastrin levels due to calcium influx related gastrin release following activation of CaSR [24]. Since the CaSR is also activated by L-type amino acids, the receptor can be considered as a nutrient sensor, which supports the modulation of gastric acid secretion and gastric luminal pH. Mice studies by Feng et al. illustrated the correlation between high luminal calcium and L-type amino acid concentrations and increase in gastrin release. The effect of gastrin release under conditions of high calcium and L-type amino acid concentration in the gastric lumen did not appear in CaSR (−/−) animals [25]. In the parietal cell, the activation of CaSR results in a H+,K+-ATPase-mediated proton secretion, thereby increasing the amount of gastric acid secretion into the gastric lumen. This effect can be achieved by the CaSR agonists such as calcium; trivalent ions, e.g., Gd^{3+}; and allosteric modifiers, L-type amino acids [26]. The intracellular signal cascade for H+,K+-ATPase is induced by intracellular calcium elevation, PLC, MAPK, and PKC [27]. Concluding, activation of CaSR on G-cells and parietal cells can either indirectly or directly increase gastric acid secretion.

Investigations of the CaSR in the intestine were mainly made on colonic crypts, where the receptor can be found on both apical and basolateral membranes. CaSR is also localized on the neuronal plexuses, Meissner and Auerbach [6]. The CaSR in the colon is considered to be an important mediator of fluid secretion. Intracellular calcium concentration increases under the exposure of crypts to CaSR agonists. Forskolin-stimulated fluid secretion, along with cholera toxin and STa toxin, can be inhibited by the activation of CaSR, thus having a reversal effect on secretory diarrhea [28].

In the bone, CaSR is expressed on osteoclasts and osteoblasts, as well as on their precursors [5].

However, the functional role of the receptor in bone is less clear. It is well known that extracellular calcium is having impact on proliferation, migration, and differentiation processes in osteoblasts and the differentiation process in osteoclasts [29]. With the introduction of CaSR (−/−) mice, the importance of the CaSR in in vivo studies is under debate [1]. Resulting in rickets but also in hyperparathyroidism, knockout of CaSR did not permit a differentiation between CaSR and high PTH concentration effects on bone turnover [30]. Besides, it seems that not only CaSR is responsible for calcium sensing in osteoblasts, since changes in extracellular calcium concentrations can still lead to a response in CaSR (−/−) osteoblasts [31]. This study led to the introduction of another GPRC, namely, the GPRC6A, which is sensitive to the calcimimetic R568. Mutation and knockout in GPRC6A results in osteopenia and mineralization defects, as well as decreased sensitivity toward extracellular calcium. This suggests GPRC6A being a potential alternative calcium-sensing pathway [32].

Nevertheless, the osteoblast-specific CaSR (−/−) model suggests that the CaSR is an important modulator of bone turnover. Deformities in the skeleton of knockout animals clearly propose that the CaSR is playing an important role in osteoblast function.

Also, the upregulation of CaSR in osteoblasts with the help of a constitutively active receptor mutant results in bone loss due to a decreased bone volume and density especially in trabecular bone, coming along with an increased number of osteoclast and unchanged osteoblast parameters. Scientists suggest RANKL of osteoblasts to be responsible for the osteoclastogenic signal, recruiting osteoclasts and inducing their own maturation via CaSR activation resulting in an increased bone turnover and osteoclast number [33].

PTH

The initial studies on the parathyroid glands back in the 1900s illustrated the appearance of tetany after performing a surgical parathyroidectomy. However, the tetany could be prevented by the application of Ca^{2+} that led scientists to conclude a correlation between parathyroid glands and serum calcium concentration.

One of the responsible hormones for calcium homeostasis next to 1,25$(OH)_2$ vitamin D and calcitonin is the 84-amino acid peptide hormone PTH synthesized in the parathyroid gland. With a very short plasma half-life less than 5 min, extracellular calcium concentration can be strictly regulated and kept between 1.1 and 1.3 mM. The PTH gene encodes for the prepro-PTH (115 amino acids), which can be transformed to mature 85-amino acid PTH within two scissions. Mature PTH is then stored in secretory vesicles, which can fuse with the membrane and release their content into the blood as an answer to low plasma Ca^{2+} concentration. Extracellular Ca^{2+} concentrations are permanently monitored by CaSR, located on the epithelial membrane of the parathyroid glands, leading to an intracellular signaling cascades following PTH secretion when Ca^{2+} is detected to be low. Besides, PTH is regulated by 1,25$(OH)_2$ vitamin D on a transcriptional level in the form of a negative feedback loop.

The receptor of PTH is represented by the parathyroid hormone receptor type 1 (PTH1R) which belongs to the family of class B GPCRs. PTH2R was found on a variety of tissues in the CNS and is not affecting Ca^{2+} homeostasis. Interestingly, the receptor was found in the nucleus as well, where its existence remains unclear.

The NH_2 domain of PTH activates PTH1R. It can bind to the α-β-βα binding fold of PTH1R and lead to the activation of the receptor. Hence, the full length of the hormone is not necessary for the activation of PTH1R. For instance, PTH (1–34) is the clinically used form as a PTH analogon and has a similar effect as the usual form PTH (1–85). In contrast, PTH (2–34) only has a partial agonizing effect on PTH1R and PTH (3–34) exhibits a complete loss of biological activity.

In an inactivated state, the receptor appears in a homodimer form at the cell surface and disintegrates after the binding of PTH. The extracellular domain of the receptor contains disulfide bond pattern including six cysteine residues, which are responsible for the stabilization of the α-β-βα binding pocket. When PTH binds to the receptor, two different types of G-proteins are activated: $G\alpha_{q/11}$ mediates intracellular Ca^{2+} store release via PLC and IP3 next to $G\alpha_s$ activating AC with an intracellular increase of cAMP. The receptor is regulated by trafficking, internalization, and direct protein interactions on the surface. Receptor-associated proteins like NHRF and β-arrestin can influence intracellular signaling. Desensitization of receptor and ligand is realized by GRK2 binding/phosphorylation and β-arrestin binding. These proteins are able to uncouple the receptor from G-protein and trigger internalization. A knockout mouse model of β-arrestin illustrated high intracellular amount of cAMP in osteoblasts upon PTH stimulation. PTH1R is supported by the protein NHERF1, which stabilizes the receptor at the cell membrane and prevents endocytosis or desensitization, thus preventing interaction between β-arrestin and

PTH1R. Interestingly, β-arrestin is not only responsible for desensitization but is also playing an important role in the activation of downstream signaling via MAPKs. NHERF is a potent modulator of the cell's response to PTH, and in the presence of NHERF2, a Ca^{2+} response via increased PLC and decreased cAMP can be found.

Another regulating mechanism of PTH1R occurs through scission of its extracellular domain by metalloproteases, leading to a receptor degradation.

The cellular effect of PTH applies to bone and kidney. In general, PTH influences the kidney in the form of phosphaturia, increase of Ca^{2+} absorption, and increase of $1,25(OH)_2$ vitamin D synthesis.

Phosphaturia is the result of decrease of expression of the NaP_i transporter type IIa (NaPi-IIa), located on the apical membrane of the proximal tubule. This occurs through activation of the basolateral PTH1R by PTH. Previous studies demonstrate that NaPi-IIa is the major phosphate transporter in the kidney, responsible for approximately 80% of the total amount of phosphate transport. Besides, the cotransporters NaPi-IIc and Pit-2 are mediated by PTH; however, the molecular pathway still remains unclear.

PTH is able to stimulate the synthesis of $1,25(OH)_2$ vitamin D in the kidney, in turn increasing renal and intestinal calcium absorption as a response to low plasma Ca^{2+} concentration, which initially led to PTH secretion. PTH affects the transcription of CYP27B1, the mitochondrial enzyme responsible for 1α-hydroxylation of 25(OH) vitamin D to $1,25(OH)_2$ vitamin D. Of note, PTH is able to upregulate renal Ca^{2+} absorption directly in the distal convoluted and connecting tubule, expressing all calcium transport proteins, such as TRPV5, calbindinD-28k, and NCX1. This effect is $1,25(OH)_2$ vitamin D independent.

In the bone, PTH can have two different contrary effects, which are dependent on exposure conditions. A pulsatile or rather intermittent PTH exposure has an anabolic effect on the bone, thereby increasing bone mass, whereas continuous secretion of PTH has a catabolic effect, thus increasing plasma Ca^{2+} levels. The retrieval of bone mass under constant PTH levels led to the introduction of clinical use of PTH administration for osteoporosis. Enhanced bone formation is caused by an increase of osteoblast differentiation and inhibition of apoptosis. These anti-apoptotic effects were investigated and concluded to be attributable to run-related transcription factor 2 (RUNX2) – mediated transcription of various survival genes and increased molecular DNA repair. A potential cofactor is represented by fibroblast growth factor 2 (FGF2).

Continuous PTH secretion is characterized by high bone turnover as a result of an increase of osteoclasts and their activation. Of note, the catabolism of bone occurs through osteoblasts, since osteoclasts don't express PTH1R. A so-called cross-talk between osteoblasts and osteoclasts takes place aided by the receptor activator of nuclear factor κB (RANK), osteoprotegerin (OPG), and RANK ligand (RANKL). RANKL plus OPG on osteoblasts causes the contrary effect on osteoclasts. RANKL binds to RANK and promotes osteoclastogenesis, which means increase of osteoclasts, improvement of bone resorption followed by an enhanced

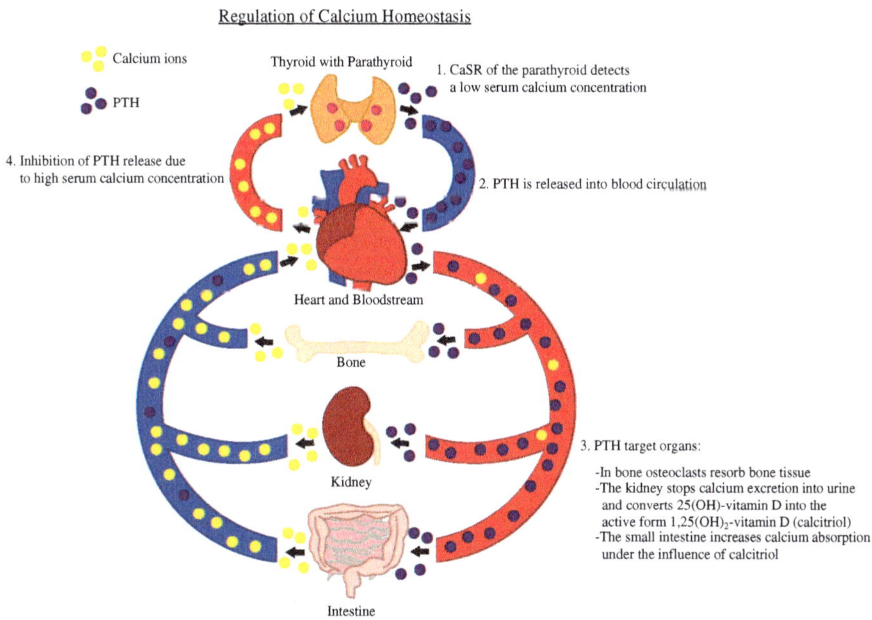

Fig. 2 The importance of PTH: Low serum calcium concentration is detected by the CaSR in parathyroid glands, which will subsequently lead to the release of the hormone PTH into systemic bloodstream. The hormone then affects the bone, kidney, and intestine. In bone, PTH causes calcium release. The kidney stops excreting calcium and the intestine increases absorption of calcium. The now normalized serum calcium level leads to CaSR deactivation and resultant termination in PTH release

serum Ca^{2+} concentration, whereas OPG tries to inhibit ligand-receptor interaction by functioning as an antagonist.

Accordingly, studies on the intestine demonstrated the interdependency of PTH and intestinal, $1,25(OH)_2$ vitamin D-irrespective Ca^{2+} absorption, showing an increase of Ca^{2+} uptake in enterocytes and intestinal loops after PTH application (Fig. 2).

Vitamin D

$1,25(OH)_2$ vitamin D (calcitriol) is considered to be the pivotal influencer of calcium homeostasis. Two different sources of $1,25(OH)_2$ vitamin D exist in our body. There is dietary vitamin D from mostly dairy products, and the skin synthesized vitamin D, whose reaction is catalyzed by UV-B light. With regard to the pathophysiology of rickets with symptoms like childhood skeletal deformities due to inadequate calcium mineralization and abstinence of sunlight, the importance of the hormone and the dependency of disease prevention on sunlight become clear [34].

The intestinal vitamin D absorption takes place in the proximal and mid-small intestine [35, 36]. The fat-soluble vitamin molecules accumulate to micelle-like structures, which can be absorbed by the enterocyte with the help of bile salts, monoglycerides, and free fatty acids. Patients with cholestasis or pancreatic insufficiency experience a vitamin D shortage, hence having a higher risk of suffering from bone diseases, such as osteomalacia or osteoporosis [37, 38]. After the absorption of vitamin D in the enterocyte, it is subsequently organized in chylomicrons or transported by the vitamin D-binding protein (DBP) and excreted into the lymph system [39, 40]. Reaching the liver, vitamin D experiences its first hydroxylation on the 25th position and becomes 25(OH) vitamin D [41].

In the skin, 7-dehydrocholesterol is converted to previtamin D by UV-B light and subsequently isomerized to vitamin D. The conversion and isomerization is slow, and a 1,25$(OH)_2$ vitamin D serum concentration peak is reached 2 days after sun exposure [42].

Like dietary vitamin D, the product of the endogenous, cutaneous vitamin D synthesis is transported to the liver via DBP and hydroxylated to 25(OH) vitamin D. The responsible enzyme for this reaction is the mitochondrial 25-hydroxylase (CYP27A1), which belongs to the family of cytochrome P-450. Accordingly, 25(OH) vitamin D is transported by DBP to the proximal tubule of the kidney, where the hormone is metabolized to 1,25$(OH)_2$ vitamin D by the enzyme 1α-hydroxylase. This vitamin is up to tenfold more active than vitamin D [43].

1α-Hydroxylase belongs to the family of cytochrome P-450 (CYP27B1). The main regulator of CYP27B1 activity, thus of 1,25$(OH)_2$ vitamin D synthesis, is PTH. Next to PTH, CYP27B1 is regulated by 1,25$(OH)_2$ vitamin D itself, creating a negative feedback to CYP27B1 expression [44]. CYP27B1 can adapt to extracellular calcium concentration in the form of a decrease of 1,25$(OH)_2$ vitamin D production if serum calcium is high and an increase of 1,25$(OH)_2$ vitamin D production if serum calcium is low, either through downregulation of CYP27B1 expression or regulation through CaSR [45]. Vitamin D and all of its metabolized products bind to DBP, the major protein of vitamin D transport next to albumin and lipoproteins [46, 47]. The mechanism of cellular uptake of vitamin D is relatively unknown. Since it is a very lipophilic molecule, passive diffusion is plausible. Another suggestion is that the 25(OH) vitamin D-DBP complex is collected by the endothelial cell in the proximal tubule via endocytosis.

1,25$(OH)_2$ vitamin D has effects on cells of the intestine, bone, and kidney. Two different ways of cellular response to 1,25$(OH)_2$ vitamin D exist: a slow genomic and a fast non-genomic effect on cells [1].

The genomic response is mediated by the binding of 1,25$(OH)_2$ vitamin D to the nuclear receptor VDR (vitamin D receptor), which has transcriptional abilities. Following the binding, the VDR couples with the retinoid C receptor (RCR) and formats a VDR-RCR complex. This complex can influence the VDRE in the 5′promotor region and therefore regulate gene activity.

In the intestine, 1,25$(OH)_2$ vitamin D can influence the transcellular pathway of calcium absorption by increasing the expression of TRPV6, calbindin-D9k, and PMCA. Additionally, 1,25$(OH)_2$ vitamin D may upregulate the paracellular mechanism of calcium absorption. Thus, $1,25(OH)_2$ vitamin D increases serum calcium by increasing the absorption of calcium. Patients suffering from $1,25(OH)_2$ vitamin D deficiency can only absorb 20% of the calcium that a healthy human is absorbing. These conditions illustrate the importance of $1,25(OH)_2$ vitamin D for calcium absorption in the intestine.

VDR was found to be located on osteoblasts, osteoclasts, and chondrocytes; thus, scientists are speculating $1,25(OH)_2$ vitamin D to have impact on bone generation and mineralization. $1,25(OH)_2$ vitamin D is affecting the differentiation of osteoblasts from mesenchymal stem cells, as well as different proteins synthesized by osteoblasts like RANKL, phosphatase, osteocalcin, osteopontin, etc. Certainly, most of $1,25(OH)_2$ vitamin D effects are operated by osteoblasts. The dependency of mineralization on $1,25(OH)_2$ vitamin D is under debate. Studies have illustrated that bone mineralization can be increased under vitamin D and vitamin K application. In contrast, VDR knockout mouse models displayed the opposite: Usually, these mice suffer from rickets and osteomalacia, since they are not able to absorb calcium in the intestine without the receptor. If animals are exposed to normal calcium concentration, interestingly, the bone can recover to a normal bone. Thereby, the influence of $1,25(OH)_2$ vitamin D as a mediator of differentiation and mineralization of bone seems to be ambiguous. A contrary result was found by another group, demonstrating a decrease of osteoblasts and bone volume under normocalcemic conditions in cyp27b/VDR (−/−) animals. They concluded that $1,25(OH)_2$ vitamin D is of high importance for bone metabolism.

In the kidney, calcium homeostasis occurs through regulating the secretion of calcium, which was absorbed in the primary urine previously. The reabsorption takes place in the proximal tubule. A very precise regulated calcium absorption takes place in the distal tubule and collecting duct. There, the proteins TRPV5 and TRPV6, calbindin-D28k, NCX1, and PMCA1b can be found, which ensure the uptake, transport, and extrusion of calcium. $1,25(OH)_2$ vitamin D can enhance these proteins and achieve a raise in calcium absorption in the kidney.

For the faster, non-genomic effects of $1,25(OH)_2$ vitamin D, the VDR is required, which is located subcellularly, in plasma membrane invaginations and caveolae. The microdomains of VDR were found to be localized on the intestine, kidney, and lung. The characteristic effects of $1,25(OH)_2$ vitamin D are modifying ion channel gating in osteoblasts, contraction of myocytes in the heart, secretion of insulin in the pancreas, and photoprotection in keratinocytes. Besides, $1,25(OH)_2$ vitamin D can bind secondary bile acid, lithocholic acid. Studies suggest that VDR functions as a secondary bile acid sensor and can activate CYP3A, which induces lithocholic acid breakdown, therefore having an autoprotective function. Secondary bile acid lithocholic acid is a carcinogenic agent and has a toxic effect on the intestinal mucosa.

Calcitonin

Calcitonin is a peptide hormone which can affect serum calcium levels. It is produced in the parafollicular cells, also called C-cells, of the thyroid gland and is released during increased serum calcium levels. A hypocalcemic effect is achieved by reducing osteoclast activity; more precisely, calcitonin can change the morphology, motility, and differentiation of osteoclasts, thus leading to less calcium mobilization from bone [48]. Calcitonin is also affecting the kidney. The role of calcitonin in the kidney is not entirely clear yet, but studies illustrated oppositional effects. On the one hand, calcitonin causes calciuria in humans, thereby decreasing serum calcium levels [49]. On the other hand, calcitonin can lead to calcium and magnesium reabsorption in the Henle loop in rats and rabbits [50, 51].

Apparently, calcitonin is only playing a minor role in calcium homeostasis compared to the hormones PTH and 1,25$(OH)_2$ vitamin D. This hypothesis is corroborated by studies on patients suffering from medullary carcinoma of the thyroid (MTC), which leads to calcitonin hypersecretion demonstrating no bone loss or increase in bone mineralization [52]. Furthermore, studies on rats with calcitonin depletion achieved by thyroidectomy showed no change in serum calcium levels [53].

Calcium Uptake

The uptake and absorption of dietary calcium occurs in the intestine. Another crucial player in regulation and calcium homeostasis is the kidney, which can excrete calcium in the urine.

The intestine is able to absorb 25–35% [54] of the total amount of the ingested dietary calcium, which is mainly extracted from dairy-based food [55]. Two different pathways are responsible for the calcium uptake into the systemic circulation: the paracellular and the transcellular pathways.

Paracellular Pathway

The paracellular pathway is dependent on a downhill concentration gradient between the luminal and extracellular space. It is a passive mechanism, which does not require ATP. The paracellular calcium uptake takes place throughout the whole small intestine in a constant manner. The absorption curve is non-saturable and constant, since it cannot be induced or activated. It has been proposed that the paracellular pathway is insensitive for 1,25$(OH)_2$ vitamin D and low calcium diet [56]. However, there is evidence that tight junctions can be mediated by 1,25$(OH)_2$

vitamin D, subsequently influencing the permeability of calcium between epithelial cells through relative leakiness of tight junctions [57, 58].

Transcellular Pathway

In contrast to the paracellular mechanism, calcium absorption via the transcellular pathway can occur toward an uphill gradient with the help of distinct calcium transport proteins. These proteins are expressed on both apical and basolateral sides of the enterocytes. It is considered as an active transport of calcium, which requires molecular energy in the form of ATP. Location of absorption is in most instances the duodenum, following the jejunum and with a decreasing capability of absorption in the more distal segments of the small intestine [59]. Since there is no dependency of any gradient, a possibility of calcium absorption at low calcium concentration in the chyme does exist. The absorption curve follows a saturable and exponential manner, due to the potential of 1,25$(OH)_2$ vitamin D to upregulate calcium absorption under conditions of restriction and diet. First evidence for 1,25$(OH)_2$ vitamin D-dependent calcium uptake against a high gradient in the small intestine were shown by the usage of calcium radioisotopes. The primary apical calcium uptake channel is the transient receptor potential vanilloid channel type 6 (TRPV6), which consists of six transmembrane and four ankyrin repeat domains, having the highest expression in the duodenum [60, 61]. The structure of the channel resembles voltage-gated cation channels, with the difference that the TRPV6 is not provided with the voltage sensor domain in the fourth α-helix and stays permanently in an open state during resting membrane potential [62]. The ankyrin domains were considered to be responsible for the formation of the tetramer; however, newer studies refute this hypothesis. The channel is highly selective for calcium and inward rectifying only. Interestingly, the TRPV6 is sensitive for intracellular calcium as well. High intracellular calcium concentrations were able to inhibit the receptor and therefore act as a negative feedback loop [63]. Latest studies from Al-Ansary et al. demonstrate the ATP dependency of TRPV6. With the direct binding of ATP, the channel undergoes a conformation change and unlocks from the inhibiting inactivated state to an open state. Next to ATP calcium, magnesium and protons can link to the channel and lead to the opponent effect, namely, inhibition and inactivation of the channel [64].

Since calcium is having a pivotal role in intracellular messaging and signaling, the regulation of intracellular calcium in enterocytes is of high importance. A number of TRPV6-associated proteins like S100A10-anexin complex, protein kinases like SGK1 and WNK3, as well as the sodium hydrogen exchanger regulating factor (NHERF4 aka PDZK2) interact in the tightly regulated and precise concentration control of intracellular calcium [65–67]. As mentioned before, intracellular calcium itself can inhibit TRPV6, creating an autoinhibitory feedback loop on the channel with a subsequent stop of transcellular calcium absorption. In addition, calmodulin (CaM), which is considered as the molecular calcium sensor, can bind in a calcium-dependent manner to the channel and therefore regulate the activity state of the channel. A correlation between TRPV6 and CaM was demonstrated by FRET studies, showing tight association when intracellular calcium is elevated and termination of the association under intracellular calcium depletion [68].

The 1,25(OH)$_2$ vitamin D-inducible binding protein calbindin-D9k, which was firstly found in chick intestines by Wassermann et al. in 1966, functions intracellularly as a calcium shuttling protein and allows the diffusion of calcium between the two poles the cell [69]. Albeit calbindin-D9k facilitates as a shuttle protein, it should be rather considered as a calcium gradient amplifier, since intracellular calcium concentrations are located in the nanomolar range and diffusion is heavily dependent on concentration gradients [70]. A linear correlation between calbindin-D9k and calcium absorption was described [70], as well as a possible inducement of calbindin-D9k by 1,25(OH)$_2$ vitamin D [71]; hence, it is not definite in which way 1,25(OH)$_2$ vitamin D can influence calbindin-D9k on a molecular level and highly controversial. With the help of various knockout mice, further investigations about 1,25(OH)$_2$ vitamin D dependency were conducted. Calbindin-D9k (−/−) mice showed neither an apparent phenotype alteration nor serum calcium shifts, but rather a similar response to 1,25(OH)$_2$ vitamin D relating to calcium absorption as wild-type animals. Scientists ascribed this phenomenon to an inducement of the expression of TRPV6 as a compensation and upregulation of calcium absorption to ensure sufficient calcium levels [72]. Next, a calbindin-D9k plus TRPV6 knockout mouse was created and also demonstrated no imbalances in calcium homeostasis [73]. These findings still remain unclear and missing explanations need further investigations.

As calcium reaches the basolateral membrane of the enterocyte, it is exported into the systemic circulation. With the release of calcium into the extracellular environment, the last step of transcellular calcium absorption is completed. The outer calcium concentration is higher than intracellular; hence, calcium needs to be transported against an uphill gradient which requires energy in the form of ATP. Calcium extrusion occurs via two different proteins: plasma membrane calcium ATPase (PMCA) and sodium calcium exchanger (NCX) [1].

PMCA belongs to the family of P-type primary ion transport and is localized on various tissue membranes. The protein is responsible for intracellular calcium homeostasis and extrudes one calcium ion into the extracellular space at the expense of one ATP [74]. The calcium affinity as well as the turnover of PMCA can be increased by CaM to a tenfold. Besides, calbindin-D9k can directly stimulate calcium extrusion by the PMCA, and 1,25(OH)$_2$ vitamin D can increase protein expression [75]. Since embryonic death occurs if PMCA is knocked out, the pivotal role of PMCA during development and organogenesis becomes clear.

The NCX needs three sodium ions to exchange one calcium ion into the extracellular space. The pump exploits the sodium gradient to extrude calcium out of the cell. Three different isoforms of the NCX are known, whereas the NCX1 is known to be localized in the intestine [76, 77]. In contrast to the PMCA, the NCX is not regulated and influenced by 1,25(OH)$_2$ vitamin D.

Moreover, two other possible calcium extruders were found on the basolateral membrane of the enterocytes in the intestine, both belonging to the potassium-dependent sodium calcium exchanger family (NCKX), namely, NCKX3 and NCKX4 [78]. Further clarification of the role of the NCKX is needed.

Gastric Acid Secretion

The mechanisms of production, secretion, and regulation of concentrated hydrochloric acid secretion from the parietal cells through the gastric glands and eventually to the end release into the stomach require a variety of highly specialized cells, both in the glands that secrete the acid and other important enzymes, to the surface cells in the stomach that are responsible for the production of a bicarbonate-rich mucus [1]. The generation and maintenance of low pH conditions through HCl production and release is necessary for sterilization and digestion of foodstuffs and is ensured by the parietal cell, also known as oxyntic cell [79]. Protection of the gastric mucosa and avoidance from tissue damage or autodigestion in the form of a mucus layer are guaranteed by mucus neck cells, which are creating a buffer gel lining across the gastric luminal epithelium. The gastric mucosa is equipped with a complex selection of endocrine cell types, which are integrated in the process of tight regulation of gastric acid secretion. Disturbances and imbalances in either regulation of gastric acid secretion or mucosal protection mechanism can lead to morphological changes of the gastric mucosa in the form of gastric ulcers [80].

H^+,K^+-ATPase

About 70–90 parietal cells can be typically found in a single gastric gland [81]. Since the parietal cell is dependent on energy in the form of ATP, it has a high abundance of mitochondria, reaching up to 40% of total cell volume, making it one of the richest mitochondrial cells [1].

Due to its structure and high homology to the Na^+,K^+-ATPases and the sarcoplasmatic reticulum Ca^{2+}-ATPase (SERCA), the gastric H^+,K^+-ATPase belongs to the family of P_2-type ATPases [81]. The structure of the H^+,K^+-ATPase is a heterodimer with an α-subunit and a β-subunit, which are arranged as $(\alpha\beta)_4$ tetramers on the surface of the parietal cell [82]. The β-subunit is heavily glycosylated and responsible for the stabilization of the α-subunit [83, 84], whereas the α-subunit has ten transmembrane domains and contains the catalytic center, which is in charge of the ion exchange. In addition, the β-subunit is capable of preventing a reversal of ion transport by a "ratchet"-like mechanism, which gives the H^+,K^+-ATPase the possibility to pump against a high proton gradient [85]. In this way, an intraluminal pH of 1–2 can be achieved. The H^+,K^+-ATPase is able to resolve an acid gradient up to six pH units, guaranteeing adequate acidification.

Two hypotheses about resting and stimulating states of the proton pumps (H^+,K^+-ATPase) in parietal cells exist. The first theory suggests that apical canaliculi extend into the parietal cell and H^+,K^+-ATPase is accumulated in microvilli tubulovesicles under resting conditions. Once hormonally or neuronally stimulated, tubulovesicles mobilize and start to fuse with the apical membrane resulting in an expansion of the canaliculi. The H^+,K^+-ATPase embeds in the membrane and releases H^+ into the gastric lumen. In the second hypothesis, basolateral ion exchangers like Na/H, Cl^-/HCO^{3-}, Na/K, and apical Cl^- channel are abounded in the cell membrane and perpetuate an electrogenic flow for Cl^- transport across the cell. In resting state, the H^+,K^+-ATPase appears in cytoplasmatic tubulovesicles. At this point, the pump does not transport H^+ because the vesicular permeability to K^+ is low. Under stimulation through a secretagogue, the H^+,K^+-ATPases containing tubulovesicles are mobilized and pumps are incorporated into the apical membrane to power gastric acid secretion so that a potassium conductance through the K^+-recycle mechanism is created. Electrogenic H^+ secretion can occur through the electroneutral H^+,K^+-ATPase. Apical Cl^- channels ensure Cl^- extrusion and electroneutrality [86]. The H^+,K^+-ATPase contains one binding site for hydrogen ion and four binding sites for potassium. When an occupation of all binding [sites] of the pump occur, a conformational change leads to a moves of cytosolic hydrogen ions into the gastric lumen at the expense of ATP. The stoichiometry is pH dependent and differs between two protons and two potassium ions for one ATP molecule at alkaline pH (>3.0) to one ion proton and one potassium for one ATP at acidic pH (<3.0) [87, 88].

The protons are capable of acidifying the gastric lumen to a pH of 1–2. To maintain the efflux of protons via the H^+,K^+-ATPase, a series of potassium channels are necessary to provide a constant efflux of K that can then be recycled by the ATPase [1]. Furthermore, in addition to protons, the parietal cells have a series of Cl secretion pathways that lead to a concurrent efflux of Cl providing the negative charge necessary to maintain the electrochemical process and the creation of HCl.

A recycling mechanism ensures the clearing of the pump after stimulation. The H^+,K^+-ATPase comes off the cell membrane by the formation of clathrin-coated pits and vesicle budding [1]. Huntingtin interacting protein 1-related (Hip1r) supports vesicle formation and membrane trafficking [89]. Investigations with Hip1r-deficient animals demonstrated a loss in tubulovesicles and decrease in acid secretion, as well as a decrease in the number of parietal cells [90].

Chloride Channels

The counterion conductance to protons is provided by chloride ions, which ensure the generation of HCl and electroneutrality. With the help of the patch-clamp technique, scientists were able to demonstrate the importance of chloride efflux and the presence of chloride conductance at the cell membrane of the parietal cell in *Necturus*, human parietal cell line HGT-1, and rabbit parietal cells [91]. Channels for apical chloride secretion in the parietal cell include the cystic fibrosis conductance regulator (CFTR), chloride channel protein 2 (ClC-2), and solute carrier 26A9 (SLC26A9) [92].

The CFTR channel is located on a variety of epithelia tissues, e.g., airways, intestine, and pancreas. A very popular correlated disease with this chloride channel is the mutation causing cystic fibrosis (CF) [92]. A plausible reason why CFTR plays an important role in gastric acid production is that secretion is significantly reduced in animals carrying the typical mutation for CF (ΔF508) and under the application of a specific CFTR inhibitor [92].

An alternative to the CFTR channel is the ClC-2, whose expression was found in rabbit gastric mucosa [93]. However, the involvement of the ClC-2 in gastric acid secretion remains unclear.

The SLC26A9 is a chloride-bicarbonate antiporter, which was found in the tubulovesicles of the parietal cell [94]. Mutation of the SLC26A9 leads to a dilatation of the parietal cell with a disappearance of tubulovesicles, which is having an immense impact on the acid production, but may be due in part to the altered morphology of the parietal cells [95].

Potassium Recycling

The H^+,K^+-ATPase is fueled by a K^+ recycling mechanism, which prevents luminal K^+ depletion [1]. K^+ is leaking into the lumen of the gland through a variety of potassium channels outside the parietal cell into the gland's lumen holding up the necessary supply for the H^+,K^+-ATPase. Responsible candidates for the K^+ extrusion are KCNQ1 (Kv7.1), KCNJ10 (K_{ir}4.1), KCNJ15 (K_{ir}4.2), KCNJ2 (K_{ir}2.1), and KCC4 [96]. Of note, the relevance of some of these channels and impact on K^+ extrusion remain under debate [1].

KCNQ1 was first identified in the heart, where its mutation is responsible for cardiac arrhythmias [1]. Later, studies with knockout mice KCNQ1 (−/−) demonstrated severe hypochlorhydria next to dilated gastric glands, vacuolated parietal cells, and gastric hyperplasia [97].

Members of the family of the inward-rectifier potassium channels (K_{ir}) were all confirmed to be on a mRNA level and immunohistochemically localized in the gastric mucosa [96]. More detailed investigations are necessary to fully understand the K_{ir} pathways and the role they play in the stomach.

Next to the potassium channels, Fujii et al. recently described the KCC4, a K-2Cl cotransporter [98]. They suggest that the KCC4 is functionally coupled to the H^+,K^+-ATPase due to the decrease in chloride and proton transport under pharmacological inhibition of the KCC4 [98]. The exact molecular pathway remains, like in all other potassium efflux hypotheses, unclear and is in need of more critical investigations.

Neuroendocrine Regulation

With regard to gut diseases caused by hypersecretion of gastric acid such as Zollinger-Ellison syndrome (ZES, gastrinoma), the importance and necessity of a very strict control of gastric acid production becomes clear. Hormonal (gastrin, somatostatin), paracrine (histamine, somatostatin), and neuronal (vagal) on-demand mechanisms are keeping the balance between low pH and tissue protection [81].

Vagus Nerve (Cholinergic Stimulation)

Before the pharmacological therapy of hypersecretion existed, surgical vagotomy was the effective treatment of hypersecretion disorders [99]. The parietal cell receives neuronal stimulation by cholinergic postganglionic enteric fibers of the enteric nervous system, which are embedded in the gastric mucosa [1]. These fibers follow orders from the efferent fraction of the vagus nerve and are triggered during the uptake of foodstuffs during the cephalic phase. The released acetylcholine (ACh) binds to muscarinic M_3 receptors of the parietal cell, leading to an intracellular increase of Ca^{2+}, mobilized from intracellular stores, in response to PLC-mediated IP_3 generation [100]. Kinases, which are involved in this transduction, are the protein kinase C (PKC) and calcium-/calmodulin-dependent protein kinase II (CaMKII). CaMKII has a clear stimulatory effect on acid secretion, whereas PKC has stimulatory and inhibitory effect on acid secretion [1]. Besides, the cholinergic stimulation activates the MAPKs, which are playing a dual role in acid secretion. On the one hand, MAPKs can lead to an acute inhibition of gastric secretion; on the other hand, they can cause chronic augmentation [1]. Next to this direct effect of ACh on the parietal cells, the vagus nerve is able to stimulate gastric acid secretion via an indirect pathway. ACh can bind muscarinic M_3 receptors on the basolateral surface of G-cells, which are in the vicinity of parietal cells [1]. This binding causes the release of the peptide hormone gastrin [81]. Gastrin binds to CCK_2 receptors, which are located on both parietal cells and ECL cells [81]. In this way the previously described intracellular transduction pathway occurs in parietal cells [1]. With gastrin binding to the surface on ECL cells, histamine is released and binds to H_2 receptors on the basolateral membrane of the parietal cell [81].

G-Cells (Gastrin)

The peptide hormone gastrin is released by G-cells found in the antral section of the stomach but also by endocrine cells located in the small and large intestine, pancreas, testis, and pituitary gland. It functions as the most important mediator of gastric acid secretion, mucosal cell growth, and calcium homeostasis in the stomach

[1]. During the gastric phase of food digestion, which is characterized by foodstuff reaching the stomach, G-cells are highly activated and secrete gastrin. G-cells can respond to a direct neuronal stimulation in the form of ACh and gastrin-releasing peptide (GRP) coming from postganglionic neurons of the ENS [101]. Other direct and food-related triggers of gastrin release are calcium, amino acids, and amines [102]. New studies suggest the CaSR to be a potent activator of the G-cells [25]. Scientists are having several explanations for this condition: The same dietary components (amino acids, amines, calcium) activating gastrin release from G-cells can also activate the CaSR. The CaSR is on both apical and basolateral surface of the G-cell and can be considered as a nutrient sensor of the gastric lumen and the bloodstream. In addition, the activation of CaSR leads to acid secretion. Furthermore, CaSR (−/−) animals cannot respond with gastrin secretion to intraluminal dietary components [25]. Gastrin release evolves from a rise of intracellular calcium with a subsequent vesicle fusion and gastrin release [81].

A negative feedback mechanism is necessary to avoid acid hypersecretion. Low intraluminal pH stops gastrin release, whereas an alkaline pH promotes gastrin release, a well-known occurrence during proton pump inhibitor (PPI) therapy. Gastric acid directly stimulates somatostatin release from ambient antral D-cells, the main inhibitor of gastric acid secretion [103]. Apart from that, gastrin release can be directly inhibited by neuronal regulation of the ENS and its neurotransmitter galanin [104].

After the release of gastrin, it binds to the cholecystokinin receptor type 2 (CCK2) located on the membrane of parietal cells and ECL cells [105]. CCK2 is a seven-transmembrane domain G-protein-coupled receptor, causing an increase of intracellular calcium concentration during activation. On ECL cells, gastrin stimulates the release of histamine, which again stimulates the parietal cell through a paracrine fashion via the H_2 receptor. This link between G-cells, ECL cells, and parietal cells is called the gastrin-histamine axis. Besides, gastrin can directly bind the CCK_2 receptor on the parietal cell and stimulate the H^+,K^+-ATPase instantaneously. Of note is that gastrin acts as a proliferate signal for the gastric mucosa and that elevated blood gastrin levels are correlated to substantial mucosal proliferation [106].

ECL Cells (Histamine)

Popielski et al. were the first scientists to describe histamine's paracrine effect on the parietal cell and its independency on the vagal nerve [1]. The amino acid histidine serves as a precursor for histamine, directly converted by the enzyme L-histidine decarboxylase (HDC) [107]. Gastrin and neuronal signals activate secretory granules of the ECL cell to fuse with the membrane and release the containing histamine. Gastrin activates the CCK_2 receptors, which leads on the one hand to an increase of HDC expression due to an increase of gene transcription via the PKC- and ERK-dependent pathway [108]. On the other hand, the activation of CCK_2

causes an increase of the transcription of the vesicle monoamine transporter type 2 (VMAT2), which is necessary for the accumulation of histamine in the secretory vesicles [109]. This happens through the same PKC- and ERK-dependent pathway as described before. Furthermore, gastrin leads to vesicle fusion with the membrane. Therefore, intracellular calcium concentration increases due to intra- and extracellular calcium mobilization: IP_3-mediated intracellular calcium store release and calcium influx from the extracellular space through L-type calcium channels [110]. The vesicle fusion is based on the confirmed expression of the core SNARE complex in the ECL cells with the proteins syntaxin, synaptobrevin, and SNAP-25.

Another stimulus for ECL cells is a neuropeptide found in the ENS of the gastric mucosa, called pituitary adenylate cyclase-activating polypeptide (PACAP). PACAP can bind to the PAC-1 receptor of ECL cells, which triggers the release of histamine into the surrounding milieu by increasing intracellular calcium concentration via L-type and ligand-gated calcium channels [1]. Furthermore, PACAP increases HDC expression and has a trophic effect on the ECL cell [111].

Inhibition of histamine release of ECL cells occurs through somatostatin produced by D-cells. Somatostatin reaches the somatostatin receptor (SST_2 and SST_5) of the ECL cell in a paracrine fashion and blocks L-type calcium channels following an inhibition of vesicle fusion and exocytosis [112]. Somatostatin can also inhibit the proliferation of ECL cells and can be seen as the global antagonist to gastrin [1]. Neuronal inhibition of ECL cells occurs through the neuropeptide galanin showing similar molecular effects as somatostatin [113]. Other inhibitors of histamine release are prostaglandin E, nitric oxide (NO), neuropeptide YY (PYY), and calcitonin gene-related peptide (CGRP); hence, the effects of PYY and CGRP need further investigations [113, 114].

Only a small amount of histamine in the surrounding of parietal cells is necessary to induce gastric acid secretion. Histamine binds to the H_2 receptor of the parietal cell, which belongs to the family of seven-transmembrane domain GPCRs. Activation of the H_2 receptor causes intracellular increase of cAMP induced by activation of the adenylate cyclase via the G_s domain of the receptor. Besides, intracellular Ca^{2+} increases as a consequence of intracellular Ca^{2+} store mobilization and extracellular Ca^{2+} influx. This signal cascade leads to an increase of gastric acid secretion following a trafficking of the H^+,K^+-ATPase-containing tubulovesicles [115]. Studies with H_2 (−/−) knockout animals highlighted the importance of the histamine-gastrin axis by complete insensitivity of the parietal cell toward gastrin and histamine with absence of acid secretion. However, carbachol was still able to induce secretion [116].

D-Cells (Somatostatin)

The peptide hormone somatostatin, which is released by D-cells, plays the opponent role toward acid secretagogues. Two different kinds of D-cells exist: The D-cell located in the gastric mucosa, which can directly inhibit parietal cells and histamine release of ECL cells, and the antral, respectively, intestinal form, which is regulating the release of gastrin from the G-cells. The release of somatostatin is regulated by a negative feedback mechanism [117, 118]. Surrounding gastrin and cholecystokinin concentrations, as well as intraluminal pH, regulate the secretion of somatostatin by D-cells. If the extracellular gastrin secretion is high, somatostatin secretion is induced, resulting in an inhibition of further gastrin release by G-cells [119]. Cholecystokinin demonstrated a stimulatory effect on somatostatin release via CKK_1. Cholecystokinin is secreted by duodenal I-cells and is known to be the classical mediator of the intestinal phase of gastric secretion [120]. Besides its inhibitory effect on gastric acid secretion, cholecystokinin is stimulating gallbladder contraction. Another important mediator of somatostatin release is the intraluminal pH. Low gastric pH correlates with a high somatostatin release. However, the sensor mechanism of D-cell remains elusive. One suggestion is that antral D-cells are capable to measure intraluminal pH permanently with the help of a distinct morphological feature facing the luminal side [121]. Yet, the oxyntic D-cells are excluded from this pH sensor mechanism, since they are embedded in the mucosa without any contact to the lumen. Another theory is that D-cells are equipped with the calcium-sensing receptor (CaSR), whose expression on gastric mucosa was confirmed [1]. Neuronal pH sensing via the neuropeptide CGRP is under debate, since CGRP receptor blocker seems to inhibit the release of somatostatin and leads to acid secretion. Knockout of the somatostatin receptor leads to a tenfold higher basal gastric acid secretion, demonstrating the pivotal role of somatostatin as an antagonist and regulator of gastric acid secretion and secretagogues. Effects of somatostatin are reached by binding to the SST_2 receptor of parietal cells, G-cells, and ECL cells [122]. In the parietal cell, somatostatin causes an activation of G_i with a subsequent decrease of adenylate cyclase, following decrease of intracellular cAMP level [123].

Other Substances

Secretin: Secretin is a peptide hormone synthesized by duodenal S-cells in response to a low duodenal pH. It influences pancreatic bicarbonate secretion and therefore buffers the gastric acid coming into the duodenal lumen. Besides, it inhibits gastric motility and acid secretion. However, the exact pathway remains unclear [124].

Oxyntomodulin: Oxyntomodulin was found in mammalian intestine and has high similarity to glucagon. On isolated gastric glands, it stimulates gastric secretion, whereas in integrated parietal cells, it decreases gastric acid [125].

Serotonin: Serotonin can be found in vesicles of the enterochromaffin cells of the antrum of the stomach and is released at vagal stimulation and low intraluminal pH. Serotonin inhibits gastric acid secretion [126].

Neurotensin: Neurotensin is secreted by N-cells in the small intestine and decreases gastric acid secretion and gastric motility [127].

Ghrelin: Recent studies demonstrate the impact of ghrelin on gastric acid secretion, appetite, and bone remodeling. It is synthesized in P/D1 cells of the gastric fundus. Investigations showed the various effects of ghrelin, including stimulation and inhibition, regardless of its effect on gastric acid secretion. Clarification in the form of further studies is needed [128].

Nitric oxide: NO is a very important intracellular signaling molecule and is a main player in vasodilatation, immune response, and acid secretion. The effect of NO on gastric acid is inhibitory, either due to direct inhibition of the parietal cell or indirect inhibition of histamine release from ECL cells. NO causes intracellular increase of cGMP and guanylate cyclase [129].

Interleukin: Parietal and ECL cells express the IL-1 receptor and are potent targets for interleukins, immune response coordinators. IL-1 showed an inhibition of acid secretion [130].

Proton Pump Inhibitors and Other Acid Suppressants

The indications for an inhibition of gastric acid secretion are all acid-related disorders such as gastroesophageal reflux disease (GERD), nonerosive reflux disease (NERD), esophagitis, and peptic ulcer disease (PUD). State-of-the-art therapy for increased gastric acid is the application of proton pump inhibitors (PPI), targeting the final step of the formation of HCl, namely, the H^+,K^+-ATPase. Next to PPIs, acid pump inhibitors (APAs) are potent inhibitors of the H^+,K^+-ATPase. H_2 antagonists, such as ranitidine or cimetidine, can also achieve a decrease of gastric acid secretion by the inhibition of the hormonal signaling pathway through the H_2 receptor. Another pharmacological way to decrease intraluminal pH is the application of antacids, which are neutralizing the secreted gastric acid [1].

Omeprazole was the first clinically prescribed PPI [131]; recently, the enantiomer esomeprazole became one of the most successfully marketed pharmaceuticals in the United States [1].

PPIs are now one of many over-the-counter medicines for the hypersecretion of acid and are available to everyone, since they were considered to be safe with only a low incidence of adverse effects [132]. However, the most recent studies demonstrate the potential longtime side effect of PPI intake and their negative influence on bone health.

After the prodrug of the PPI reaches the parietal cell via the bloodstream and passes the cell membrane, the intracellular acidic conditions of the secretory canaliculus convert it to the active form called cyclic sulfenamide [133]. The pK_a value of 4, consequently the preferred accumulation and conversion at low pH environments of the prodrug, identifies PPIs as highly specific drugs [1]. In their activated form, PPIs are binding irreversibly and covalently to the H^+,K^+-ATPase by forming disulfide bonds, thus blocking its capability to pump out protons [134]. The members of the PPI family are distinguished from each other in terms of the involved reaction of cysteine residues; however, cysteine-813 interacts with all of them [134]. Cysteine-892 is typical for omeprazole, cysteine-321 for lansoprazole, and cysteine-822 for pantoprazole and tenatoprazole [1]. The effect of the PPI lasts usually 0.5–2 h (well beyond their half-life time), until they are metabolized by the cytochrome P-450 in the liver [135] (Fig. 3).

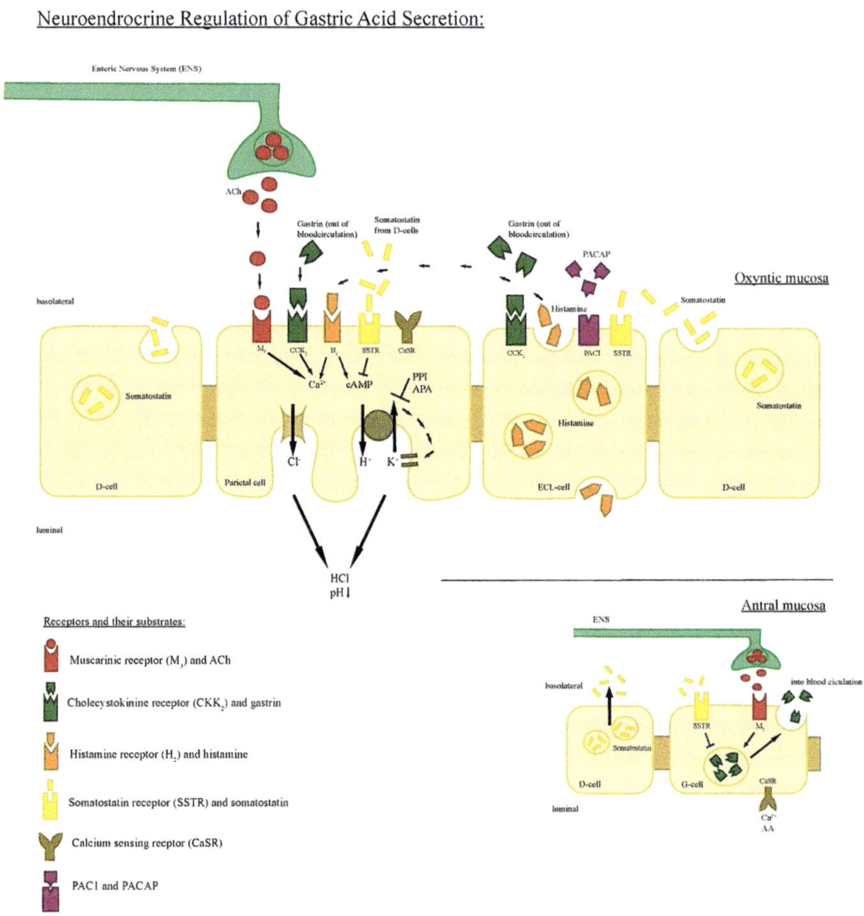

Fig. 3 Neuronal and endocrine regulation of gastric acid secretion: The parietal cell receives stimuli directly from neurons, which are releasing ACh molecules, and, in an indirect, paracrine fashion from vicinal D-cells and ECL cells. Gastrin secretion occurs in the antral part of the stomach and reaches the oxyntic mucosa through the bloodstream. Gastrin can either directly stimulate the parietal cell or indirectly by binding to the CCK_2 receptors on ECL cells, which will lead to a subsequent histamine release. The gastrin-mediated histamine release incorporates the gastrin-histamine axis. All of these pathways lead to an increase of gastric acid secretion. Gastrin release is pH dependent and stops when intragastric pH becomes more acidic. The negative feedback loop is tied to somatostatin release via D-cells.

Stomach, Calcium, and Bone: The Link

After exemplifying and reviewing the physiology of bone calcium homeostasis, gastric acid secretion, intestinal calcium absorption, and their hormonal and neuronal regulation, this section will highlight the functional intersections between the stomach, calcium, and bone. We include a discussion of normal and pathophysiology disease patterns and clinical conditions, which have an impact on the calcium homeostasis cycle.

Inherited CaSR Mutations

Clinical pathologies which arose from CaSR mutations illustrate the pivotal role of the receptor in calcium homeostasis. The threshold of the receptor is changed in either direction. Examples of CaSR pathologies are the familial hypocalciuric hypercalcemia (FHH, OMIM 145980) and neonatal severe hyperparathyroidism (NSHPT, OMIM 239200), typical consequences of loss-of-function mutations. The patients suffer from hypercalcemia due to reduced calcium sensing and decreased calcium excretion. The pathology of autosomal dominant hypocalcemia (OMIM 601198) is the result of an activating mutation. Here, the increased sensitivity of the receptor causes hypocalcemia next to an increase in calcium excretion via the kidneys. More than 300 mutations are reported to date, yet the majority of the mutations are characterized by inactivation of the receptor [136] (Fig. 4).

Gain-of-function mutation of the receptor can cause autosomal dominant hypocalcemia (ADH) with symptoms of decreased serum calcium and decreased calcium resorption in kidneys.

PPI and Bone Fracture

Proton pump inhibitors (PPIs), such as esomeprazole and lansoprazole, are used in state-of-the-art therapy of gastroesophageal reflux disease (GERD), esophagitis, and gastric ulcer disease (GUD) in the consequence of prolonged gastric acid exposure. Other ways to suppress hypersecretion of gastric acid are buffer agents (bicarbonate and carbonate) and H_2 blockers.

PPIs realize their medication effect by inhibiting the release of the acid component of HCl in the form of blocking the H^+-K^+-ATPase irreversibly. The widespread usage of PPI can be accounted to the easy access to drug and over-the-counter policy. Unfortunately, PPI users and physicians often do not deliberate enough about whether benefits are worth the potential risks PPIs can cause. Possible adverse

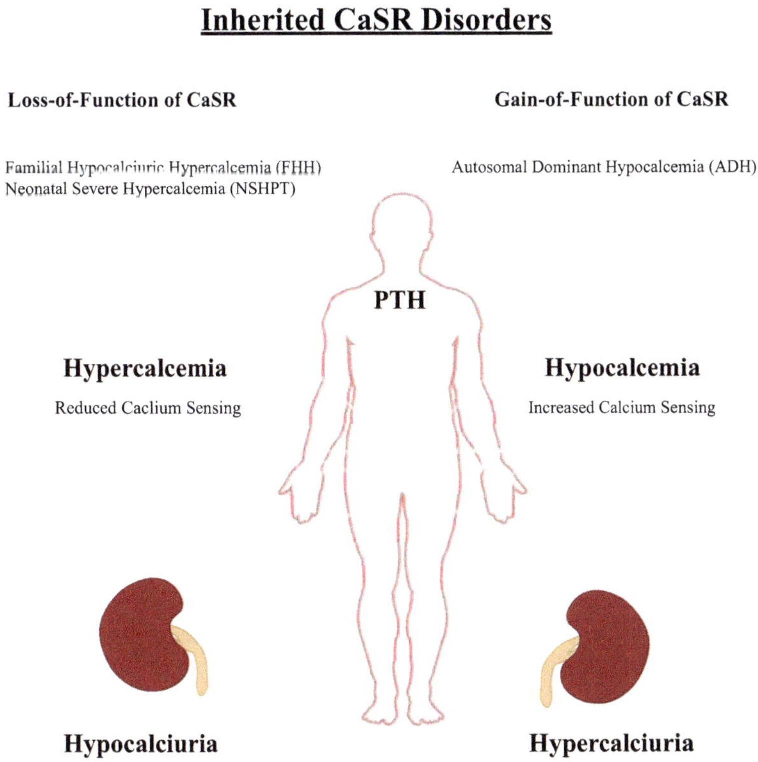

Fig. 4 Mutations of the CaSR may lead to abnormalities in the receptor in the form of loss-of-function or gain-of-function. Typical loss-of-function mutations lead to the diseases familial hypocalciuric hypercalcemia (FHH) and neonatal severe hypercalcemia (NSHPT), which are characterized by elevated serum calcium levels and decreased calcium excretion via the urine

effects after the intake of PPIs are higher risks of osteoporosis, enteric infections, and community-acquired pneumonia [137]. In 2010 the Food and Drug Administration (FDA) released a safety warning about the usage of PPIs emphasizing the possible increased risk of hip, wrist, and spine fractures. One of the largest studies about the correlation between increased risk of fractures and PPI intake is made by Yang et al. with the conclusion that long-term intake of PPIs for more than 1 year is associated with a higher incidence of hip fractures in humans over the age of 50 compared to a control group [138]. The ongoing debate about an overly prescription of PPIs, next to the fact that they are the third commonly prescribed medication in the United States, and the establishment of an OTC availability in Northern America should concern physicians and patients.

Gastrectomy and Vagotomy

The observation of fractures under PPI intake leads to the hypothesis that a reduction in gastric acid through inhibition of the main player of acid production, namely, the H^+-K^+-ATPase, prevents an adequate calcium absorption, thus leading to decreased bone density. This assumption is justified by a high number of animal studies as well as patient observations. Next to PPIs, gastrectomies and vagotomies, which were former therapies of gastric acid control, result in the development of osteoporosis, as well [139]. Differentiating between the effect of PPIs on gastric acid secretion and a total gastrectomy, the latter has an additional effect on stomach emptying, emulsification of foodstuffs, and food habits. Furthermore, there are differences in the procedure of gastrectomy; some sustain the duodenum (Billroth I) and others bypass it (Billroth II, Roux-en-Y, total gastrectomy). Gastrectomized patients showed to have a low 25(OH)-vitamin D and high 1,25$(OH)_2$-vitamin D [140]. The pathophysiology of this observation is not entirely clear yet. Scientists argue for the decrease in 25(OH)-vitamin D with an impaired uptake rate of vitamin D as a result of the surgery, yet the broad agreement is that vitamin D absorption is not impaired in these patients but might be a consequence of improper nutrition, fat and milk intolerance with insufficient intake of fat-soluble vitamin D [38, 141]. However, Billroth I and Billroth II demonstrate same bone loss rate, although a bypass of the duodenum is linked to a stronger fat malabsorption [142]. Another study suggests an impairment of calcium absorption being responsible for the decrease of bone density, high amounts of 1,25$(OH)_2$-vitamin D promote a decline of 25(OH)-vitamin D, and the increase in 1,25$(OH)_2$-vitamin D level is more likely due to an upregulation consequence of insufficient calcium absorption [143]. Still, the pathophysiology of bone loss after gastrectomy remains elusive, and an exact attribution of vitamin D shortage, calcium absorption impairment, or malnutrition following a diet is not possible. Secondary hyperparathyroidism does manifest in gastrectomized patients [1].

Studies of vagotomized patients demonstrated an abolishment of parasympathetic effects on the stomach in the form of a decreased gastric acid secretion. These patients showed the same constellation of vitamin D parameters as gastrectomized patients, low 25(OH)-vitamin D level accompanied with a high 1,25$(OH)_2$-vitamin D rate, assuming a positive feedback to compensate the low calcium absorption. In addition, serum calcium concentration is decreased, and due to an elevated 1,25$(OH)_2$-vitamin D, intestinal calcium absorption increased [144].

An interesting study by Granzinani et al. highlighted the effects of PPI on mineral metabolism and bone health. A group of eight healthy patients under PPI treatment (3x/d 20 mg) showed no elevation of serum calcium concentration after food intake together with a decreased calcium excretion in their urine. In contrast, the control group showed a rise in postprandial serum calcium concentration [145]. Another more recent study illustrated the impairment of calcium absorption under PPI regime with the help of radiolabeled calcium isotopes. A group of elderly and fasted women experienced a significant reduction of calcium absorption after 7 days of 20 mg PPI intake once a day. This did not occur to the control, placebo group [146]. However, this and other studies are under debate, since the experimental techniques for measuring calcium absorption vary a lot and radioactive tracers, as well as the use of different calcium salts, may manipulate calcium uptake results. In conclusion, it is difficult to find the correlation between a decrease in gastric acid secretion and calcium absorption on an experimental basis, and further investigation is needed to unmask this link.

Next to human test studies, many investigations were conducted on animals to demonstrate osteopenia as a consequence of calcium uptake impairment under gastric acid hyposecretion.

In an elegant study by Axelson et al., serum calcium concentration was measured in rats which had undergone several types of surgeries. In the first set of experiments, serum calcium level was measured in parathyroidectomized rats, which was reduced. In the next set, rats had undergone gastric operations, such as gastrectomy, antrectomy, and vagotomy, which had only a little to no effect on serum calcium levels. However, calcium uptake rates in these rats seem to be elevated, most likely as a consequence of the upregulation of PTH secretion. The third set represented parathyroidectomized rats combined with a gastric surgery, which resulted in a massive depletion of calcium and animal death after just a few days. This observation let the scientists conclude that acid secretion is a pivotal requirement for a sustained calcium homeostasis [147].

Several studies were made on vagotomized rats, showing that vagotomy alone has no effect on calcium absorption. When rats underwent vagotomy and parathyroidectomy together, intestinal calcium absorption was heavily affected compared to animals after vagotomy or parathyroidectomy alone [148].

In a lately published study by Schinke et al., $CCK_2(-/-)$ knockout mice were used to represent a complete achlorhydric state. These mice demonstrated an extreme calcium deficiency, osteoporosis, and secondary hyperparathyroidism. As mentioned before, these rats were not able to produce hydrochloric acid while the stomach morphology remained intact [149].

Conclusion

This chapter should elucidate the role of the stomach in the regulation of bone density by the modulation of calcium absorption and the ever-changing needs in bone to maintain a healthy skeletal environment. Furthermore, we have discussed the effects of mutations, diseases, and pharmaceutical influence on bone health. The intertwined mechanisms between gastric and intestinal physiology and calcium absorption are having great impact on the maintenance of bone health and are in need of a strict regulation. With the combination of receptors, channels, and transport proteins, a tight control of serum calcium concentration can be ensured for a physiological bioavailability to bone.

References

1. Kopic S, Geibel JP. Gastric acid, calcium absorption, and their impact on bone health. Physiol Rev. 2013;93(1):189–268.
2. Bushinsky DA, Lechleider RJ. Mechanism of proton-induced bone calcium release: calcium carbonate-dissolution. Am J Phys. 1987;253(5 Pt 2):F998–1005.
3. Jouret F, Wu J, Hull M, Rajendran V, Mayr B, Schofl C, et al. Activation of the Ca(2)+-sensing receptor induces deposition of tight junction components to the epithelial cell plasma membrane. J Cell Sci. 2013;126(Pt 22):5132–42.
4. Dufner MM, Kirchhoff P, Remy C, Hafner P, Muller MK, Cheng SX, et al. The calcium-sensing receptor acts as a modulator of gastric acid secretion in freshly isolated human gastric glands. Am J Physiol Gastrointest Liver Physiol. 2005;289(6):G1084–90.
5. Alfadda TI, Saleh AM, Houillier P, Geibel JP. Calcium-sensing receptor 20 years later. Am J Phys Cell Physiol. 2014;307(3):C221–31.
6. Geibel JP, Hebert SC. The functions and roles of the extracellular Ca2+-sensing receptor along the gastrointestinal tract. Annu Rev Physiol. 2009;71:205–17.
7. Marie PJ. The calcium-sensing receptor in bone cells: a potential therapeutic target in osteoporosis. Bone. 2010;46(3):571–6.
8. Riccardi D, Brown EM. Physiology and pathophysiology of the calcium-sensing receptor in the kidney. Am J Physiol Renal Physiol. 2010;298(3):F485–99.
9. Bai M, Trivedi S, Brown EM. Dimerization of the extracellular calcium-sensing receptor (CaR) on the cell surface of CaR-transfected HEK293 cells. J Biol Chem. 1998;273(36):23605–10.
10. Ray K, Clapp P, Goldsmith PK, Spiegel AM. Identification of the sites of N-linked glycosylation on the human calcium receptor and assessment of their role in cell surface expression and signal transduction. J Biol Chem. 1998;273(51):34558–67.
11. Zhuang X, Adipietro KA, Datta S, Northup JK, Ray K. Rab1 small GTP-binding protein regulates cell surface trafficking of the human calcium-sensing receptor. Endocrinology. 2010;151(11):5114–23.
12. Huang Y, Niwa J, Sobue G, Breitwieser GE. Calcium-sensing receptor ubiquitination and degradation mediated by the E3 ubiquitin ligase dorfin. J Biol Chem. 2006;281(17):11610–7.
13. Poon G. Cinacalcet hydrochloride (Sensipar). Proc (Baylor Univ Med Cent). 2005;18(2):181–4.
14. Bourdeau A, Souberbielle JC, Bonnet P, Herviaux P, Sachs C, Lieberherr M. Phospholipase-A2 action and arachidonic acid metabolism in calcium-mediated parathyroid hormone secretion. Endocrinology. 1992;130(3):1339–44.
15. Russell J, Bar A, Sherwood LM, Hurwitz S. Interaction between calcium and 1,25-dihydroxyvitamin D3 in the regulation of preproparathyroid hormone and vitamin D receptor messenger ribonucleic acid in avian parathyroids. Endocrinology. 1993;132(6):2639–44.
16. Imanishi Y, Kawata T, Kenko T, Wada M, Nagano N, Miki T, et al. Cinacalcet HCl suppresses Cyclin D1 oncogene-derived parathyroid cell proliferation in a murine model for primary hyperparathyroidism. Calcif Tissue Int. 2011;89(1):29–35.
17. Pi M, Chen L, Huang M, Luo Q, Quarles LD. Parathyroid-specific interaction of the calcium-sensing receptor and G alpha q. Kidney Int. 2008;74(12):1548–56.
18. Ba J, Brown D, Friedman PA. Calcium-sensing receptor regulation of PTH-inhibitable proximal tubule phosphate transport. Am J Physiol Renal Physiol. 2003;285(6):F1233–43.
19. Shareghi GR, Stoner LC. Calcium transport across segments of the rabbit distal nephron in vitro. Am J Phys. 1978;235(4):F367–75.
20. Loffing J, Loffing-Cueni D, Valderrabano V, Klausli L, Hebert SC, Rossier BC, et al. Distribution of transcellular calcium and sodium transport pathways along mouse distal nephron. Am J Physiol Renal Physiol. 2001;281(6):F1021–7.
21. Topala CN, Schoeber JP, Searchfield LE, Riccardi D, Hoenderop JG, Bindels RJ. Activation of the Ca2+−sensing receptor stimulates the activity of the epithelial Ca2+ channel TRPV5. Cell Calcium. 2009;45(4):331–9.

22. Renkema KY, Velic A, Dijkman HB, Verkaart S, van der Kemp AW, Nowik M, et al. The calcium-sensing receptor promotes urinary acidification to prevent nephrolithiasis. J Am Soc Nephrol. 2009;20(8):1705–13.
23. Sands JM, Flores FX, Kato A, Baum MA, Brown EM, Ward DT, et al. Vasopressin-elicited water and urea permeabilities are altered in IMCD in hypercalcemic rats. Am J Phys. 1998;274(5 Pt 2):F978–85.
24. Buchan AM, Squires PE, Ring M, Meloche RM. Mechanism of action of the calcium-sensing receptor in human antral gastrin cells. Gastroenterology. 2001;120(5):1128–39.
25. Feng J, Petersen CD, Coy DH, Jiang JK, Thomas CJ, Pollak MR, et al. Calcium-sensing receptor is a physiologic multimodal chemosensor regulating gastric G-cell growth and gastrin secretion. Proc Natl Acad Sci U S A. 2010;107(41):17791–6.
26. Busque SM, Kerstetter JE, Geibel JP, Insogna K. L-type amino acids stimulate gastric acid secretion by activation of the calcium-sensing receptor in parietal cells. Am J Physiol Gastrointest Liver Physiol. 2005;289(4):G664–9.
27. Remy C, Kirchhoff P, Hafner P, Busque SM, Mueller MK, Geibel JP, et al. Stimulatory pathways of the calcium-sensing receptor on acid secretion in freshly isolated human gastric glands. Cell Physiol Biochem. 2007;19(1–4):33–42.
28. Cheng I, Qureshi I, Chattopadhyay N, Qureshi A, Butters RR, Hall AE, et al. Expression of an extracellular calcium-sensing receptor in rat stomach. Gastroenterology. 1999;116(1):118–26.
29. Chattopadhyay N, Yano S, Tfelt-Hansen J, Rooney P, Kanuparthi D, Bandyopadhyay S, et al. Mitogenic action of calcium-sensing receptor on rat calvarial osteoblasts. Endocrinology. 2004;145(7):3451–62.
30. Garner SC, Pi M, Tu Q, Quarles LD. Rickets in cation-sensing receptor-deficient mice: an unexpected skeletal phenotype. Endocrinology. 2001;142(9):3996–4005.
31. Pi M, Garner SC, Flannery P, Spurney RF, Quarles LD. Sensing of extracellular cations in CasR-deficient osteoblasts. Evidence for a novel cation-sensing mechanism. J Biol Chem. 2000;275(5):3256–63.
32. Pi M, Chen L, Huang MZ, Zhu W, Ringhofer B, Luo J, et al. GPRC6A null mice exhibit osteopenia, feminization and metabolic syndrome. PLoS One. 2008;3(12):e3858.
33. Dvorak MM, Chen TH, Orwoll B, Garvey C, Chang W, Bikle DD, et al. Constitutive activity of the osteoblast Ca2+-sensing receptor promotes loss of cancellous bone. Endocrinology. 2007;148(7):3156–63.
34. Hess AF. The prevention and cure of rickets by sunlight. Am J Public Health (N Y). 1922;12(2):104–7.
35. Hollander D, Muralidhara KS, Zimmerman A. Vitamin D-3 intestinal absorption in vivo: influence of fatty acids, bile salts, and perfusate pH on absorption. Gut. 1978;19(4):267–72.
36. Hollander D, Truscott TC. Mechanism and site of small intestinal uptake of vitamin D3 in pharmacological concentrations. Am J Clin Nutr. 1976;29(9):970–5.
37. Rautureau M, Rambaud JC. Aqueous solubilisation of vitamin D3 in normal man. Gut. 1981;22(5):393–7.
38. Davies M, Mawer EB, Krawitt EL. Comparative absorption of vitamin D3 and 25-hydroxyvitamin D3 in intestinal disease. Gut. 1980;21(4):287–92.
39. Dueland S, Pedersen JI, Helgerud P, Drevon CA. Transport of vitamin D3 from rat intestine. Evidence for transfer of vitamin D3 from chylomicrons to alpha-globulins. J Biol Chem. 1982;257(1):146–50.
40. Dueland S, Pedersen JI, Helgerud P, Drevon CA. Absorption, distribution, and transport of vitamin D3 and 25-hydroxyvitamin D3 in the rat. Am J Phys. 1983;245(5 Pt 1):E463–7.
41. Dueland S, Helgerud P, Pedersen JI, Berg T, Drevon CA. Plasma clearance, transfer, and distribution of vitamin D3 from intestinal lymph. Am J Phys. 1983;245(4):E326–31.
42. Adams JS, Clemens TL, Parrish JA, Holick MF. Vitamin-D synthesis and metabolism after ultraviolet irradiation of normal and vitamin-D-deficient subjects. N Engl J Med. 1982;306(12):722–5.

43. Holick MF, Schnoes HK, DeLuca HF, Suda T, Cousins RJ. Isolation and identification of 1,25-dihydroxycholecalciferol. A metabolite of vitamin D active in intestine. Biochemistry. 1971;10(14):2799–804.
44. Booth BE, Tsai HC, Morris RC Jr. Vitamin D status regulates 25-hydroxyvitamin D3-1 alpha-hydroxylase and its responsiveness to parathyroid hormone in the chick. J Clin Invest. 1985;75(1):155–61.
45. Bland R, Walker EA, Hughes SV, Stewart PM, Hewison M. Constitutive expression of 25-hydroxyvitamin D3-1alpha-hydroxylase in a transformed human proximal tubule cell line: evidence for direct regulation of vitamin D metabolism by calcium. Endocrinology. 1999;140(5):2027–34.
46. Haddad JG Jr, Walgate J. 25-hydroxyvitamin D transport in human plasma. Isolation and partial characterization of calcifidiol-binding protein. J Biol Chem. 1976;251(16):4803–9.
47. Silver J, Fainaru M. Transport of vitamin D sterols in human plasma: effect of excess vitamin D, 25 hydroxyvitamin D and 1,25 dihydroxyvitamin D. Eur J Clin Investig. 1979;9(6):433–8.
48. Friedman J, Raisz LG. Thyrocalcitonin: inhibitor of bone resorption in tissue culture. Science. 1965;150(3702):1465–7.
49. Ardaillou R, Vuagnat P, Milhaud G, Richet G. Effects of thyrocalcitonin on the renal excretion of phosphates, calcium and hydrogen ions in man. Nephron. 1967;4(5):298–314.
50. Carney SL. Calcitonin and human renal calcium and electrolyte transport. Miner Electrolyte Metab. 1997;23(1):43–7.
51. Berndt TJ, Knox FG. Effects of parathyroid hormone and calcitonin on electrolyte excretion in the rabbit. Kidney Int. 1980;17(4):473–8.
52. Wuster C, Raue F, Meyer C, Bergmann M, Ziegler R. Long-term excess of endogenous calcitonin in patients with medullary thyroid carcinoma does not affect bone mineral density. J Endocrinol. 1992;134(1):141–7.
53. Cooper CW, Hirsch PF, Munson PL. Importance of endogenous thyrocalcitonin for protection against hypercalcemia in the rat. Endocrinology. 1970;86(2):406–15.
54. Yergey AL, Abrams SA, Vieira NE, Aldroubi A, Marini J, Sidbury JB. Determination of fractional absorption of dietary calcium in humans. J Nutr. 1994;124(5):674–82.
55. Institute of Medicine Committee to Review Dietary Reference Intakes for Vitamin D, Calcium. The National Academies collection: reports funded by National Institutes of Health. In: Ross AC, Taylor CL, Yaktine AL, Del Valle HB, editors. Dietary reference intakes for calcium and vitamin D. Washington, DC: National Academies Press (US) National Academy of Sciences; 2011.
56. Pansu D, Bellaton C, Bronner F. Effect of Ca intake on saturable and nonsaturable components of duodenal Ca transport. Am J Phys. 1981;240(1):G32–7.
57. Tudpor K, Teerapornpuntakit J, Jantarajit W, Krishnamra N, Charoenphandhu N. 1,25-dihydroxyvitamin D(3) rapidly stimulates the solvent drag-induced paracellular calcium transport in the duodenum of female rats. J Physiol Sci. 2008;58(5):297–307.
58. Chirayath MV, Gajdzik L, Hulla W, Graf J, Cross HS, Peterlik M. Vitamin D increases tight-junction conductance and paracellular Ca2+ transport in Caco-2 cell cultures. Am J Phys. 1998;274(2 Pt 1):G389–96.
59. Schachter D, Dowdle EB, Schenker H. Active transport of calcium by the small intestine of the rat. Am J Phys. 1960;198:263–8.
60. Miller A 3rd, Bronner F. Calcium uptake in isolated brush-border vesicles from rat small intestine. Biochem J. 1981;196(2):391–401.
61. Peng JB, Chen XZ, Berger UV, Vassilev PM, Tsukaguchi H, Brown EM, et al. Molecular cloning and characterization of a channel-like transporter mediating intestinal calcium absorption. J Biol Chem. 1999;274(32):22739–46.
62. Hoenderop JG, Voets T, Hoefs S, Weidema F, Prenen J, Nilius B, et al. Homo- and heterotetrameric architecture of the epithelial Ca2+ channels TRPV5 and TRPV6. EMBO J. 2003;22(4):776–85.

63. Hoenderop JG, Vennekens R, Muller D, Prenen J, Droogmans G, Bindels RJ, et al. Function and expression of the epithelial Ca(2+) channel family: comparison of mammalian ECaC1 and 2. J Physiol. 2001;537(Pt 3):747–61.
64. Al-Ansary D, Bogeski I, Disteldorf BM, Becherer U, Niemeyer BA. ATP modulates Ca2+ uptake by TRPV6 and is counteracted by isoform-specific phosphorylation. FASEB J. 2010;24(2):425–35.
65. van de Graaf SF, Hoenderop JG, Gkika D, Lamers D, Prenen J, Rescher U, et al. Functional expression of the epithelial Ca(2+) channels (TRPV5 and TRPV6) requires association of the S100A10-annexin 2 complex. EMBO J. 2003;22(7):1478–87.
66. Bohmer C, Palmada M, Kenngott C, Lindner R, Klaus F, Laufer J, et al. Regulation of the epithelial calcium channel TRPV6 by the serum and glucocorticoid-inducible kinase isoforms SGK1 and SGK3. FEBS Lett. 2007;581(29):5586–90.
67. Kim HJ, Yang DK, So I. PDZ domain-containing protein as a physiological modulator of TRPV6. Biochem Biophys Res Commun. 2007;361(2):433–8.
68. Derler I, Hofbauer M, Kahr H, Fritsch R, Muik M, Kepplinger K, et al. Dynamic but not constitutive association of calmodulin with rat TRPV6 channels enables fine tuning of Ca2+-dependent inactivation. J Physiol. 2006;577(Pt 1):31–44.
69. Wasserman RH, Taylor AN. Vitamin d3-induced calcium-binding protein in chick intestinal mucosa. Science. 1966;152(3723):791–3.
70. Bronner F, Pansu D, Stein WD. An analysis of intestinal calcium transport across the rat intestine. Am J Phys. 1986;250(5 Pt 1):G561–9.
71. Duflos C, Bellaton C, Baghdassarian N, Gadoux M, Pansu D, Bronner F. 1,25-Dihydroxycholecalciferol regulates rat intestinal calbindin D9k posttranscriptionally. J Nutr. 1996;126(4):834–41.
72. Kutuzova GD, Akhter S, Christakos S, Vanhooke J, Kimmel-Jehan C, Deluca HF. Calbindin D(9k) knockout mice are indistinguishable from wild-type mice in phenotype and serum calcium level. Proc Natl Acad Sci U S A. 2006;103(33):12377–81.
73. Benn BS, Ajibade D, Porta A, Dhawan P, Hediger M, Peng JB, et al. Active intestinal calcium transport in the absence of transient receptor potential vanilloid type 6 and calbindin-D9k. Endocrinology. 2008;149(6):3196–205.
74. Keeton TP, Burk SE, Shull GE. Alternative splicing of exons encoding the calmodulin-binding domains and C termini of plasma membrane Ca(2+)-ATPase isoforms 1, 2, 3, and 4. J Biol Chem. 1993;268(4):2740–8.
75. Walters JR. Calbindin-D9k stimulates the calcium pump in rat enterocyte basolateral membranes. Am J Phys. 1989;256(1 Pt 1):G124–8.
76. Lytton J. Na+/Ca2+ exchangers: three mammalian gene families control Ca2+ transport. Biochem J. 2007;406(3):365–82.
77. Bers DM, Ginsburg KS. Na:Ca stoichiometry and cytosolic Ca-dependent activation of NCX in intact cardiomyocytes. Ann N Y Acad Sci. 2007;1099:326–38.
78. Li XF, Kraev AS, Lytton J. Molecular cloning of a fourth member of the potassium-dependent sodium-calcium exchanger gene family, NCKX4. J Biol Chem. 2002;277(50):48410–7.
79. Prinz C, Kajimura M, Scott D, Helander H, Shin J, Besancon M, et al. Acid secretion and the H,K ATPase of stomach. Yale J Biol Med. 1992;65(6):577–96.
80. Sachs G, Shin JM, Hunt R. Novel approaches to inhibition of gastric acid secretion. Curr Gastroenterol Rep. 2010;12(6):437–47.
81. Kopic S, Murek M, Geibel JP. Revisiting the parietal cell. Am J Phys Cell Physiol. 2010;298(1):C1–c10.
82. Abe K, Kaya S, Taniguchi K, Hayashi Y, Imagawa T, Kikumoto M, et al. Evidence for a relationship between activity and the tetraprotomeric assembly of solubilized pig gastric H/K--ATPase. J Biochem. 2005;138(3):293–301.
83. Asano S, Kawada K, Kimura T, Grishin AV, Caplan MJ, Takeguchi N. The roles of carbohydrate chains of the beta-subunit on the functional expression of gastric H(+),K(+)-ATPase. J Biol Chem. 2000;275(12):8324–30.

84. Vagin O, Denevich S, Sachs G. Plasma membrane delivery of the gastric H,K-ATPase: the role of beta-subunit glycosylation. Am J Phys Cell Physiol. 2003;285(4):C968–76.
85. Abe K, Tani K, Nishizawa T, Fujiyoshi Y. Inter-subunit interaction of gastric H+,K+-ATPase prevents reverse reaction of the transport cycle. EMBO J. 2009;28(11):1637–43.
86. Forte JG, Zhu L. Apical recycling of the gastric parietal cell H,K-ATPase. Annu Rev Physiol. 2010;72:273–96.
87. Kopic S, Geibel JP. Update on the mechanisms of gastric acid secretion. Curr Gastroenterol Rep. 2010;12(6):458–64.
88. Shin JM, Munson K, Vagin O, Sachs G. The gastric HK-ATPase: structure, function, and inhibition. Pflugers Archiv Eur J Physiol. 2009;457:609. 2011;461(3):399-
89. Engqvist-Goldstein AE, Warren RA, Kessels MM, Keen JH, Heuser J, Drubin DG. The actin-binding protein Hip1R associates with clathrin during early stages of endocytosis and promotes clathrin assembly in vitro. J Cell Biol. 2001;154(6):1209–23.
90. Keeley TM, Samuelson LC. Cytodifferentiation of the postnatal mouse stomach in normal and Huntingtin-interacting protein 1-related-deficient mice. Am J Physiol Gastrointest Liver Physiol. 2010;299(6):G1241–51.
91. Supplisson S, Loo DD, Sachs G. Whole-cell currents in isolated resting Necturus gastric oxynticopeptic cells. J Physiol. 1993;463:57–82.
92. Sidani SM, Kirchhoff P, Socrates T, Stelter L, Ferreira E, Caputo C, et al. DeltaF508 mutation results in impaired gastric acid secretion. J Biol Chem. 2007;282(9):6068–74.
93. Malinowska DH, Kupert EY, Bahinski A, Sherry AM, Cuppoletti J. Cloning, functional expression, and characterization of a PKA-activated gastric Cl- channel. Am J Phys. 1995;268(1 Pt 1):C191–200.
94. Petrovic S, Wang Z, Ma L, Seidler U, Forte JG, Shull GE, et al. Colocalization of the apical Cl-/HCO3- exchanger PAT1 and gastric H-K-ATPase in stomach parietal cells. Am J Physiol Gastrointest Liver Physiol. 2002;283(5):G1207–16.
95. Xu J, Song P, Miller ML, Borgese F, Barone S, Riederer B, et al. Deletion of the chloride transporter Slc26a9 causes loss of tubulovesicles in parietal cells and impairs acid secretion in the stomach. Proc Natl Acad Sci U S A. 2008;105(46):17955–60.
96. Vucic E, Alfadda T, MacGregor GG, Dong K, Wang T, Geibel JP. Kir1.1 (ROMK) and Kv7.1 (KCNQ1/KvLQT1) are essential for normal gastric acid secretion: importance of functional Kir1.1. Pflugers Arch Eur J Physiol. 2015;467(7):1457–68.
97. Lee MP, Ravenel JD, Hu RJ, Lustig LR, Tomaselli G, Berger RD, et al. Targeted disruption of the Kvlqt1 gene causes deafness and gastric hyperplasia in mice. J Clin Invest. 2000;106(12):1447–55.
98. Fujii T, Takahashi Y, Ikari A, Morii M, Tabuchi Y, Tsukada K, et al. Functional association between K+-Cl- cotransporter-4 and H+,K+-ATPase in the apical canalicular membrane of gastric parietal cells. J Biol Chem. 2009;284(1):619–29.
99. Edwards LW, Herrington JL Jr. Vagotomy and gastro-enterostomy; vagotomy and conservative gastrectomy; a comparative study. Ann Surg. 1953;137(6):873–83.
100. Chew CS, Brown MR. Release of intracellular Ca2+ and elevation of inositol trisphosphate by secretagogues in parietal and chief cells isolated from rabbit gastric mucosa. Biochim Biophys Acta. 1986;888(1):116–25.
101. Berthoud HR. Morphological analysis of vagal input to gastrin releasing peptide and vasoactive intestinal peptide containing neurons in the rat glandular stomach. J Comp Neurol. 1996;370(1):61–70.
102. DelValle J, Yamada T. Amino acids and amines stimulate gastrin release from canine antral G-cells via different pathways. J Clin Invest. 1990;85(1):139–43.
103. Larsson LI, Goltermann N, de Magistris L, Rehfeld JF, Schwartz TW. Somatostatin cell processes as pathways for paracrine secretion. Science. 1979;205(4413):1393–5.
104. Madaus S, Schusdziarra V, Seufferlein T, Classen M. Effect of galanin on gastrin and somatostatin release from the rat stomach. Life Sci. 1988;42(23):2381–7.

105. Kazumori H, Ishihara S, Kawashima K, Fukuda R, Chiba T, Kinoshita Y. Analysis of gastrin receptor gene expression in proliferating cells in the neck zone of gastric fundic glands using laser capture microdissection. FEBS Lett. 2001;489(2–3):208–14.
106. Hakanson R, Blom H, Carlsson E, Larsson H, Ryberg B, Sundler F. Hypergastrinaemia produces trophic effects in stomach but not in pancreas and intestines. Regul Pept. 1986;13(3–4):225–33.
107. Schayer RW. Formation and binding of histamine by rat tissues in vitro. Am J Phys. 1956;187(1):63–5.
108. Dimaline R, Sandvik AK. Histidine decarboxylase gene expression in rat fundus is regulated by gastrin. FEBS Lett. 1991;281(1–2):20–2.
109. Gerhard M, Neumayer N, Presecan-Siedel E, Zanner R, Lengyel E, Cramer T, et al. Gastrin induces expression and promoter activity of the vesicular monoamine transporter subtype 2. Endocrinology. 2001;142(8):3663–72.
110. Zanner R, Hapfelmeier G, Gratzl M, Prinz C. Intracellular signal transduction during gastrin-induced histamine secretion in rat gastric ECL cells. Am J Phys Cell Physiol. 2002;282(2):C374–82.
111. Miampamba M, Germano PM, Arli S, Wong HH, Scott D, Tache Y, et al. Expression of pituitary adenylate cyclase-activating polypeptide and PACAP type 1 receptor in the rat gastric and colonic myenteric neurons. Regul Pept. 2002;105(3):145–54.
112. Kidd M, Modlin IM, Black JW, Boyce M, Culler M. A comparison of the effects of gastrin, somatostatin and dopamine receptor ligands on rat gastric enterochromaffin-like cell secretion and proliferation. Regul Pept. 2007;143(1–3):109–17.
113. Bjorkqvist M, Bernsand M, Eliasson L, Hakanson R, Lindstrom E. Somatostatin, misoprostol and galanin inhibit gastrin- and PACAP-stimulated secretion of histamine and pancreastatin from ECL cells by blocking specific Ca2+ channels. Regul Pept. 2005;130(1–2):81–90.
114. Lindstrom E, Bjorkqvist M, Boketoft A, Chen D, Zhao CM, Kimura K, et al. Neurohormonal regulation of histamine and pancreastatin secretion from isolated rat stomach ECL cells. Regul Pept. 1997;71(2):73–86.
115. Wang LD, Hoeltzel M, Gantz I, Hunter R, Del Valle J. Characterization of the histamine H2 receptor structural components involved in dual signaling. J Pharmacol Exp Ther. 1998;285(2):573–8.
116. Kobayashi T, Tonai S, Ishihara Y, Koga R, Okabe S, Watanabe T. Abnormal functional and morphological regulation of the gastric mucosa in histamine H2 receptor-deficient mice. J Clin Invest. 2000;105(12):1741–9.
117. Baldissera FG, Nielsen OV, Holst JJ. The intestinal mucosa preferentially releases somatostatin-28 in pigs. Regul Pept. 1985;11(3):251–62.
118. Alumets J, Ekelund M, El Munshid HA, Hakanson R, Loren I, Sundler F. Topography of somatostatin cells in the stomach of the rat: possible functional significance. Cell Tissue Res. 1979;202(2):177–88.
119. Schubert ML, Makhlouf GM. Gastrin secretion induced by distention is mediated by gastric cholinergic and vasoactive intestinal peptide neurons in rats. Gastroenterology. 1993;104(3):834–9.
120. Konturek SJ, Bilski J, Cieszkowski M. Role of cholecystokinin in the intestinal fat- and acid-induced inhibition of gastric secretion. Regul Pept. 1992;42(1–2):97–109.
121. Lamberts R, Stumps D, Plumpe L, Creutzfeldt W. Somatostatin cells in rat antral mucosa: qualitative and quantitative ultrastructural analyses in different states of gastric acid secretion. Histochemistry. 1991;95(4):373–82.
122. Martinez V, Curi AP, Torkian B, Schaeffer JM, Wilkinson HA, Walsh JH, et al. High basal gastric acid secretion in somatostatin receptor subtype 2 knockout mice. Gastroenterology. 1998;114(6):1125–32.
123. Park J, Chiba T, Yamada T. Mechanisms for direct inhibition of canine gastric parietal cells by somatostatin. J Biol Chem. 1987;262(29):14190–6.

124. Chey WY, Hitanant S, Hendricks J, Lorber SH. Effect of secretin and cholecystokinin on gastric emptying and gastric secretion in man. Gastroenterology. 1970;58(6):820–7.
125. Bado A, Cloarec D, Moizo L, Laigneau JP, Bataille D, Lewin MJ. Neurotensin and oxyntomodulin-(30-37) potentiate PYY regulation of gastric acid and somatostatin secretions. Am J Phys. 1993;265(1 Pt 1):G113–7.
126. Black JW, Fisher EW, Smith AN. The effects of 5-hydroxytryptamine on gastric secretion in anaesthetized dogs. J Physiol. 1958;141(1):27–34.
127. Andersson S, Chang D, Folkers K, Rosell S. Inhibition of gastric acid secretion in dogs by neurotensin. Life Sci. 1976;19(3):367–70.
128. Dornonville de la Cour C, Lindstrom E, Norlen P, Hakanson R. Ghrelin stimulates gastric emptying but is without effect on acid secretion and gastric endocrine cells. Regul Pept. 2004;120(1–3):23–32.
129. Berg A, Redeen S, Grenegard M, Ericson AC, Sjostrand SE. Nitric oxide inhibits gastric acid secretion by increasing intraparietal cell levels of cGMP in isolated human gastric glands. Am J Physiol Gastrointest Liver Physiol. 2005;289(6):G1061–6.
130. Beales IL, Calam J. Inhibition of carbachol stimulated acid secretion by interleukin 1beta in rabbit parietal cells requires protein kinase C. Gut. 2001;48(6):782–9.
131. Fellenius E, Berglindh T, Sachs G, Olbe L, Elander B, Sjostrand SE, et al. Substituted benzimidazoles inhibit gastric acid secretion by blocking (H+ + K+)ATPase. Nature. 1981;290(5802):159–61.
132. Ferstl FS, Kitay AM, Trattnig RM, Alsaihati A, Geibel JP. Secretagogue-dependent and -independent transport of zinc hydration forms in rat parietal cells. Pflugers Arch. 2016;468(11–12):1877–83.
133. Lindberg P, Nordberg P, Alminger T, Brandstrom A, Wallmark B. The mechanism of action of the gastric acid secretion inhibitor omeprazole. J Med Chem. 1986;29(8):1327–9.
134. Besancon M, Shin JM, Mercier F, Munson K, Miller M, Hersey S, et al. Membrane topology and omeprazole labeling of the gastric H+,K(+)-adenosinetriphosphatase. Biochemistry. 1993;32(9):2345–55.
135. Klotz U, Schwab M, Treiber G. CYP2C19 polymorphism and proton pump inhibitors. Basic Clin Pharmacol Toxicol. 2004;95(1):2–8.
136. Pidasheva S, D'Souza-Li L, Canaff L, Cole DE, Hendy GN. CASRdb: calcium-sensing receptor locus-specific database for mutations causing familial (benign) hypocalciuric hypercalcemia, neonatal severe hyperparathyroidism, and autosomal dominant hypocalcemia. Hum Mutat. 2004;24(2):107–11.
137. Abramowitz J, Thakkar P, Isa A, Truong A, Park C, Rosenfeld RM. Adverse event reporting for proton pump inhibitor therapy: an overview of systematic reviews. Otolaryngol Head Neck Surg Off J Am Acad Otolaryngol Head Neck Surg. 2016;155(4):547–54.
138. Yang YX, Lewis JD, Epstein S, Metz DC. Long-term proton pump inhibitor therapy and risk of hip fracture. JAMA. 2006;296(24):2947–53.
139. Baird IM, Oleesky S. Osteomalacia following gastric surgery. Gastroenterology. 1957;33(2):284–92.
140. Bisballe S, Eriksen EF, Melsen F, Mosekilde L, Sorensen OH, Hessov I. Osteopenia and osteomalacia after gastrectomy: interrelations between biochemical markers of bone remodelling, vitamin D metabolites, and bone histomorphometry. Gut. 1991;32(11):1303–7.
141. Gertner JM, Lilburn M, Domenech M. 25-Hydroxycholecalciferol absorption in steatorrhoea and postgastrectomy osteomalacia. Br Med J. 1977;1(6072):1310–2.
142. Blichert-Toft M, Beck A, Christiansen C, Transbol I. Effects of gastric resection and vagotomy on blood and bone mineral content. World J Surg. 1979;3(1):99–102. 33–5
143. Davies M, Heys SE, Selby PL, Berry JL, Mawer EB. Increased catabolism of 25-hydroxyvitamin D in patients with partial gastrectomy and elevated 1,25-dihydroxyvitamin D levels. Implications for metabolic bone disease. J Clin Endocrinol Metab. 1997;82(1):209–12.
144. Nilas L, Christiansen C. Vitamin D deficiency after highly selective vagotomy. Acta Med Scand. 1987;221(3):303–6.

145. Graziani G, Como G, Badalamenti S, Finazzi S, Malesci A, Gallieni M, et al. Effect of gastric acid secretion on intestinal phosphate and calcium absorption in normal subjects. Nephrol Dial Transplant. 1995;10(8):1376–80.
146. O'Connell MB, Madden DM, Murray AM, Heaney RP, Kerzner LJ. Effects of proton pump inhibitors on calcium carbonate absorption in women: a randomized crossover trial. Am J Med. 2005;118(7):778–81.
147. Axelson J, Persson P, Gagnemo-Persson R, Hakanson R. Importance of the stomach in maintaining calcium homoeostasis in the rat. Gut. 1991;32(11):1298–302.
148. Engelhardt W, Grohmann C, Schwille PO, Geus A. Calcium absorption in the rat as influenced by highly selective vagotomy with special regard to endogenous gastrin. Res Exp Med (Berl). 1982;180(1):1–9.
149. Schinke T, Schilling AF, Baranowsky A, Seitz S, Marshall RP, Linn T, et al. Impaired gastric acidification negatively affects calcium homeostasis and bone mass. Nat Med. 2009;15(6):674–81.

Inflammatory Bowel Disease: Effects on Bone and Mechanisms

Francisco A. Sylvester

Introduction

Inflammatory bowel disease (IBD) encompasses a spectrum of chronic inflammatory disorders of the intestinal tract, including Crohn's disease and ulcerative colitis. Crohn's disease can affect any portion of the gastrointestinal tract, and the inflammation affects the full thickness of the intestinal wall. Ulcerative colitis affects the superficial layers of the lining of the colon. Both bone mass and bone architecture are significantly affected in IBD. The skeleton of pediatric patients with Crohn's disease is more severely affected than those with ulcerative colitis. This is probably because Crohn's disease inhibits linear growth in children more frequently than ulcerative colitis [1]. In adults with IBD, Crohn's disease also has a larger effect on bone mass than ulcerative colitis [2], which may be secondary to the increased inflammatory burden in adult Crohn's disease vs. ulcerative colitis (Figs. 1 and 2).

In this chapter we will discuss studies in human and mice that describe the effects of IBD on bone mineral density, bone geometry, and bone architecture. We will also present several hypotheses of mechanisms that link the inflammatory events in the intestine with bone cell function.

F.A. Sylvester, MD (✉)
Division Chief of Pediatric Gastroenterology, The University of North Carolina at Chapel Hil, 333 South Columbia Street, MacNider Hall 247, Chapel Hill, NC 27599-7229, USA
e-mail: fsylvester@unc.edu

L.R. McCabe, N. Parameswaran (eds.), *Understanding the Gut-Bone Signaling Axis*, Advances in Experimental Medicine and Biology 1033,
DOI 10.1007/978-3-319-66653-2_7

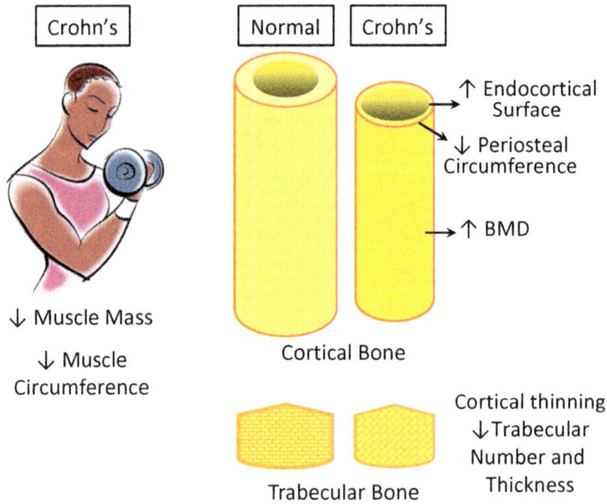

Fig. 1 The muscle-bone unit in IBD. Patients with IBD are at risk for decreased bone mineral density and reduced skeletal muscle mass. Longitudinal studies in inception cohorts of children with Crohn's disease have demonstrated sarcopenia by both DXA and pQCT. Sarcopenia can persist despite absence of symptoms of active Crohn's disease. Long bones in children with Crohn's disease are shorter and thinner. Cortical BMD is increased, endocortical surface is expanded, and the periosteal circumference is lower at diagnosis. These architectural abnormalities are reversed by anti-inflammatory therapy, improved nutrition, and increased weight-bearing activity

The Bone Phenotype in Human IBD: The Importance of the Bone-Muscle Unit

The effects of IBD on bone mineral density (BMD) in patients with IBD have been studied primarily with dual x-ray absorptiometry (DXA). DXA scanners are widely available and measure BMD accurately. DXA calculates BMD by dividing bone mineral content by the bone area in a region of interest. Therefore, DXA BMD measurements (expressed in g/cm^2) are dependent on the size of the subject (a larger patient will appear to have denser bones than a smaller patient, even when the actual density of the bone in g/cm^3 is identical in the two patients). This is an important consideration for individuals with IBD that have growth retardation, which is common in children with IBD (especially Crohn's disease). The reduction of BMD in children with IBD may be attributed in part to decreased bone size due to growth delay. For this reason, it is recommended that BMD be reported as the number of standard deviations from the mean for age- and sex-matched individuals (Z-score) that is further adjusted to height Z-score [3]. DXA studies have been conducted in both incident and prevalent cohorts of patients with IBD. The advantage of studying incident cohorts of patients with IBD from the time of diagnosis is that the results are reflective of disease factors that affect BMD rather than treatment factors such as corticosteroids or biologics (monoclonal antibodies directed against cytokines or adhesion molecules) [4–6]. Studies of BMD in patients with IBD have been either

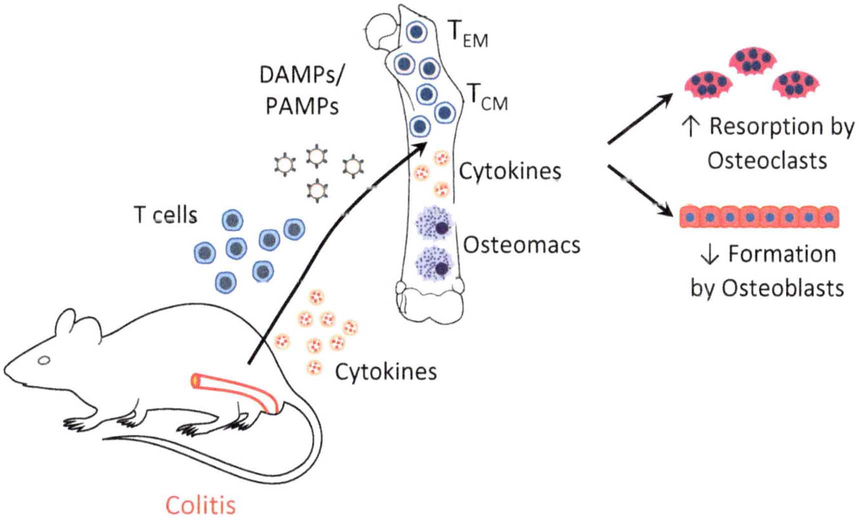

Fig. 2 Possible mechanisms of bone loss in IBD. Experiments in mice with colitis suggest that IBD can affect the activity of bone cells by circulating factors (cytokines, DAMPs, PAMPs) and inflammatory cells. The bone marrow is a reservoir for T effector (T_{EM}) and T central memory cells (T_{CM}) that can react to circulating stimuli. Macrophages (Osteomacs) form a canopy over osteoblasts. Activated T cells and Osteomacs can then influence the activity of osteoblasts, osteoclasts, and osteocytes

longitudinal or cross-sectional. These reports suggest that patients with IBD have significant reductions in BMD, especially in patients with delayed growth and sexual maturation and active disease and those with decreased lean tissue mass [4, 5, 7, 8]. Patients with low body mass index, low serum albumin, and active severe IBD appear to be at particular risk for decreased BMD [5]. The role of corticosteroids on BMD in pediatric Crohn disease however is not clear. It is possible that the negative effects of corticosteroids on bone formation are offset by its anti-inflammatory effects in IBD [9, 10]. Nonetheless, long-term corticosteroid therapy may have adverse effects on bone mass in patients with IBD, and therefore, prolonged corticosteroid use should be avoided in these patients.

Studies of bone architecture and geometry have been performed primarily in children with incident Crohn's disease using peripheral quantitated tomography (pQCT) [5, 6]. As opposed to DXA, which gives a two-dimensional projection of bone, pQCT analyzes bone in three dimensions. Collectively, this work has revealed that trabecular bone volume in long bones is significantly reduced in Crohn's disease at the time of diagnosis. The endocortical surface (bone marrow cavity) is expanded, probably due to increased bone resorption. The periosteal circumference is diminished, likely secondary to reduced bone formation in the periosteal envelope [5, 11]. This results in a thinner bone cortex. Histomorphometry of transiliac bone biopsies of 20 children with newly diagnosed Crohn disease showed that both bone formation and resorption are reduced at diagnosis. In addition, there is cortical thinning, but trabecular thickness and number were unaffected [12]. At the time of diagnosis of Crohn's disease, cortical bone density of long bones is higher than normal, probably due to decreased bone remodeling [5, 10]. Cortical bone density decreases in parallel with clinical response to anti-inflammatory therapy, suggesting restoration of cortical bone remodeling. The mechanism responsible for increased cortical bone density in Crohn's disease is unknown. It is possible that elevated serum osteoprotegerin (OPG) in children with active IBD may inhibit osteoclast activity in cortical bone. Serum OPG decreases when inflammation is under control. In parallel, soluble serum receptor activator of nuclear factor κB is decreased in patients with active Crohn's disease, which may also contribute to decrease remodeling in cortical bone [13]. Therefore, there are regional differences in how IBD affects bone mass and bone architecture in humans. Trabecular BMD and cortical dimensions are likely to improve in patients with well-controlled inflammation and limited exposure to corticosteroids and in children who exhibit catch-up growth [10].

The skeletal muscle mass is significantly affected in IBD [14]. Large skeletal forces are important in the development of the bone. The development of muscle mass precedes the sharp rise in bone mass that occurs during adolescence [15]. At diagnosis sarcopenia is well documented in pediatric Crohn's disease [8]. There is also a reduction in grip strength, suggesting that the reduced muscle mass affects muscle function [6]. Sarcopenia persists even in patients who are in clinical remission secondary to anti-inflammatory therapy and who are well nourished, as judged by a normal body mass index [8]. Therefore, it is possible that IBD affects the regenerative capacity of skeletal muscle. It is also possible that low level of inflammation in patients who are in clinical remission has negative effects on muscle regeneration. Muscle mass can also be adversely affected by pubertal delay, physical inactivity, and corticosteroids that are used to treat IBD. Over time, the skeleton adapts to the lower muscle cross-sectional area in IBD, resulting in lower bone mass [16].

Reduction in BMD, alterations in bone geometry, and bone remodeling are well established in IBD, but it is not yet known whether IBD increases the risk of fractures. Even in the setting of skeletal fragility, physical inactivity may shelter patients with IBD from activities that can result in fractures. The risk of clinically apparent fractures appears to be modestly increased [17] or not elevated [18] in population-based studies of adults with IBD. Asymptomatic vertebral fractures however may be prevalent in adults with Crohn's disease, even in patients with normal bone density by DXA [19]. Vertebral fractures have been reported in children as well [20]. There is no detectable increase in the overall frequency of fractures in children with IBD however [21, 22]. Fractures are common in children (especially of long bones), notably in teenage boys, and therefore, it might be difficult to detect a significant increase in fracture risk in adolescents with IBD. A study of sufficient sample size to detect such an effect size may be impractical due its large population size and associated costs.

In summary, the muscle-bone unit is subject to multiple negative influences during active human IBD. This results in significant reductions in muscle and bone mass and a disturbed bone geometry. Whether these abnormalities in muscle and bone mass and bone architecture result in increase in fracture risk is yet not clear however.

The Bone Phenotype in Laboratory Models of IBD

Patients with IBD can have significant decreases in BMD and muscle mass and changes in bone architecture. However, it is difficult to study disease mechanisms in humans. To examine mechanistic questions, animal models of IBD have been developed, primarily in laboratory mice. Generally speaking, mouse models of IBD are generated by knocking out genes (e.g., IL-10, IL-2), by transferring T effector cells into immune-deficient mice (e.g., $CD45RB^{Hi}$ cells into $Rag^{-/-}$), or by administering chemical irritants orally or rectally (TNBS, DSS).

Dressner-Pollak et al. compared the skeleton of 8- and 12-week-old $IL\text{-}10^{-/-}$ mice with colitis to wild-type mice without colitis of the same age and sex. $IL\text{-}10^{-/-}$ mice with colitis had decreased bone mass secondary to decreased bone formation; bone resorption was not increased [23]. Long bones were more fragile, and ash weight (bone mineral content) was reduced in $IL\text{-}10^{-/-}$ with colitis. Due to the lack of a control of $IL\text{-}10^{-/-}$ mice without colitis, it is not clear if some of the changes in the skeleton in $IL\text{-}10^{-/-}$ mice with colitis were due to deficiency of IL-10. Ciucci et al. more recently compared bone of $IL\text{-}10^{-/-}$ mice with colitis with $IL\text{-}10^{-/-}$ mice without colitis by histomorphometry. $IL\text{-}10^{-/-}$ mice with colitis had significant reductions in trabecular thickness, trabecular number, and bone surface density and in the trabecular bone volume per tissue volume [24]. These findings suggest that colitis (and not IL-10 deficiency) is responsible for the bone phenotype in this model. Cytokine-producing $CD4^+$ T cells are present in the bone marrow of $IL\text{-}10^{-/-}$ mice with colitis. IL-17 and TNF-α produced by these cells and bone marrow stromal cells may attract osteoclasts, increase bone resorption, and suppress bone formation [24].

$IL\text{-}2^{-/-}$ mice spontaneously develop colitis and have other features of autoinflammation. Ashcroft et al. showed that $IL\text{-}2^{-/-}$ with colitis have decreased trabecular bone at 7 and 9 weeks of age [25]. $IL\text{-}2^{-/-}$ $CD3^+$ cells transferred into C57BL/6-$Rag1^{-/-}$ resulted in significantly lower femoral BMD and percent trabecular volume 6–8 weeks post-grafting compared to recipients of $IL\text{-}2^{+/+}$ $CD3^+$ cells. The number of osteoclasts was significantly higher in C57BL/6-$Rag1^{-/-}$ engrafted with $IL\text{-}2^{-/-}$ $CD3^+$ cells. This suggests that activated T cells may play a role in inducing bone loss in the setting of colitis by inducing osteoclast formation. Exogenous OPG improved BMD. However, this does not necessarily prove that increased bone resorption is responsible for decreasing BMD in this model, since exogenous OPG increases BMD even in the absence of colitis.

Byrne et al. used a transfer model of colitis, transplanting $CD4^+CD45RB^{Hi}$ or $CD4^+CD45RB^{Lo}$ from CB6F1 mice to C.B.17 *scid/scid* mice [26]. As expected, $CD4^+CD45RB^{Hi}$ transfer induced colitis in recipient mice, but not transfer of

$CD4^+CD45RB^{Lo}$ cells. Mice with colitis had lower bone mineral density in the femur/tibia. Engraftment of $CD4^+CD45RB^{Hi}$ was associated with an inflammatory infiltrate in the bone marrow containing TNF-α-producing cells [26], suggesting that activated T cells in the bone marrow may affect bone cell function. This is significant because even low expression of TNF-α may inhibit bone formation [27]. Fc-OPG 3.4–5 mg/kg SC three times weekly for 34 days significantly improved BMD in mice with colitis. Similarly to the finding in $IL\text{-}2^{-/-}$ mice with colitis, this only demonstrates that exogenous OPG can suppress osteoclast activity, but does not imply necessarily that bone resorption is primarily responsible for the bone loss in this model.

Colitis induced by irritants such as TNBS and DSS is associated with reduced bone mass. Lin et al. observed significant decreases in bone mass in rats 3 weeks after a single TNBS enema that induced colitis. Rats with colitis had a 33% loss of trabecular bone in the tibia compared with age-matched, pair-fed control animals. Trabecular bone formation rate was markedly reduced. Bone formation reactivated when the colitis healed, resulting in normalization of bone mass [28]. Hamdani et al. observed that mice with DSS colitis have lower femoral trabecular bone mass compared to controls secondary to reduced bone formation. No changes were observed in cortical bone indices [29]. Harris et al. reported that in mice with DSS-induced colitis, there were significant decreases in trabecular bone mineral density, bone volume, and bone thickness. Cortical bone thickness, outer perimeter, and density were also decreased, whereas inner perimeter and marrow area were increased. There was significant inhibition of bone formation. Bone mass recovered after resolution of DSS-induced colitis in mice [30].

In summary, experimental colitis in rodents induced in three different ways reduces bone mass. Depending on the laboratory model, reduction in bone mass is secondary to decreased bone formation, increased bone resorption, or both. Gender may affect how colitis affects bone mass in laboratory animals [31].

Mechanisms by Which IBD May Affect Bone Cells

Malnutrition

IBD is associated with reduced nutrient intake and affects nutrient utilization. Children with IBD maintain a normal basal metabolic rate that does not adjust to the degree of malnutrition [32]. Patients with IBD who undergo intestinal resection may have malabsorption of macro- and micronutrients that are important for bone formation. IBD is associated with nutrient deficiencies that are important for bone development [33]. Vitamin D levels may be low due to the lack of unprotected exposure to sunlight due to inactivity and the application of sunscreen to protect against nonmelanoma skin cancers in patients with IBD on immunosuppression [34]. Vitamin K, which is synthesized by enteric bacteria, may be affected by the changes in the gut microbiota that accompany IBD [35]. Patients with active IBD lose protein through the intestine, which may affect proteins that are necessary for adequate bone mineralization.

Malnutrition can have direct effects on bone formation. Malnutrition can affect the availability of substrate for mineralization due to decrease intake of protein and calcium in the diet. Patients with IBD commonly complain of lactose intolerance and as a result curtail their intake of dairy products that contain calcium. Vitamin D deficiency in IBD impairs calcium absorption. The correction of vitamin D deficiency in patients with IBD requires high oral doses [36]. Moreover, IBD may affect calcium utilization, similar to patients with cystic fibrosis, who actively lose calcium in their colon [37]. Low vitamin D level can lead to hypocalcemia and secondary hyperparathyroidism, which results in urinary losses of phosphate. Malnutrition can also have indirect effects on bone. Insulin-like growth factor 1 (IGF-1), which is anabolic to bone, is sensitive to the nutritional state, and it lowers in patients with IBD who are malnourished [38]. After nutritional rehabilitation, serum IGF-1 returns to normal. Active inflammation can also lower serum IGF-1 due to effects in the liver, which is the main source of IGF-1 [39].

A possible role for vitamin D in the pathogenesis of IBD is being actively investigated [40]. In addition to its role in calcium metabolism, vitamin D also enhances the activity of the innate immune system and in general suppresses the adoptive immune system. Since some forms of IBD are associated with defects in innate barrier function in the intestine, it is possible that vitamin D deficiency predisposes to the development of IBD. There is epidemiological evidence that supports this hypothesis. IBD is more common in latitudes where availability of sunlight of sufficient energy to induce the synthesis of vitamin D in the skin is limited. IBD is more likely to appear in individuals with low vitamin D levels. Mice with the absence of vitamin D or who receive a vitamin D-deficient diet are susceptible to developing worse experimental colitis [41–44]. Mice with intestinal epithelium that overexpresses a vitamin D receptor on the other hand are less susceptible to colitis [45]. Therefore, vitamin D may become an adjuvant therapy for patients with IBD. This needs to be confirmed in double blind, placebo-controlled trials.

Circulating Factors

It is possible that factors generated in the inflamed intestine can reach the bone via circulation and affect bone cell activity. Serum from newly diagnosed children with Crohn's disease decreases markers of osteoblastic activity in bone explants [46] and in osteoblasts [47] in vitro, while indicators of bone resorption are not increased in this model. Neutralization of IL-6, a pro-inflammatory cytokine that is increased in the circulation of patients with active IBD, reverses the effects of serum on bone explants [48], suggesting an important role of IL-6 on the effects of Crohn's disease on bone formation.

Tumor necrosis factor (TNF)-α can be elevated in the circulation of patients with active IBD. TNF-α is an important therapeutic target in IBD. Monoclonal anti-TNF-α antibodies are among the most effective agents to treat IBD that are resistant to other therapies. TNF-α has effects on both osteoblasts and osteoclasts. In osteoblasts, TNF-α induces the degradation of Runx2, a critical transcription factor in osteoblast development [49]. TNF-α suppresses both canonical Wnt [50] and bone morphogenetic protein-2 signaling [51–53], thus interfering with osteoblast differentiation and activity. TNF-α also can induce apoptosis in pre-osteoblasts [54]. TNF-α decreases the expression of *Phex* in osteoblasts, which impairs bone mineralization [55, 56]. The response to monoclonal anti-TNF-α antibodies in Crohn's disease is characterized by a brisk increase in biomarkers of bone formation, suggesting an activation of bone modeling and linear growth [57, 58]. TNF-α can induce osteoclastic formation both directly and indirectly through the increase in RANKL. Several cytokines relevant to the pathogenesis of IBD inhibit osteoclast differentiation, including IFN-γ [59], IL-10 [60, 61], and IL-12 [62, 63]. The effects of IL-17 on osteoclasts are complex and context specific. IL-17 can both inhibit [64, 65] and induce osteoclast formation in vitro [24, 64, 66–69] and in vivo [24].

RANKL and OPG in IBD

RANKL is a potent stimulator of osteoclast formation. RANKL is secreted by osteoblasts, stromal cells and activated T cells. Osteoprotegerin (OPG) is a soluble decoy receptor for RANKL and inhibits osteoclast development and activity [70]. Both OPG and RANKL also regulate immune responses. RANKL is involved in normal dendritic cell function and survival and plays a role in the early development of B and T cells [71–73]. In addition, RANKL/RANK contributes to intestinal mucosal tolerance [73]. OPG is secreted by both B cells and dendritic cells, in a process regulated by the CD40 receptor [74]. In vitro, dendritic cells isolated from $OPG^{-/-}$ mice more efficiently present antigen and secrete more inflammatory cytokines when stimulated with bacterial products or soluble RANKL [75]. Serum OPG is increased in patients with active IBD, whereas soluble RANKL is decreased [13, 76]. This occurs in the setting of a marked decrease in bone metabolic activity in IBD, suggesting that the increase in serum OPG is due to the production of OPG outside of the bone. One possibility is that serum OPG is produced in the liver in response to systemic exposure to LPS in IBD [77]. It is also possible that the increase in serum OPG is secondary to its production in the inflamed intestine [78]. An increase in serum OPG paired with a decrease in soluble RANKL may be responsible for the increased cortical bone density observed in children with incident Crohn's disease [5]. The inflamed intestine secretes OPG into its lumen. Elevated fecal concentrations of OPG are associated with resistance to corticosteroids and to infliximab in patients with IBD [79, 80]. The concentration of OPG in feces significantly decreases in children with IBD in remission [76]. Collectively, this evidence suggests that RANKL/RANK/OPG plays an important role in the regulation of the immune response in IBD and that OPG may be a biomarker of IBD severity.

Effects of IBD on the Bone Microenvironment

The bone marrow of mice with colitis harbors $CD4^+$ T central memory cells and T effector memory cells [24, 81]. T effector memory cell populations are maintained by IL-7 produced by bone marrow stromal cells [82]. It is possible that microbial antigens that translocate into the circulation from the intestine may activate memory T cells. Another possibility is that T cells may be activated by damage-associated molecular patterns (DAMPs) in IBD. As a result, T cells residing in the bone marrow can produce cytokines that affect osteoblasts and osteoclasts. In animal models of arthritis and periodontal disease T cells induce bone resorption [83]. Activated $CD4^+$ and $CD8^+$ T cells produce both soluble and membrane-bound RANKL [84]. In $IL\text{-}2^{-/-}$ mice with intestinal inflammation, activated T cells accumulate in the bone marrow and produce RANKL [25]. The bone marrow of $IL\text{-}10^{-/-}$ mice with colitis (but not from $IL\text{-}10^{-/-}$ without colitis or wild-type mice) contains $CD4^+$ T cells that produce IL-17 and TNF-α. This T-cell subset expresses membrane-bound RANKL, secretes M-CSF, and induces osteoclast formation in vitro without requiring exogenous RANKL/M-CSF [24]. $CD4^+$ Th17 T cells in the bone marrow [64] secrete cytokines that stimulate osteoclast formation and activity [85], upregulate RANK in osteoclast precursors [86], and increase expression of RANKL in osteocytes [87]. This is significant given the importance of Th17 cells in the pathogenesis of IBD [88]. In the steady state, T regulatory cells (Treg) populate the bone marrow as well [89, 90] and may inhibit bone resorption [91]. T cells may also regulate bone formation by osteoblasts. Intermittent parathyroid hormone can stimulate bone marrow $CD8^+$ T cells and activate anabolic canonical Wnt signaling in pre-osteoblasts [92].

In conclusion, in IBD there is a crosstalk between the inflamed intestines in the bone microenvironment that is mediated by bone marrow T cells. Activated T cells from the intestine may reach the bone marrow through the circulatory system. Memory T cells may be activated by PAMPs or DAMPs that originate in the gut and influence bone cells [93]. Other immune cells in the bone marrow such as macrophages that overlay osteoblasts may also respond to systemic and local signals and affect osteoblast function [94]. LPS AND DAMPs can activate circulating CD14+ or bone marrow CD11b+ monocytes/macrophages and induce oncostatin M, resulting in increased osteoblast differentiation from human mesenchymal stem cells and matrix mineralization [95].

IBD and Skeletal Muscle

An important stimulus for bone formation is mechanical stress by skeletal muscle. Muscle mass expands rapidly during puberty, and gains in muscle mass antecede increases in bone mass [96]. The mechanical stress from large muscle forces is anabolic to bone. Children with IBD often present with malnutrition, with significant losses in both the fat and lean tissue compartments and decreased body mass index. With treatment and clinical improvement, children gain weight, but deficits in lean body mass often persist [8, 38, 97]. This may result in decreased mechanical loading and be a reason for decreased bone formation in children. In addition, children with IBD may be less active than their peers, which may also affect gains in muscle and bone mass over time. In theory, it is possible that inflammation may affect the regenerative capacity of muscle satellite cells, which are responsible for muscle regeneration after injury. Exclusive enteral nutrition with defined commercially available formulas and anti-TNF-α antibodies offers promise to reconstitute skeletal muscle mass [97].

Endocrine Effects of IBD

Puberty is commonly delayed in children with IBD, especially in Crohn's disease. Active inflammation and malnutrition are probably responsible. Pubertal delays are associated with a relative decrease in sex steroids, both estrogen and testosterone that are important for bone formation. Effective anti-inflammatory therapy and nutritional reconstitution can put pubertal progression back on track [98].

Insulin-like growth factor-1 (IGF-1) is secreted by the liver in response to stimulation by growth hormone and enhances the expression of the mature osteoblast phenotype [99]. Serum IGF-1 is frequently reduced in children with active IBD due to growth hormone insensitivity in the liver and malnutrition [39]. Consequently, relative IGF-1 deficiency in children with IBD may negatively affect osteoblast differentiation and function and bone formation.

IBD Genetics and Bone

Genome-wide association studies (GWAS) performed in large cohorts of patients with IBD have uncovered over 300 susceptibility genes and the importance of previously unrecognized pathogenic pathways [100, 101]. GWAS have pointed to endoplasmic reticulum (ER) stress, the unfolded protein response (UPR), autophagy, the intestinal epithelial barrier, innate immune function, and interactions between the host and intestinal microbes as relevant systems in IBD [100].

ER stress is triggered by an overabundance of unfolded and misfolded proteins in its vesicles. The UPR then decreases transcription and protein synthesis, degrades proteins inside the ER and shuttles proteins away from the ER with chaperones, or induces apoptosis. ER stress and UPR are particularly active in highly secretory cells [102]. Therefore, it is possible that the same genetic defects in ER stress and UPR that compromise the function of intestinal Paneth cells and goblets cells may also affect osteoblasts and osteoclasts in bone [103], but this has not yet been studied in IBD. ER stress transducers are stimulated by bone morphogenetic protein-2 (BMP-2), a stimulator of osteoblast development and activity [104, 105]. BMP-2 also activates the UPR during osteogenesis [105, 106], which induces the synthesis of RANKL [107]. ER stress transducers are also important in osteoclast development [107, 108]. Therefore, defects in the UPR and ER stress present in IBD may affect the development and activity of both osteoblasts and osteoclasts.

Autophagy is a process by which cells recycle old proteins, damaged organelles, and other cellular debris [109]. Autophagy also is important in the uptake and clearance of bacteria. Autophagy can be defective in IBD, allowing persistence of intracellular live bacteria and triggering inflammation [110]. Autophagy regulates bone cell function. Induction of autophagy in osteoclasts is associated with decreased bone resorption [111]. However, during hypoxia and microgravity, autophagy induces osteoclast formation [112, 113]. Therefore, the effects of autophagy on osteoclastogenesis are context-dependent. Autophagy is important for osteoblast differentiation [114, 115] and bone mineralization [116]. Altered autophagy in IBD may therefore impair normal bone matrix mineralization. Autophagy may regulate bone mineral density in humans [117]. In summary, novel pathogenic pathways involved in IBD may also affect bone cell function.

Gut Microbiome and Bone

Depending on the mouse strain, fecal microbial transplant into gnotobiotic mice can either decrease or increase bone mass [118]. The intestinal microbiome in IBD is less diverse compared to heathy individuals and contains blooms of pathogenic bacteria (dysbiosis) [119]. The gut microbiome probably plays an important role in the pathogenesis of IBD [119, 120]. Therefore, the IBD microbiome may have unique effects on bone. However, the relationship between the enteric microbiome and the bone-muscle unit in IBD has not yet been investigated. Recent work by Ohlsson et al. suggests that the effects of the gut microbiota on bone depends on NOD1 and NOD2 signaling, suggesting that innate immune mechanisms are involved [121]. It is possible that activated immune cells in the gut circulate into the bone marrow and affect bone cell function. Metabolites generated by intestinal microbes may also reach the bone and have effects on bone cells. Research focused in this field should uncover the role of specific populations of intestinal microbes that can be modified to help restore normal bone mass and architecture in patients with IBD.

References

1. Sylvester FA, Gordon CM, Thayu M, Burnham JM, Denson LA, Essers J, et al. Report of the CCFA pediatric bone, growth and muscle health workshop, New York City, November 11–12, 2011, with updates. Inflamm Bowel Dis. 2013;19(13):2919–26.
2. Targownik LE, Bernstein CN, Leslie WD. Risk factors and management of osteoporosis in inflammatory bowel disease. Curr Opin Gastroenterol. 2014;30:168–74.
3. Bianchi ML, Leonard MB, Bechtold S, Hogler W, Mughal MZ, Schonau E, et al. Bone health in children and adolescents with chronic diseases that may affect the skeleton: the 2013 ISCD pediatric official positions. J Clin Densitom. 2014;17(2):281–94.
4. Sylvester FA, Wyzga N, Hyams JS, Davis PM, Lerer T, Vance K, et al. Natural history of bone metabolism and bone mineral density in children with inflammatory bowel disease. Inflamm Bowel Dis. 2007;13(1):42–50.
5. Dubner SE, Shults J, Baldassano RN, Zemel BS, Thayu M, Burnham JM, et al. Longitudinal assessment of bone density and structure in an incident cohort of children with Crohn's disease. Gastroenterology. 2009;136(1):123–30.
6. Werkstetter KJ, Pozza SB, Filipiak-Pittroff B, Schatz SB, Prell C, Bufler P, et al. Long-term development of bone geometry and muscle in pediatric inflammatory bowel disease. Am J Gastroenterol. 2011;106:988–98.
7. Thayu M, Shults J, Burnham JM, Zemel BS, Baldassano RN, Leonard MB. Gender differences in body composition deficits at diagnosis in children and adolescents with Crohn's disease. Inflamm Bowel Dis. 2007;13(9):1121–8.
8. Sylvester FA, Leopold S, Lincoln M, Hyams JS, Griffiths AM, Lerer T. A two-year longitudinal study of persistent lean tissue deficits in children with Crohn's disease. Clin Gastroenterol Hepatol. 2009;7(4):452–5.
9. Leonard MB. Glucocorticoid-induced osteoporosis in children: impact of the underlying disease. Pediatrics. 2007;119(Suppl 2):S166–74.
10. Tsampalieros A, Lam CK, Spencer JC, Thayu M, Shults J, Zemel BS, et al. Long-term inflammation and glucocorticoid therapy impair skeletal modeling during growth in childhood Crohn disease. J Clin Endocrinol Metab. 2013;98(8):3438–45.
11. Bechtold S, Alberer M, Arenz T, Putzker S, Filipiak-Pittroff B, Schwarz HP, et al. Reduced muscle mass and bone size in pediatric patients with inflammatory bowel disease. Inflamm Bowel Dis. 2010;16(2):216–25.
12. Ward LM, Rauch F, Matzinger MA, Benchimol EI, Boland M, Mack DR. Iliac bone histomorphometry in children with newly diagnosed inflammatory bowel disease. Osteoporos Int. 2010;21:331–7.
13. Sylvester FA, Davis PM, Wyzga N, Hyams JS, Lerer T. Are activated T cells regulators of bone metabolism in children with Crohn disease? J Pediatr. 2006;148(4):461–6.
14. Evans WJ. Skeletal muscle loss: cachexia, sarcopenia, and inactivity. Am J Clin Nutr. 2010;91(4):1123S–7S.
15. Rauch F, Bailey DA, Baxter-Jones A, Mirwald R, Faulkner R. The 'muscle-bone unit' during the pubertal growth spurt. Bone. 2004;34(5):771–5.
16. Bechtold S, Alberer M, Arenz T, Putzker S, Filipiak-Pittroff B, Schwarz HP, et al. Reduced muscle mass and bone size in pediatric patients with inflammatory bowel disease. Inflamm Bowel Dis. 2009;16:216–25.
17. Bernstein CN, Blanchard JF, Leslie W, Wajda A, Yu BN. The incidence of fracture among patients with inflammatory bowel disease. A population-based cohort study. Ann Intern Med. 2000;133(10):795–9.
18. Loftus EJ, Crowson CS, Sandborn WJ, Tremaine WJ, O'Fallon WM, Melton LJ 3rd. Long-term fracture risk in patients with Crohn's disease: a population-based study in Olmsted County, Minnesota. Gastroenterology. 2002;123(2):468–75.
19. Siffledeen JS, Siminoski K, Jen H, Fedorak RN. Vertebral fractures and role of low bone mineral density in Crohn's disease. Clin Gastroenterol Hepatol. 2007;5(6):721–8.

20. Semeao EJ, Stallings VA, Peck SN, Piccoli DA. Vertebral compression fractures in pediatric patients with Crohn's disease. Gastroenterology. 1997;112(5):1710–3.
21. Persad R, Jaffer I, Issenman RM. The prevalence of long bone fractures in pediatric inflammatory bowel disease. J Pediatr Gastroenterol Nutr. 2006;43(5):597–602.
22. Kappelman MD, Galanko JA, Porter CQ, Sandler RS. Risk of diagnosed fractures in children with inflammatory bowel diseases. Inflamm Bowel Dis. 2011;17:1125–30.
23. Dresner-Pollak R, Gelb N, Rachmilewitz D, Karmeli F, Weinreb M. Interleukin 10-deficient mice develop osteopenia, decreased bone formation, and mechanical fragility of long bones. Gastroenterology. 2004;127(3):792–801.
24. Ciucci T, Ibanez L, Boucoiran A, Birgy-Barelli E, Pene J, Abou-Ezzi G, et al. Bone marrow Th17 TNFalpha cells induce osteoclast differentiation, and link bone destruction to IBD. Gut. 2015;64(7):1072–81.
25. Ashcroft AJ, Cruickshank SM, Croucher PI, Perry MJ, Rollinson S, Lippitt JM, et al. Colonic dendritic cells, intestinal inflammation, and T cell-mediated bone destruction are modulated by recombinant osteoprotegerin. Immunity. 2003;19(6):849–61.
26. Byrne FR, Morony S, Warmington K, Geng Z, Brown HL, Flores SA, et al. CD4+CD45RBHi T cell transfer induced colitis in mice is accompanied by osteopenia which is treatable with recombinant human osteoprotegerin. Gut. 2005;54(1):78–86.
27. Gilbert L, He X, Farmer P, Boden S, Kozlowski M, Rubin J, et al. Inhibition of osteoblast differentiation by tumor necrosis factor-alpha. Endocrinology. 2000;141(11):3956–64.
28. Lin CL, Moniz C, Chambers TJ, Chow JW. Colitis causes bone loss in rats through suppression of bone formation. Gastroenterology. 1996;111(5):1263–71.
29. Hamdani G, Gabet Y, Rachmilewitz D, Karmeli F, Bab I, Dresner-Pollak R. Dextran sodium sulfate-induced colitis causes rapid bone loss in mice. Bone. 2008;43(5):945–50.
30. Harris L, Senagore P, Young VB, McCabe LR. Inflammatory bowel disease causes reversible suppression of osteoblast and chondrocyte function in mice. Am J Physiol Gastrointest Liver Physiol. 2009;296(5):G1020–9.
31. Irwin R, Lee T, Young VB, Parameswaran N, McCabe LR. Colitis-induced bone loss is gender dependent and associated with increased inflammation. Inflamm Bowel Dis. 2013;19:1586–97.
32. Azcue M, Rashid M, Griffiths A, Pencharz PB. Energy expenditure and body composition in children with Crohn's disease: effect of enteral nutrition and treatment with prednisolone. Gut. 1997;41(2):203–8.
33. Owczarek D, Rodacki T, Domagala-Rodacka R, Cibor D, Mach T. Diet and nutritional factors in inflammatory bowel diseases. World J Gastroenterol. 2016;22(3):895–905.
34. Kabbani TA, Koutroubakis IE, Schoen RE, Ramos-Rivers C, Shah N, Swoger J, et al. Association of vitamin D level with clinical status in inflammatory bowel disease: a 5-year longitudinal study. Am J Gastroenterol. 2016;111:712–9.
35. Nakajima S, Iijima H, Egawa S, Shinzaki S, Kondo J, Inoue T, et al. Association of vitamin K deficiency with bone metabolism and clinical disease activity in inflammatory bowel disease. Nutrition. 2011;27(10):1023–8.
36. Simek RZ, Prince J, Syed S, Sauer CG, Martineau B, Hofmekler T, et al. Pilot study evaluating efficacy of two regimens for hypovitaminosis D repletion in pediatric inflammatory bowel disease. J Pediatr Gastroenterol Nutr. 2016;62:252–8.
37. Schulze KJ, O'Brien KO, Germain-Lee EL, Booth SL, Leonard A, Rosenstein BJ. Calcium kinetics are altered in clinically stable girls with cystic fibrosis. J Clin Endocrinol Metab. 2004;89(7):3385–91.
38. Thayu M, Denson LA, Shults J, Zemel BS, Burnham JM, Baldassano RN, et al. Determinants of changes in linear growth and body composition in incident pediatric Crohn's disease. Gastroenterology. 2010;139:430–8.
39. Difedele LM, He J, Bonkowski EL, Han X, Held MA, Bohan A, et al. Tumor necrosis factor-a blockade restores growth hormone signaling in murine colitis. Gastroenterology. 2005;128(5):1278–91.

40. Reich KM, Fedorak RN, Madsen K, Kroeker KI. Vitamin D improves inflammatory bowel disease outcomes: basic science and clinical review. World J Gastroenterol WJG. 2014;20(17):4934–47.
41. Zhu Y, Mahon BD, Froicu M, Cantorna MT. Calcium and 1 alpha,25-dihydroxyvitamin D3 target the TNF-alpha pathway to suppress experimental inflammatory bowel disease. Eur J Immunol. 2005;35(1):217–24.
42. Froicu M, Weaver V, Wynn TA, McDowell MA, Welsh JE, Cantorna MT. A crucial role for the vitamin D receptor in experimental inflammatory bowel diseases. Mol Endocrinol. 2003;17(12):2386–92.
43. Cantorna MT, Munsick C, Bemiss C, Mahon BD. 1,25-Dihydroxycholecalciferol prevents and ameliorates symptoms of experimental murine inflammatory bowel disease. J Nutr. 2000;130(11):2648–52.
44. Wang F, Johnson RL, DeSmet ML, Snyder PW, Fairfax KC, Fleet JC. Vitamin D receptor-dependent signaling protects mice from dextran sulfate sodium-induced colitis. Endocrinology. 2017;
45. Liu W, Chen Y, Golan MA, Annunziata ML, Du J, Dougherty U, et al. Intestinal epithelial vitamin D receptor signaling inhibits experimental colitis. J Clin Invest. 2013;123:3983–96.
46. Hyams JS, Wyzga N, Kreutzer DL, Justinich CJ, Gronowicz GA. Alterations in bone metabolism in children with inflammatory bowel disease: an in vitro study. J Pediatr Gastroenterol Nutr. 1997;24(3):289–95.
47. Varghese S, Wyzga N, Griffiths AM, Sylvester FA. Effects of serum from children with newly diagnosed Crohn disease on primary cultures of rat osteoblasts. J Pediatr Gastroenterol Nutr. 2002;35(5):641–8.
48. Sylvester FA, Wyzga N, Hyams JS, Gronowicz GA. Effect of Crohn's disease on bone metabolism in vitro: a role for interleukin-6. J Bone Miner Res. 2002;17(4):695–702.
49. Kaneki H, Guo R, Chen D, Yao Z, Schwarz EM, Zhang YE, et al. Tumor necrosis factor promotes Runx2 degradation through up-regulation of Smurf1 and Smurf2 in osteoblasts. J Biol Chem. 2006;281(7):4326–33.
50. Gilbert LC, Chen H, Lu X, Nanes MS. Chronic low dose tumor necrosis factor-alpha (TNF) suppresses early bone accrual in young mice by inhibiting osteoblasts without affecting osteoclasts. Bone. 2013;56(1):174–83.
51. Yamazaki M, Fukushima H, Shin M, Katagiri T, Doi T, Takahashi T, et al. Tumor necrosis factor alpha represses bone morphogenetic protein (BMP) signaling by interfering with the DNA binding of Smads through the activation of NF-kappaB. J Biol Chem. 2009;284(51):35987–95.
52. Lee HL, Yi T, Woo KM, Ryoo HM, Kim GS, Baek JH. Msx2 mediates the inhibitory action of TNF-alpha on osteoblast differentiation. Exp Mol Med. 2010;42(6):437–45.
53. Jang WG, Jeong BC, Kim EJ, Choi H, Oh SH, Kim DK, et al. Cyclic AMP response element-binding protein H (CREBH) mediates the inhibitory actions of tumor necrosis factor alpha in osteoblast differentiation by stimulating Smad1 degradation. J Biol Chem. 2015;290(21):13556–66.
54. Shen F, Ruddy MJ, Plamondon P, Gaffen SL. Cytokines link osteoblasts and inflammation: microarray analysis of interleukin-17- and TNF-a-induced genes in bone cells. J Leukoc Biol. 2005;77(3):388–99.
55. Uno JK, Kolek OI, Hines ER, Xu H, Timmermann BN, Kiela PR, et al. The role of tumor necrosis factor-a in down-regulation of osteoblast Phex gene expression in experimental murine colitis. Gastroenterology. 2006;131(2):497–509.
56. Majewski PM, Thurston RD, Ramalingam R, Kiela PR, Ghishan FK. Cooperative role of NF-{kappa}B and poly(ADP-ribose) polymerase 1 (PARP-1) in the TNF-induced inhibition of PHEX expression in osteoblasts. J Biol Chem. 2010;285(45):34828–38.
57. Thayu M, Leonard MB, Hyams JS, Crandall WV, Kugathasan S, Otley AR, et al. Improvement in biomarkers of bone formation during infliximab therapy in pediatric Crohn's disease: results of the REACH study. Clin Gastroenterol Hepatol. 2008;6(12):1378–84.

58. Griffin LM, Thayu M, Baldassano RN, DeBoer MD, Zemel BS, Denburg MR, et al. Improvements in bone density and structure during anti-TNF-alpha therapy in pediatric Crohn's disease. J Clin Endocrinol Metab. 2015;100(7):2630–9.
59. Takayanagi H, Ogasawara K, Hida S, Chiba T, Murata S, Sato K, et al. T-cell-mediated regulation of osteoclastogenesis by signalling cross-talk between RANKL and IFN-gamma. Nature. 2000;408(6812):600–5.
60. Sasaki H, Hou L, Belani A, Wang CY, Uchiyama T, Muller R, et al. IL-10, but not IL-4, suppresses infection-stimulated bone resorption in vivo. J Immunol. 2000;165(7):3626–30.
61. Park-Min KH, Ji JD, Antoniv T, Reid AC, Silver RB, Humphrey MB, et al. IL-10 suppresses calcium-mediated costimulation of receptor activator NF-kappa B signaling during human osteoclast differentiation by inhibiting TREM-2 expression. J Immunol. 2009;183(4):2444–55.
62. Yoshimatsu M, Kitaura H, Fujimura Y, Eguchi T, Kohara H, Morita Y, et al. IL-12 inhibits TNF-alpha induced osteoclastogenesis via a T cell-independent mechanism in vivo. Bone. 2009;45(5):1010–6.
63. Kitaura H, Fujimura Y, Yoshimatsu M, Kohara H, Morita Y, Aonuma T, et al. IL-12- and IL-18-mediated, nitric oxide-induced apoptosis in TNF-alpha-mediated osteoclastogenesis of bone marrow cells. Calcif Tissue Int. 2011;89(1):65–73.
64. Sato K, Suematsu A, Okamoto K, Yamaguchi A, Morishita Y, Kadono Y, et al. Th17 functions as an osteoclastogenic helper T cell subset that links T cell activation and bone destruction. J Exp Med. 2006;203(12):2673–82.
65. Tyagi AM, Mansoori MN, Srivastava K, Khan MP, Kureel J, Dixit M, et al. Enhanced immunoprotective effects by anti-IL-17 antibody translates to improved skeletal parameters under estrogen deficiency compared with anti-RANKL and anti-TNF-alpha antibodies. J Bone Miner Res. 2014;29(9):1981–92.
66. Van Bezooijen RL, Farih-Sips HC, Papapoulos SE, Lowik CW. Interleukin-17: a new bone acting cytokine in vitro. J Bone Miner Res. 1999;14(9):1513–21.
67. Yago T, Nanke Y, Ichikawa N, Kobashigawa T, Mogi M, Kamatani N, et al. IL-17 induces osteoclastogenesis from human monocytes alone in the absence of osteoblasts, which is potently inhibited by anti-TNF-alpha antibody: a novel mechanism of osteoclastogenesis by IL-17. J Cell Biochem. 2009;108(4):947–55.
68. Balani D, Aeberli D, Hofstetter W, Seitz M. Interleukin-17A stimulates granulocyte-macrophage colony-stimulating factor release by murine osteoblasts in the presence of 1,25-dihydroxyvitamin D(3) and inhibits murine osteoclast development in vitro. Arthritis Rheum. 2013;65(2):436–46.
69. Kotake S, Udagawa N, Takahashi N, Matsuzaki K, Itoh K, Ishiyama S, et al. IL-17 in synovial fluids from patients with rheumatoid arthritis is a potent stimulator of osteoclastogenesis. J Clin Invest. 1999;103(9):1345–52.
70. Simonet WS, Lacey DL, Dunstan CR, Kelley M, Chang MS, Luthy R, et al. Osteoprotegerin: a novel secreted protein involved in the regulation of bone density. Cell. 1997;89(2):309–19.
71. Anderson DM, Maraskovsky E, Billingsley WL, Dougall WC, Tometsko ME, Roux ER, et al. A homologue of the TNF receptor and its ligand enhance T-cell growth and dendritic-cell function. Nature. 1997;390(6656):175–9.
72. Kong YY, Yoshida H, Sarosi I, Tan HL, Timms E, Capparelli C, et al. OPGL is a key regulator of osteoclastogenesis, lymphocyte development and lymph-node organogenesis. Nature. 1999;397(6717):315–23.
73. Summers deLuca L, Gommerman JL. Fine-tuning of dendritic cell biology by the TNF superfamily. Nat Rev Immunol. 2012;12(5):339–51.
74. Yun TJ, Chaudhary PM, Shu GL, Frazer JK, Ewings MK, Schwartz SM, et al. OPG/FDCR-1, a TNF receptor family member, is expressed in lymphoid cells and is up-regulated by ligating CD40. J Immunol. 1998;161(11):6113–21.
75. Chino T, Draves KE, Clark EA. Regulation of dendritic cell survival and cytokine production by osteoprotegerin. J Leukoc Biol. 2009;86(4):933–40.

76. Nahidi L, Leach ST, Sidler MA, Levin A, Lemberg DA, Day AS. Osteoprotegerin in pediatric Crohn's disease and the effects of exclusive enteral nutrition. Inflamm Bowel Dis. 2011;17(2):516–23.
77. Maruyama K, Takada Y, Ray N, Kishimoto Y, Penninger JM, Yasuda H, et al. Receptor activator of NF-kappa B ligand and osteoprotegerin regulate proinflammatory cytokine production in mice. J Immunol. 2006;177(6):3799–805.
78. Franchimont N, Reenaers C, Lambert C, Belaiche J, Bours V, Malaise M, et al. Increased expression of receptor activator of NF-kappaB ligand (RANKL), its receptor RANK and its decoy receptor osteoprotegerin in the colon of Crohn's disease patients. Clin Exp Immunol. 2004;138(3):491–8.
79. Arijs I, Li K, Toedter G, Quintens R, Van Lommel L, Van Steen K, et al. Mucosal gene signatures to predict response to infliximab in patients with ulcerative colitis. Gut. 2009;58:1612–9.
80. Sylvester FA, Turner D, Draghi A 2nd, Uuosoe K, McLernon R, Koproske K, et al. Fecal osteoprotegerin may guide the introduction of second-line therapy in hospitalized children with ulcerative colitis. Inflamm Bowel Dis. 2011;17(8):1726–30.
81. Nemoto Y, Kanai T, Makita S, Okamoto R, Totsuka T, Takeda K, et al. Bone marrow retaining colitogenic CD4+ T cells may be a pathogenic reservoir for chronic colitis. Gastroenterology. 2007;132(1):176–89.
82. Nemoto Y, Kanai T, Takahara M, Oshima S, Nakamura T, Okamoto R, et al. Bone marrow-mesenchymal stem cells are a major source of interleukin-7 and sustain colitis by forming the niche for colitogenic CD4 memory T cells. Gut. 2013;62(8):1142–52.
83. Takayanagi H. Osteoimmunology in 2014: two-faced immunology-from osteogenesis to bone resorption. Nat Rev Rheumatol. 2015;11(2):74–6.
84. Kong YY, Feige U, Sarosi I, Bolon B, Tafuri A, Morony S, et al. Activated T cells regulate bone loss and joint destruction in adjuvant arthritis through osteoprotegerin ligand. Nature. 1999;402(6759):304–9.
85. Jovanovic DV, Di Battista JA, Martel-Pelletier J, Jolicoeur FC, He Y, Zhang M, et al. IL-17 stimulates the production and expression of proinflammatory cytokines, IL-beta and TNF-alpha, by human macrophages. J Immunol. 1998;160(7):3513–21.
86. Adamopoulos IE, Chao CC, Geissler R, Laface D, Blumenschein W, Iwakura Y, et al. Interleukin-17A upregulates receptor activator of NF-kappaB on osteoclast precursors. Arthritis Res Ther. 2010;12(1):R29.
87. Li JY, D'Amelio P, Robinson J, Walker LD, Vaccaro C, Luo T, et al. IL-17A is increased in humans with primary hyperparathyroidism and mediates PTH-induced bone loss in mice. Cell Metab. 2015;22(5):799–810.
88. Harbour SN, Maynard CL, Zindl CL, Schoeb TR, Weaver CT. Th17 cells give rise to Th1 cells that are required for the pathogenesis of colitis. Proc Natl Acad Sci U S A. 2015;112(22):7061–6.
89. Zou L, Barnett B, Safah H, Larussa VF, Evdemon-Hogan M, Mottram P, et al. Bone marrow is a reservoir for CD4+CD25+ regulatory T cells that traffic through CXCL12/CXCR4 signals. Cancer Res. 2004;64(22):8451–5.
90. Tokoyoda K, Zehentmeier S, Hegazy AN, Albrecht I, Grun JR, Lohning M, et al. Professional memory CD4+ T lymphocytes preferentially reside and rest in the bone marrow. Immunity. 2009;30(5):721–30.
91. Kelchtermans H, Geboes L, Mitera T, Huskens D, Leclercq G, Matthys P. Activated CD4+CD25+ regulatory T cells inhibit osteoclastogenesis and collagen-induced arthritis. Ann Rheum Dis. 2009;68(5):744–50.
92. Terauchi M, Li JY, Bedi B, Baek KH, Tawfeek H, Galley S, et al. T lymphocytes amplify the anabolic activity of parathyroid hormone through Wnt10b signaling. Cell Metab. 2009;10(3):229–40.
93. Yang S, Takahashi N, Yamashita T, Sato N, Takahashi M, Mogi M, et al. Muramyl dipeptide enhances osteoclast formation induced by lipopolysaccharide, IL-1a, and TNF-a through

nucleotide-binding oligomerization domain 2-mediated signaling in osteoblasts. J Immunol. 2005;175(3):1956–64.
94. Chang MK, Raggatt LJ, Alexander KA, Kuliwaba JS, Fazzalari NL, Schroder K, et al. Osteal tissue macrophages are intercalated throughout human and mouse bone lining tissues and regulate osteoblast function in vitro and in vivo. J Immunol. 2008;181(2):1232–44.
95. Guihard P, Danger Y, Brounais B, David E, Brion R, Delecrin J, et al. Induction of osteogenesis in mesenchymal stem cells by activated monocytes/macrophages depends on oncostatin M signaling. Stem Cells. 2012;30(4):762–72.
96. Laurent MR, Dubois V, Claessens F, Verschueren SM, Vanderschueren D, Gielen E, et al. Muscle-bone interactions: from experimental models to the clinic? A critical update. Mol Cell Endocrinol. 2015;
97. Werkstetter KJ, Ullrich J, Schatz SB, Prell C, Koletzko B, Koletzko S. Lean body mass, physical activity and quality of life in paediatric patients with inflammatory bowel disease and in healthy controls. J Crohns Colitis. 2012;
98. DeBoer MD, Denson LA. Delays in puberty, growth, and accrual of bone mineral density in pediatric Crohn's disease: despite temporal changes in disease severity, the need for monitoring remains. J Pediatr. 2013;163(1):17–22.
99. Zhao G, Monier-Faugere MC, Langub MC, Geng Z, Nakayama T, Pike JW, et al. Targeted overexpression of insulin-like growth factor I to osteoblasts of transgenic mice: increased trabecular bone volume without increased osteoblast proliferation. Endocrinology. 2000;141(7):2674–82.
100. Jostins L, Ripke S, Weersma RK, Duerr RH, McGovern DP, Hui KY, et al. Host-microbe interactions have shaped the genetic architecture of inflammatory bowel disease. Nature. 2012;491(7422):119–24.
101. Cleynen I, Boucher G, Jostins L, Schumm LP, Zeissig S, Ahmad T, et al. Inherited determinants of Crohn's disease and ulcerative colitis phenotypes: a genetic association study. Lancet. 2015;387:156–67.
102. Adolph TE, Tomczak MF, Niederreiter L, Ko HJ, Bock J, Martinez-Naves E, et al. Paneth cells as a site of origin for intestinal inflammation. Nature. 2013;503(7475):272–6.
103. Wu Y, Yang M, Fan J, Peng Y, Deng L, Ding Y, et al. Deficiency of osteoblastic Arl6ip5 impaired osteoblast differentiation and enhanced osteoclastogenesis via disturbance of ER calcium homeostasis and induction of ER stress-mediated apoptosis. Cell Death Dis. 2014;5:e1464.
104. Murakami T, Saito A, Hino S, Kondo S, Kanemoto S, Chihara K, et al. Signalling mediated by the endoplasmic reticulum stress transducer OASIS is involved in bone formation. Nat Cell Biol. 2009;11(10):1205–11.
105. Saito A, Ochiai K, Kondo S, Tsumagari K, Murakami T, Cavener DR, et al. Endoplasmic reticulum stress response mediated by the PERK-eIF2(alpha)-ATF4 pathway is involved in osteoblast differentiation induced by BMP2. J Biol Chem. 2011;286(6):4809–18.
106. Tohmonda T, Miyauchi Y, Ghosh R, Yoda M, Uchikawa S, Takito J, et al. The IRE1alpha-XBP1 pathway is essential for osteoblast differentiation through promoting transcription of Osterix. EMBO Rep. 2011;12(5):451–7.
107. Tohmonda T, Yoda M, Mizuochi H, Morioka H, Matsumoto M, Urano F, et al. The IRE1alpha-XBP1 pathway positively regulates parathyroid hormone (PTH)/PTH-related peptide receptor expression and is involved in pth-induced osteoclastogenesis. J Biol Chem. 2013;288(3):1691–5.
108. Tohmonda T, Yoda M, Iwawaki T, Matsumoto M, Nakamura M, Mikoshiba K, et al. IRE1alpha/XBP1-mediated branch of the unfolded protein response regulates osteoclastogenesis. J Clin Invest. 2015;125(8):3269–79.
109. Cao SS, Luo KL, Shi L. Endoplasmic reticulum stress interacts with inflammation in human diseases. J Cell Physiol. 2016;231(2):288–94.
110. Baxt LA, Xavier RJ. Role of autophagy in the maintenance of intestinal homeostasis. Gastroenterology. 2015;

111. Kneissel M, Luong-Nguyen NH, Baptist M, Cortesi R, Zumstein-Mecker S, Kossida S, et al. Everolimus suppresses cancellous bone loss, bone resorption, and cathepsin K expression by osteoclasts. Bone. 2004;35(5):1144–56.
112. Zhao Y, Chen G, Zhang W, Xu N, Zhu JY, Jia J, et al. Autophagy regulates hypoxia-induced osteoclastogenesis through the HIF-1alpha/BNIP3 signaling pathway. J Cell Physiol. 2012;227(2):639–48.
113. Sambandam Y, Townsend MT, Pierce JJ, Lipman CM, Haque A, Bateman TA, et al. Microgravity control of autophagy modulates osteoclastogenesis. Bone. 2014;61:125–31.
114. Liu F, Fang F, Yuan H, Yang D, Chen Y, Williams L, et al. Suppression of autophagy by FIP200 deletion leads to osteopenia in mice through the inhibition of osteoblast terminal differentiation. J Bone Miner Res. 2013;28(11):2414–30.
115. Pantovic A, Krstic A, Janjetovic K, Kocic J, Harhaji-Trajkovic L, Bugarski D, et al. Coordinated time-dependent modulation of AMPK/Akt/mTOR signaling and autophagy controls osteogenic differentiation of human mesenchymal stem cells. Bone. 2013;52(1):524–31.
116. Nollet M, Santucci-Darmanin S, Breuil V, Al-Sahlanee R, Cros C, Topi M, et al. Autophagy in osteoblasts is involved in mineralization and bone homeostasis. Autophagy. 2014;10(11):1965–77.
117. Zhang L, Guo YF, Liu YZ, Liu YJ, Xiong DH, Liu XG, et al. Pathway-based genome-wide association analysis identified the importance of regulation-of-autophagy pathway for ultradistal radius BMD. J Bone Miner Res. 2010;25(7):1572–80.
118. Hernandez CJ, Guss JD, Luna M, Goldring SR. Links between the microbiome and bone. J Bone Miner Res. 2016;
119. Gevers D, Kugathasan S, Denson LA, Vazquez-Baeza Y, Van Treuren W, Ren B, et al. The treatment-naive microbiome in new-onset Crohn's disease. Cell Host Microbe. 2014;15(3):382–92.
120. Knights D, Silverberg MS, Weersma RK, Gevers D, Dijkstra G, Huang H, et al. Complex host genetics influence the microbiome in inflammatory bowel disease. Genome Med. 2014;6(12):107.
121. Ohlsson C, Nigro G, Boneca IG, Backhed F, Sansonetti P, Sjogren K. Regulation of bone mass by the gut microbiota is dependent on NOD1 and NOD2 signaling. Cell Immunol. 2017;317:55–8.

Epithelial Barrier Function in Gut-Bone Signaling

Naiomy Deliz Rios-Arce, Fraser L. Collins, Jonathan D. Schepper, Michael D. Steury, Sandi Raehtz, Heather Mallin, Danny T. Schoenherr, Narayanan Parameswaran, and Laura R. McCabe

Introduction

The gastrointestinal (GI) epithelium plays an essential role in maintaining host health through its ability to digest and absorb nutrients. At the same time, it is essential for providing a selective barrier that prevents translocation of harmful substances as well as pathogens and their products from the external environment to the bloodstream. The intestinal epithelium is composed of a continuous single layer of intestinal epithelial cells (IECs) that are sealed together by tight junction (TJ) proteins. This epithelial layer allows the movement of materials from the mucosal side of the epithelium to the serosal side via transcellular and paracellular pathways. A mucus layer, secreted by specialized epithelial cells (goblet cells), is located on the surface of the epithelium and is important for limiting the ability of gut bacteria and pathogens to access host cells. The lumen of the GI tract also harbors a variety of

---

Laura R. McCabe and Narayanan Parameswaran equally contributed to this work and are co-senior authors.

N.D. Rios-Arce • N. Parameswaran (✉)
Comparative Medicine and Integrative Biology Program, East Lansing, MI, USA

Department of Physiology, Michigan State University, East Lansing, MI, USA
e-mail: narap@msu.edu

F.L. Collins • J.D. Schepper • M.D. Steury • S. Raehtz • H. Mallin • D.T. Schoenherr
Department of Physiology, Michigan State University, East Lansing, MI, USA

L.R. McCabe (✉)
Department of Physiology and Department of Radiology, Biomedical Imaging Research Centre, Michigan State University, East Lansing, MI, USA
e-mail: mccabel@msu.edu

L.R. McCabe, N. Parameswaran (eds.), *Understanding the Gut-Bone Signaling Axis*, Advances in Experimental Medicine and Biology 1033,
DOI 10.1007/978-3-319-66653-2_8

commensal microorganisms referred to as the gut microbiota which accounts for 90% of the cells in the human body, approximately 10^{14} total bacteria. The intestine also secretes immunoglobulins, defensins, and other antimicrobial products that contribute to maintaining a healthy environment. Beneath the epithelial layer is the lamina propria which contains immune cells, fibroblasts, and plasma cells.

Disruption of the epithelial barrier can (1) affect efficient nutrient absorption, (2) facilitate pathogen translocation into the bloodstream and cause systemic inflammation, and (3) alter gut microbiota composition [1]. As a consequence, barrier disruption can trigger the development of GI diseases such as inflammatory bowel disease (IBD), celiac disease, and colon cancer [2–5]. Other systemic and metabolic diseases such as type I diabetes can also be influenced by barrier changes [6, 7]. However, whether barrier dysfunction is causal or consequence of these systemic and metabolic diseases is controversial. Recent studies from our lab and others demonstrate that GI barrier dysregulation can critically affect bone health [8, 9]. In this chapter, we will review several important aspects of intestinal epithelial barrier function including tight junction protein composition, the mucus layer, epithelial barrier integrity measurements, barrier alterations associated with disease processes, and barrier dysregulation-induced bone loss during aging, dysbiosis, and metabolic diseases.

Pathophysiology of Tight Junction Proteins

Tight junction (TJ) proteins connect adjacent epithelial cells on their apical side and therefore are critical for controlling paracellular permeability by selectively regulating the flow of ions, solutes, and small molecules across the epithelium. TJ proteins respond to a variety of stimuli including changes in diet, dysbiosis, viruses, inflammation, antibiotic treatment, and/or humoral or neuronal signals [1, 4, 10]. Stimuli can have positive or adverse effects on paracellular permeability depending on the physiological status of the host [1, 11–13].

TJ protein complexes are composed of junctional adhesion molecules (JAM), occludins, desmosomes, claudins, and cytoskeletal linker proteins such as zonula occludens (ZO) (1–3) (Fig. 1). The ZO is a family of proteins (ZO-1, ZO-2, ZO-3) that link the TJ proteins to the actin cytoskeleton. This interaction, between the TJ and the actin cytoskeleton, is essential to maintain TJ structure and cytoskeletal regulation of the epithelial barrier. Desmosomes do not directly connect adjacent epithelial cells. Instead, they provide the adhesive force to ensure the integrity of the epithelial layer [1, 14]. Alterations of the TJ complexes can increase paracellular permeability and pathogen translocation that can induce sustained activation of the mucosal immune system and tissue damage.

Several cytokines can modulate TJ complexes and affect intestinal permeability. For example, the pro-inflammatory cytokine tumor necrosis factor alpha (TNFα) can directly increase intestinal permeability in cultured intestinal epithelial cells and mouse epithelium by dysregulating TJ proteins. In vitro, in a colon epithelial cell

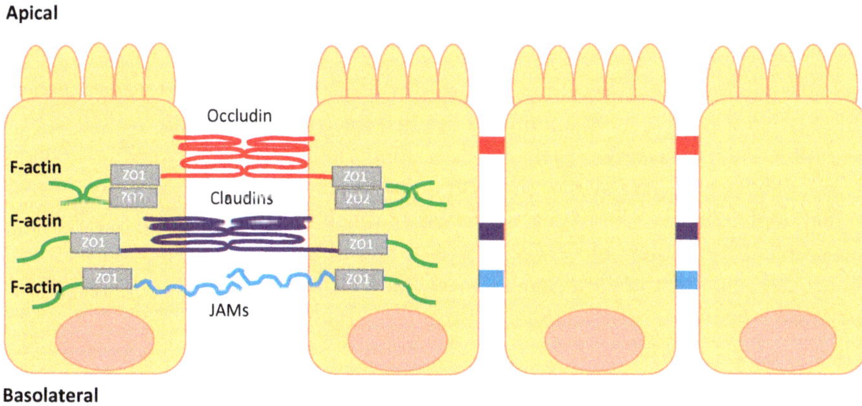

Fig. 1 Schematic representation of the intestinal tight junction proteins and their location. Tight junctions are protein complexes that span between epithelial cells to form a tight barrier. They are comprised of transmembrane proteins, such as occludin (*red*) and claudins (*purple*), and they are connected to the actin cytoskeleton via a zona occludens (ZO-1 and ZO-2 (*gray*)). The transmembrane receptor JAM (junctional adhesion molecule (*blue*)) is also found at tight junction complexes. Abbreviations: *JAM* junctional adhesion molecule, *ZO* zona occludens

line (Caco-2), TNFα-induced increases in TJ permeability were associated with increased nuclear factor kappa beta (NF-κB) activation and nuclear translocation of NF-κB p65 [13]. TNFα has also been shown to increase the expression of the myosin light-chain kinase (MLCK), and inhibition of MLCK can prevent TNFα-induced increases in permeability in intestinal epithelial cells [15, 16]. Similarly, interferon-γ (IFN-γ) increases paracellular permeability in T84 colonic epithelial cells. IFN-γ decreases ZO-1 protein synthesis, increases internalization of the TJ proteins, and rearranges the actin cytoskeleton [17]. Interleukin-1β (IL-1β) can also regulate transepithelial permeability in vitro. IL-1β caused a progressive time-dependent increase in transepithelial permeability in Caco-2 cells. This increase in permeability was attributed to the rapid activation of the NF-κB by IL-1β [12]. On the other hand, the role of the anti-inflammatory cytokine interleukin-10 (IL-10) in TJ regulation has been demonstrated in IL-10 knockout mice. IL-10 knockout mice present with an increase in ileal and colonic permeability at 2 weeks of age. However, this effect was ablated in IL-10 gene-deficient mice raised under germ-free conditions, suggesting a role of the microbiota in intestinal permeability in IL-10 knockout mice [18].

Studies in IBD patients indicate that TNFα levels are increased in the serum, stool, and intestinal mucosa [19], and correspondingly, patients display increased intestinal paracellular permeability that is characterized by suppression and redistribution of occludins and claudin-5 and claudin-8, whereas claudin-2 expression was upregulated [20]. Interestingly, people at high risk of developing Crohn's disease exhibit increases in small intestinal permeability [21]. Together, these studies demonstrate that TJ proteins are regulated by cytokines, modify intestinal permeability, and are linked to disease pathogenesis.

GI Mucus Layer

The GI epithelium is covered by a layer of mucus that creates a physical barrier. This barrier prevents the interaction of luminal microorganism with the surface of the epithelium. It also serves as the first line of host defense and allows the exchange of water, nutrients, and gases with the underlying epithelium. This layer is formed by high molecular weight glycoproteins called mucins (MUC) that are synthesized and secreted by goblet cells. Mucins are produced and stored in granules in the goblet cell and then are transported and secreted into the lumen. They can be secreted by continuous fusion (constitutive/basal secretion) or by exocytosis (exocytosis/regulated secretion). There are two groups of mucins: secreted mucins and transmembrane mucins. MUC2, MUC5AC, MUC5B, and MUC6 constitute the secreted mucins and are responsible for the formation of the mucus layer. The transmembrane mucins (MUC1, MUC4, MUC13, and MUC16) don't play a role in mucus production. Under normal physiological conditions, goblet cells continually produce mucins; however, factors such as cytokines, toxins, microbes, and microbial products can negatively or positively regulate this process [22]. Disruption of this process has been associated with GI diseases.

Inflammatory cytokines such as TNFα, IL-1β, and IL-6 are known to be major regulators of mucin synthesis and exocytosis. In vitro, in colon cells, TNFα and IL-6 increased the expression of the secreted gel-forming mucins (MUC2, MUC5AC, MUC5B, and MUC6) [23]. TNFα enhanced MUC2 transcription through the activation of the NF-κB pathway in colon cells [24]. Similarly, the anti-inflammatory cytokine IL-10 enhances MUC2 expression in goblet cells [25].

The gut microbiota depends on the mucus layer which serves as an energy source for bacteria. The mucus layer also serves as a matrix for commensal bacteria attachment and colonization that ultimately prevents opportunistic pathogenic bacteria from binding/growing within the mucus. Cross talk between the intestinal microbiota and mucus layer contributes to the regulated production of mucin by goblet cells [26–28]. Studies in germ-free mice demonstrate that microbiota deficiency leads to fewer goblet cells and thinner mucus layer in comparison with conventionally raised mice, supporting the role of the microbiota and/or microbial products in modulation of mucin synthesis and secretion [26]. Correspondingly, treatment of intestinal epithelial cells (HT-29) with the probiotic *Lactobacillus plantarum* enhanced MUC2 and MUC3 mRNA expression levels [29]. Similarly, commensal microbiota break down nondigestible carbohydrates into short-chain fatty acids (SCFAs) such as acetate, propionate, and especially butyrate, which at low concentration can increase mucus production and secretion [27, 28, 30]. Other microbial products, such as lipopolysaccharides and flagellin, also increase mucin synthesis [31]. If mucin production is chronically stimulated, goblet cells become depleted of mucin, and the lack of mucus secretion can lead to increased permeability and disease [32]. Taken together, mucus secretion is highly regulated by a variety of conditions/factors and, along with tight junctions, contributes to intestinal barrier integrity.

Epithelial Barrier Integrity Measurements

The intestinal epithelial barrier plays an essential role in host health as well as in GI diseases. Several tests have been developed to improve the diagnosis of GI diseases such as IBD. These tests (discussed below and in Table 1) include measures of serum, fecal, and urine biomarkers that are altered as a consequence of intestinal barrier dysfunction (i.e., inflammation) or that directly assess barrier permeability (i.e., endotoxin).

Serum

Serologic markers for epithelial barrier integrity in IBD patients includes C-reactive protein (CRP) measurements [33, 34]. CRP levels are elevated in the serum of patients with acute IBD, with almost 100% of patients with CD showing an increase of this protein in the serum [33, 35]. Serum measurements of inflammatory markers such as cytokines and neutrophils can also be performed. However, plasma/serum levels of inflammatory markers, including CRP, are not specific for gut inflammation since they can also be increased in other inflammatory conditions.

Breakdown of the epithelial barrier can lead to the translocation of the microbiota or their toxic products. Markers of bacterial antibodies for *Escherichia coli* and *Pseudomonas fluorescens* have been used to identify children with IBD with 67% sensitivity and 76% specificity [36]. Bacterial metabolic products, such as D-lactate or cell wall components such as LPS/endotoxin, are also commonly measured.

Table 1 Methods for the assessment of intestinal epithelial barrier integrity

Test	Measured in	Advantages	Disadvantages
Ex vivo			
Ussing chamber	Small and large intestine	Site specific	Invasive
In vivo			
C-reactive protein (CRP)	Serum	Noninvasive	Low sensitivity
Inflammatory markers (e.g., cytokines)	Serum	Noninvasive	Nonspecific
Bacteria metabolic products (e.g., LPS)	Serum	Noninvasive	High sensitivity
Fatty acid binding proteins (FABP)	Serum/urine	Noninvasive	
Calprotectin and lactoferrin	Feces	Noninvasive	Nonspecific
Dual sugar test (e.g., mannitol, lactulose)	Urine	Noninvasive	Time consuming
Polyethylene glycols (PEG)	Urine	Noninvasive Small and large intestine	Time consuming

Baseline levels of these markers are low in healthy individuals, whereas increased circulating LPS/endotoxin levels are related to an impaired mucosal barrier and increased levels of D-lactate are correlated with intestinal injury [37].

To estimate enterocyte damage, measurement of the fatty acid binding proteins (FABP) can be performed in the urine or plasma. FABP is located on top of the villi, and an increase in FABP in the blood or urine can be used as a marker of early stage intestinal diseases [38–40]. Levels of FABP rise rapidly after GI inflammation and have been correlated with the histological status of the epithelium after GI inflammation.

Feces

Invasive methods to test gut epithelial barrier inflammation in humans are not feasible. However, fecal proteins such as calprotectin and lactoferrin are specific markers for mucosal inflammation in intestinal diseases [41] and can identify patients with IBD, assess disease activity, and predict relapse [42, 43]. Calprotectin is a 36 kDa calcium- and zinc-binding protein. Approximately 60% of the cytosolic protein content in the neutrophils is made up of calprotectin. During intestinal inflammation, neutrophils migrate to the mucosa, and any break in the mucosal barrier results in the leakage of neutrophils into the lumen. Hence, the presence of calprotectin in the feces indicates the migration of neutrophils to the intestinal mucosa and potential leakage of these cells into the lumen. Calprotectin is stable in feces, and its concentration represents an indirect measure of neutrophil infiltration and barrier breaks. Lactoferrin is an iron-binding protein that is also found in neutrophils, specifically neutrophil granules. Lactoferrin is secreted during inflammation; when the intestine is inflamed and neutrophils are present, lactoferrin levels increase in the lumen and stool [41, 44].

Urine

Intestinal permeability can also be assessed by using small- to large-sized probe molecules [45, 46]. This approach involves oral ingestion of sugars, such as mannitol and lactulose, and measuring their subsequent concentration in the urine over a period of time (usually 5 h). Mannitol is a monosaccharide with a molecular weight (MW) of 182 Da and a molecular radius of $\leq$0.4 nm. Lactulose is a disaccharide with an MW of 342 Da and a molecular radius of 0.42 nm. The different sizes of the molecules allow for the measurement of transcellular and paracellular routes of permeability across the epithelia [44]. Large molecules, such as lactulose, are thought to traverse the epithelium by paracellular pathways. Small molecules, such as mannitol, cross the epithelium predominantly by the transcellular pathways. Neither sugar should be fermented by bacteria nor metabolized in the body. Thus, the ratio of urinary excretion of the relatively large molecule is compared with that of the relatively small molecule, and permeability is expressed as the ratio [45].

One concern with this approach is that many factors can influence the uptake of these sugars by epithelial cells, including (1) GI motility, (2) the use of medications such as nonsteroidal anti-inflammatory drugs, (3) intestinal transit time and surface area, (4) mucosal blood flow, and (5) renal clearance; these effects can potentially yield false-positive results. However, when both the large and small molecules are combined in the test solution at a fixed concentration ratio, the effects of variables, such as gastric emptying, intestinal transit time, and renal clearance, will apply equally to both. Thus, the urinary excretion ratio will be influenced only by the difference in gut permeability for each molecule.

Polyethylene glycols (PEG) have also been used to test intestinal barrier function. PEGs that have a molecular weight of 400–4,000 Da can only cross the intestinal mucosa under conditions of barrier integrity loss. PEG can be used to measure both small and large intestinal permeability and are not degraded by bacteria. PEG have been used to test changes in permeability in IBD and Crohn's patients [47, 48].

Barrier Pathophysiology in Development

Pediatrics

The intestine at birth is not fully developed, and many factors, such as diet, stress, and microbiota, have been implicated in influencing its permeability during development [49–51]. Increased intestinal permeability in infancy may lead to diseases that persist throughout childhood as well as those that appear later in life, such as IBD. Immune system involvement, which is developed alongside microbiota and diet changes, is also a significant indicator of the health of the intestinal barrier. Because of the fundamental differences in development between children and adults, it is important to consider pediatric intestinal barrier physiology separately.

The intestinal mucosal barrier significantly matures after birth, coinciding with changes in microbial composition and diet changes. At birth, the child is introduced to microbes, traditionally through contact with the birth canal, which colonize the intestine. It has been shown that vaginally born infants have higher numbers of *Bifidobacteria* and *Bacteroides* when compared with infants born through cesarean section [52]. In addition, the presence of *Bifidobacteria* in breast-fed infants corresponds with breast-fed infants having lower intestinal permeability than cow's milk formula-fed infants [53]. Other studies have shown that *B. infantis* promotes intestinal barrier function by regulating tight junctions. Infant mice treated with *B. infantis* exhibited decreased internalization of claudin-4 and occludin, which effectively decreased the incidence of necrotizing enterocolitis [54]. Mucin production also contributes to barrier integrity due to its importance in building the mucosal layer. In mice, the maturation and production of these glycoproteins occur after weaning, signifying the role of diet as well as hormonal and other age-associated factors in barrier development [55].

Increased intestinal permeability is associated with a variety of intestinal as well as extra-intestinal diseases, many of which persist or manifest in adulthood. Since the intestinal microbiota takes approximately 2.5 years to become functionally mature, the clinical impact of any large shifts in microbial composition during this developmental period can significantly impact intestinal permeability [56]. It has been suggested that traumatic GI events in early infancy, during the period of barrier maturation, are more powerful indicators of eventual disease than events occurring outside this "critical window." As evidence for the possibility of an early life disturbance creating lasting effects on barrier function, the trauma of maternal separation during weaning has been shown to predispose adult rats to enhanced intestinal permeability in response to stress [50].

The intestinal immune system in infants develops upon antigen exposure, directing attention to the importance of gut microbial colonization in relation to successful barrier function. Immunoglobulin A (IgA) is a class of antibody first received in breast milk and is then produced by the gut mucosa. Interestingly, by 24 months, both mono- and dizygotic twins had IgA responses comparable to unrelated children although significant differences were observed at older ages, suggesting a level of maturation acquired by age two [57]. This roughly coincides with the stabilization of the makeup of the microbiome at 2.5 years. The interplay among microbial colonization, diet, immune system, and the intestinal barrier is fundamental to the health of the gastrointestinal tract. Insults to the intestinal barrier at a young age can induce diseases such as IBD that appear in childhood and possibly persist through adulthood.

Early life disturbances such as premature birth, which can increase intestinal permeability, can also affect bone density [58]. In fact, 16–40% of very low birth weight (VLBW, <1.500 kg) and extremely low birth weight (ELBW, <1.000 kg) infants are estimated to develop bone metabolic diseases [59]. Examination of growth and bone mineralization among children born prematurely (birth weight less than 1.5 kg) indicates reduced lumbar bone mineral density and content compared to full-term children [59]. During this early period of life, the intestinal barrier is particularly permeable to allow antibodies in the mother's colostrum to cross into the infant's blood. This increased permeability in neonates can cause intestinal inflammation and can lead to necrotizing enterocolitis (NEC) [60]. While a direct link between early changes in bone density with reduced barrier function in neonates and children has yet to be proven, studies in inflammatory bowel patients and animal models support this link (see section on "IBD" below). For example, chemically increasing barrier permeability in young (5 week old) growing mice causes reduced bone density and stunted growth compared to control mice [61]. It is noteworthy that when the inflammatory insult is removed, the young mice are able to fully regain bone density and length 5 weeks later [61]. Taken together, the data support the need for more studies to understand the role of the gut epithelial barrier in early life disturbances in bone physiology.

Aging

Several studies have shown that aging can have profound effects on the GI tract. Approximately 35–40% of elderly patients report having at least one GI tract complication during a routine medical exam [62]. Effects of aging in the GI tract include changes in permeability, motility, inflammation, and disruption of the gut microbiota. However, the mechanisms by which aging contribute to shifts in any of these effects and their influence on epithelial barrier and bone health are poorly understood. This is in part because it is difficult to discern if changes are due to normal aging and common age-related disorders or result from disease treatments. This section will discuss the effects of aging on intestinal permeability, inflammation, and microbiome, and while no papers directly link barrier changes with age-related bone loss, we will discuss potential connections.

Changes in Permeability

Several studies have shown that aging can have detrimental effects on intestinal barrier permeability. A study looking at 34- vs. 133-week-old rats demonstrated that the younger rats excreted 34.3% of the administrated PEG 400 in the urine, while 43.6% was excreted by the older rats [63]. Similarly, another study showed that as rats age (12–112 weeks), intestinal permeability to PEG 400 and mannitol increased [64]. Colonic mucosal biopsies from young and old nonhuman primates (baboons) demonstrate significant differences in permeability and TJ proteins. The older baboons displayed a significant decrease in ZO-1, occludin, and JAM-A proteins, and an increase in claudin-2 expression, all of which correlated with increased permeability in the old-aged group [65]. Another study in monkeys found that old monkeys have an increase in FITC dextran flux compared to young monkeys [66].

A cross-sectional study of nonsmoking healthy adults, between 60 and 85 years old, showed no difference in the permeability index (PI = lactulose/mannitol) between young and older humans. However, the study had some limitations such as that the older age group consisted of predominantly males, and sex difference may play a role in intestinal permeability [67]. In an ex vivo assay by Man et al. [68], the authors demonstrated that the transepithelial electric resistance (TEER) is affected in the aged humans. In their study, the effects of age on TEER were tested using ileal biopsies from healthy humans, young (7–12 years), adult (20–40 years), and aging (67–77 years). The TEER was significantly reduced in the aging biopsies, whereas no difference was observed between the two younger groups. The increase in permeability in the aging group appeared to be restricted to solutes since the permeability to macromolecules was not affected by aging [68]. There were no changes in mRNA expression of ZO-1, occludin, and JAMA-1 in the aging group compared with adults and young individuals. On the other hand, they observed that levels of claudin-2 were significantly increased in the aging group and not in the adult group, suggesting that claudin-2 play an important role in intestinal permeability [68]. It has also been proposed that the changes seen in intestinal permeability in aging people can be due to an increase in inflammation and/or disruption of the gut microbiota.

Changes in Mucosal Immune System

Aging is associated with a decline in the immune response [69, 70]. About 50% of the older age group are affected by low-grade chronic inflammation known as "inflammaging" [71]. In a steady-state situation, the IECs communicate with the intestinal immune system to regulate intestinal homeostasis. IECs regulate the intestinal immune homeostasis through the secretion of cytokines that control dendritic cells and T-regulatory cells. This interaction helps to discriminate between invasive pathogenic organisms and harmless antigens. Several studies have reported a dysregulation in the intestinal immune homeostasis in different models of aging. Specifically, it has been shown that the function of the gut-associated immune system is impaired in elderly humans [66, 68, 72, 73].

Using monkeys as a model of human aging, researchers found that old monkeys have greater systemic inflammation as compared to young monkeys. This increase in inflammation was attributed to an increase in serum CRP [66]. In a similar study using baboons, it was found that aged baboons have a significant increase in IFN-γ, IL-6, and IL-β in colonic biopsies. In the same study, the old animals presented an increase in colonic permeability [74]. These results suggest that dysregulation of the immune system can alter intestinal permeability.

Several studies in humans have also confirmed the effects of aging on the intestinal immune response. Ileal biopsies from young (7–12 years), adult (20–40 years), and aging (67–77 years) individuals were assessed for inflammatory cytokine levels. They noticed an increase in the expression of IL-6, but not IFN-γ, TNFα, and IL-1β, in the aging group. The increase in IL-6 was attributed to an increase in dendritic cells. They also demonstrated a correlation between IL-6, claudin-2, and permeability [68]. Many studies indicate that IL-6 expression is induced with aging. Animal studies showed that the decline in the production of IL-1 β, TNFα, and IL-12, in response to LPS in aging, is restored in aging IL-6-deficient knockout mice, suggesting that IL-6 is responsible for the changes in the mucosal immune system during aging [75]. In conclusion, the pro-inflammatory state observed in aging populations may be related to dysfunction of the intestinal barrier.

Changes in Intestinal Microbiota

In a healthy intestinal tract, the microbiota and the gut immune system interact to maintain a homeostatic equilibrium. Perturbation of this homeostatic equilibrium has been strongly associated with many human diseases such as obesity and IBD [76, 77]. The intestinal microbiota supplies nutrients as well as protects the intestinal barrier against pathogens [78]. A variety of factors including the host and microbiological, dietary, and environmental factors can disrupt the gut microbiota. Because of the crucial role of the intestinal microbiota in host homeostasis, it is important to study the age-related differences in microbiota and how it influences intestinal function.

It has been demonstrated that the human intestinal microbiota undergoes maturation from birth to adulthood and is further altered with aging. A study looking at age-related differences in the gut microbiota composition among young (average 31-years old), elderly (average 72 years old), and centenarian humans (average 100 years old) demonstrated that the composition and diversity of the gut microbiota do not differ between young adults and elderly groups. However, there was a significant difference between the elderly and centenaries [79]. These differences were attributed to an increase in facultative anaerobes, mostly belonging to *Proteobacteria* and *Bacilli* in the centenarian group. The *Firmicutes/Bacteroidetes* ratios did not differ between the young and centenarian groups. In the same study, measurements of inflammatory cytokines were performed. An increase in IL-6 and IL-8 was observed, but not TNFα. They also found a positive correlation between bacteria belonging to the phylum *Proteobacteria* with IL-6 and IL-8 [79]. Species diversity was found to change with age in bacteria isolated from fecal samples from healthy young and elderly adults. On the other hand, the overall numbers of organisms were similar at the genus level [80]. In a different study, the results showed a change in bacterial genera with age and a reduction in the numbers *Bacteroides* and *Bifidobacteria* in the elderly group. These reductions were accompanied by reduced species diversity [81]. The *Firmicutes/Bacteroidetes* ratio of the human microbiota increased with age [82]. This shift in microbiota composition might result in a greater susceptibility to diseases by altering intestinal permeability among other consequences.

Intestinal Changes and Bone Health

While we do not know of any study directly linking the effect of aging on gut barrier-to-bone signaling, the intestinal changes that occur with aging are associated with bone loss in other conditions. Specifically, the strongest link between gut permeability and bone loss comes from colitis studies where barrier disruption in adult animal models leads to bone loss, even without weight loss [61, 83, 84]. Thus, as animals and humans age, intestinal permeability increases [61, 63–66, 68] and could contribute to age-related bone loss. In addition, increased permeability likely promotes low-grade chronic inflammation, termed "inflammaging" [66, 71, 74], and many studies link low-grade inflammation with bone loss [61, 83, 84]. In the future, studies are need to test if a direct link exists between barrier function and bone health in the elderly, since this could be a promising target for new therapeutics.

Menopause

Menopause is the natural cessation of menstruation and decline in reproductive hormones. One of the most significant reproductive hormones in females is estrogen, produced primarily in the ovaries. It is also produced at extra-gonadal sites including adipose tissue, skin, osteoblasts, osteoclasts, aorta, and the brain. After menopause, adipose tissue is the main source of estrogen. There are several forms of estrogen, 17β-estradiol being the most prevalent circulating estrogen. Only a small amount in the plasma is free and active, most is bound to globulin or albumin. The two primary receptors for estrogen are estrogen receptor α (ERα) and estrogen receptor β (ERβ), both of which are nuclear receptors. ERα is typically associated with secondary sex characteristics and regulation of the menstrual cycle in females and sperm maturation in males [85]. ERβ has less of a role in the classical estrogen target tissues and has been found to be more dominant in the brain, cardiovascular system, and the colon [86, 87]. A decrease in estrogen levels during menopause has been attributed to osteoporosis that occurs in postmenopausal women, but the role of declining estrogen in intestinal permeability is only beginning to be understood. In ovariectomized Wistar rats, colonic paracellular permeability was increased significantly, and this was reversed by estrogen treatment (estradiol benzoate) [88]. Consistent with this, colonic paracellular permeability decreases during the estrus phase (high levels of estrogen) of the rat when compared to the diestrus phase (low levels of estrogen) [88]. Although estrogen treatment has been shown to predispose ovariectomized rats to development of ulcerative colitis-induced tumor development [89], most studies to date have shown that estrogen treatment decreases colonic paracellular permeability and reduces IBD symptom severity [88, 90, 91]. ERβ is the predominant estrogen receptor in the intestinal tract. Whole body ERβ knockout mice display altered intestinal cell proliferation, decreased apoptosis, and abnormal villus/crypt architecture throughout the intestine [92]. One potential mechanism that could account for estrogen effects on the intestine is through its alterations in TJ and adhesion molecules which would alter intestinal permeability [93]. In models of IBD (IL-10-deficient mice and HLA-B27 rats), ERβ mRNA levels were decreased and colonic permeability increased [91]. Similarly, treatment of cell culture models of intestinal epithelial layers (HT-29, T84, Caco-2) with estrogen receptor antagonists increases permeability, while estrogen treatment prevents this outcome [88, 91]. Our lab demonstrated a significant increase in intestinal permeability 1 week post-surgery in the absence of estrogen (OVX model) in mice. Section-specific changes in permeability were also measured ex vivo by Ussing chambers which demonstrated that the ileum had the most dynamic changes. This study indicates that estrogen deficiency induces region-specific effects on intestinal permeability [93]. Thus, estrogen appears to predominantly inhibit increases in intestinal permeability, and therefore an increase in permeability during menopause could lead to increase in systemic and bone inflammation that could contribute to bone loss.

In addition to altering epithelial barrier function, estrogen has also been shown to impact calcium absorption in the intestine, which is important for bone maintenance. Several studies identified decreases in intestinal and renal calcium absorption following estrogen deficiency [94–98]. Though the exact mechanism is not well understood, it is thought that estrogen deficiency leads to downregulation of the expression of transcellular calcium transport protein plasma membrane calcium pump 1b (PMCA1b), transient receptor potential cation channel subfamily V member 5 (TRPV5), and calbindin-D-28K (CaBP28k) [94]. Furthermore, estrogen has been found to increase vitamin D receptor (VDR) gene and protein expression as well as $1,25(OH)_2D_3$ activity in the colon, which leads to increased intestinal calcium absorption [99, 100]. Taken together, these studies suggest that estrogen could modulate bone health via multiple mechanisms that depend on intestinal barrier function (permeability and calcium/vitamin D metabolism).

Given that one out of two postmenopausal women will fracture a bone [101], the potential for using the gut as a therapeutic target to treat osteoporosis has increased research in this area. Recent studies support a role for intestinal health in the prevention of bone loss in ovariectomy (Ovx) mice [8, 102]. Decreasing intestinal inflammation or altering the gut microbiome leads to the prevention of bone loss [8, 102]. Our lab has shown that treatment with the probiotic *Lactobacillus reuteri* significantly protected Ovx mice from bone loss. This prevention of bone loss by *Lactobacillus reuteri* was attributed to a decrease in osteoclastogenesis and an increase in bone marrow CD4+ T-lymphocytes. *Lactobacillus reuteri* also modifies microbial communities in the Ovx mouse gut [102]. In a different study, researchers found an increase in gut permeability and cytokines (TNFα and IL-17) in the small intestine of Ovx mice. Surprisingly, in the germ-free mice, the effect of estrogen deficiency in gut permeability and cytokine dysregulation was ablated, suggesting a role of the gut microbiota in Ovx-induced bone loss. Treatment with the probiotic *Lactobacillus rhamnosus* GG (LGG) or the probiotic supplement VSL#3 reduces gut permeability and intestinal inflammation and completely protects against bone loss induced by estrogen deficiency [8]. Together, these data highlight the role that of the gut epithelial barrier and microbiota in bone loss induced by estrogen deficiency.

Barrier Pathophysiology in Disease

Dysbiosis

The intestinal microbiota has been described as a virtual organ that exhibits a complex bidirectional cross talk with the environment and other systems throughout the body [103, 104]. The intestinal barrier acts as a wall between the intestinal microbiota and the host's immune system. Under normal conditions, the intestinal epithelium has numerous adaptations such as antimicrobial peptides and mucins that keep the intestinal microbiota away from the gut epithelial layer [105–108]. The TJ also impede microbial invasion into the host tissue [109]. This intestinal epithelial barrier and its adaptations are not static but can be regulated by a variety of external factors such as alteration to the gut microbiota (i.e., dysbiosis).

A number of factors can alter intestinal microbial composition. These include medications such as antibiotics, psychological and physical stress, radiation, altered peristalsis, and dietary changes [110–114]. This can lead to alterations in bacterial metabolism as well as overgrowth of potential pathogenic bacteria [115]. Changes to the gut microbiota during dysbiosis have now been linked to a myriad of diseases such as IBD, irritable bowel syndrome (IBS), obesity, and rheumatoid arthritis [116–119]. Importantly, dysbiosis can also lead to disruption of epithelial barrier leading to unwanted consequences [49] (Fig. 2).

Altered gut microbiota can signal through pattern recognition receptors on gut epithelial cells, activating the NF-κB pathway and leading to changes in gut homeostasis [120]. Epithelial cell NF-κB activation increases pro-inflammatory cytokines such as TNFα, IL-1, and INFγ [121]. An increase in gut INFγ and TNFα protein levels has been shown to increase intestinal permeability [122, 123]. This altered protein composition decreases barrier properties and leads to leaky gut properties. The exact mechanism behind these effects is not well characterized [19]. Dysbiosis has also been linked to increase in IL-1β, which has also been shown to increase permeability by decreasing TJ protein occludin expression [12].

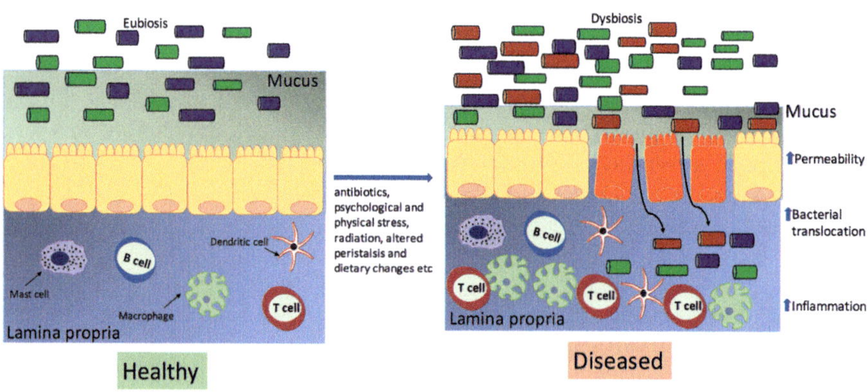

Fig. 2 Schematic representation of the gut epithelial layer in healthy gut vs. dysbiosis. In the normal state, the mucus layer prevents the interaction between the gut microbiota and the intestinal epithelial barrier. Underneath the epithelial layer is the lamina propria. The lamina propria is composed of connective tissue and cells of the innate and adaptive immune system: mast cells, macrophages, dendritic cells, and lymphocytes (T and B). The composition of the intestinal epithelial layer can be influenced by many factors including antibiotic treatment, psychological and physical stress, radiation, age, and diet. This can lead to alterations in bacterial metabolism as well as overgrowth of potential pathogenic bacteria. This dysbiosis is associated with increased levels of permeability, bacterial translocation, and inflammation

Additionally, dysbiosis of the gut microbiota also influences other adaptations such as mucin production which in turn influences gut barrier function. Intestinal mucins can inhibit bacterial adhesion to the intestinal epithelial cells, limiting immune responses and maintaining barrier function. The commensal microbiota has been shown to regulate production of intestinal mucins [124]. The abundance of mucolytic bacterium, which has the ability to degrade mucins, has been shown to increase 100-fold during dysbiosis observed in IBD [125]. In addition, under dysbiosis, pathogens can secrete proteases that have been shown to cleave MUC2, the main mucin component, thereby decreasing mucin levels [126]. Decreases in the mucin layers have been shown to compromise the intestinal barrier function leading to increases in intestinal permeability and microbe penetration [127]. All of the effects of dysbiosis on the gut barrier function can also lead to systemic changes in the body including bone loss [128–130].

Intestinal dysbiosis has also been shown to affect bone density [8, 102, 131–133]. The role of the microbiome in regulating bone remodeling was shown in germ-free mice (in C57BL/6 background) which have increased femoral bone density (both trabecular and cortical) when compared to conventionally raised mice [130]. This increase in bone density was attributed to a decrease in osteoclast as well as inflammatory cytokines in the bone and bone marrow in the germ-free mice vs. conventionally raised mice [130]. However, the effects of microbiome on bone density, as determined by studies using germ-free mice, are not consistent across mouse strains and/or sex [8, 128, 130, 134]. In addition, the impact of the microbiota changes on the epithelial barrier was not been fully examined. It is possible that changes in the microbiome that promote greater barrier function could benefit bone density, while a more pro-inflammatory microbiome could cause bone loss, thereby explaining the inconsistencies between studies.

The previous work has demonstrated that gut dysbiosis promotes inflammation in the bone marrow that correlates with bone loss [83]. It has been hypothesized that dysbiosis disrupts barrier function leading to increases in inflammation and activates T cells leading to enhanced expression of TNFα in the bone marrow [135, 136]. The increase in TNFα stimulates osteoclastogenesis and/or enhances osteoblast apoptosis, thus disrupting normal bone homeostasis leading to bone loss [135, 136]. The mechanism by which activated T cells are increased in the bone marrow in response to changes in the gut microbiota is not completely understood; T-cell activation could be due to gut antigens crossing the intestinal barrier consequent to dysbiosis [137].

The finding that dysbiosis can alter bone density led to studies investigating the role of both pre- and probiotics in bone health. Prebiotics are nondigestible fermentable nutrients which promote the growth of beneficial microorganisms [138]. In vitro studies indicate that prebiotics can enhance intestinal epithelial barrier function and increase tight junction protein expression [139]. Under healthy and estrogen-deficient conditions, prebiotics (such as fructo-oligosaccharides (FOS) and inulin) also increase bone health parameters [140–143]. In addition, probiotics (live microorganisms which have a beneficial effect on the host) have been shown to increase barrier function and bone health. The probiotic *Lactobacillus reuteri* has anti-TNFα properties, reduces gut inflammation, and strengthens gut barrier function in vitro [102, 132, 144]. When given to mice, *L. reuteri* treatment was found to increase bone density in healthy male mice in addition to preventing bone loss in both female ovariectomized mice and type 1 diabetic male mice [102, 133, 132, 145]. Taken together, these data demonstrate the role of the microbiome and intestine in maintaining bone density.

Colitis/IBD

Inflammatory bowel disease is characterized by damage to the intestinal epithelial barrier resulting in increased permeability and the resultant dissemination of the commensal microbiota. This translocation of the luminal contents into the lamina propria persistently stimulates the immune system leading to its hyper-activation and eventual damage to the intestine. IBD can occur in two different forms, through either ulcerative colitis, which affects only the large intestine, or Crohn's disease, which can occur anywhere in the gastrointestinal tract. This idea that IBD is caused by the improper localization of the microbiota and other luminal contents is largely supported through animal models of intestinal inflammation in that it is difficult to elicit these diseases in germ-free conditions [146]. In animal models, decreased epithelial resistance has been shown to precede microscopic inflammation [147]. This highlights the importance of maintaining a healthy epithelial barrier to protect and regulate the permeability and translocation of the microbiota.

An important element in maintaining this healthy barrier is the constant maintenance and restoration of the epithelial cells comprising this barrier as these cells age and eventually undergo apoptosis (approximately every week). To maintain a healthy barrier, the epithelial cells are constantly in a balance of proliferation, migration, and differentiation, migrating from the base of the crypts to the crypt surface or villous tip. Once their journey is complete, these epithelial cells are removed through shedding/apoptosis that does not result in inflammation, normally associated with mass apoptosis. In a disease state, such as IBD, this apoptosis is greatly upregulated resulting in damage and increased permeability in the epithelial barrier and impairment of its basic functions.

As mentioned before, one of the key factors maintaining the integrity and permeability of the epithelial barrier are the TJ. Additionally, inflammatory conditions can influence this regulation resulting in alterations in the mucosal barrier. Increases in pro-inflammatory cytokine such as TNFα, IL-4, IL-6, and IL-13 have all been shown to increase epithelial permeability and have been tied to increased expression of claudin-2 in animal and human models as well as decreased expression of JAM-A and occludin [148, 149]. For example, TNFα is responsible for the removal of claudin-1 from tight junctions. TNFα also induces occludin degradation while promoting MLCK phosphorylation thus resulting in augmented paracellular permeability [150, 151]. Not only can the expression of these TJ proteins be influenced but also their localization within the cell can be dysregulated. Claudin-3, claudin-5, and claudin-8 as well as occludin and JAM-A have all been observed to be internalized rather than expressed on the membrane in biopsies from patients with colitis [152]. These effects on the TJ proteins by inflammatory cytokines are in part mediated by myosin light-chain phosphorylation through myosin light-chain kinase (MLCK). This phosphorylation induces actomyosin contraction that can lead to openings in the junctional gap. In fact, mice continuously expressing MLCK are more susceptible to experimental colitis [153]. Furthermore, improper activation of protein kinase C and Rho can modulate the actin cytoskeleton and influence tight junction regulation and function [154].

In addition to causing a leaky barrier, IBD and ulcerative colitis negatively impact bone [61, 83, 155, 156]. Patients with IBD have a 40% higher risk of developing osteoporosis than the general population [157]. Inflammatory cytokines also increase in IBD and it is known that they can have negative effects in the bone [158]. Although it is not well known whether loss of intestinal barrier per se in IBD patients is causal to bone loss in these patients (see chapter on IBD and bone for further details), our lab has shown that intestinal inflammation without weight loss in an IBD model can lead to significant bone loss suggesting a link [84]. Despite these results, more studies need to be performed to further understand the role of intestinal disruption in IBD effects on bone.

Type 1 Diabetes

Type 1 diabetes (T1D) is characterized by hyperglycemia and hypoinsulinemia and requires treatment with exogenous insulin therapy. Intestinal health has been shown to play a key role in the development of T1D [159–161]. Additionally, T1D-induced changes in intestinal health and function have been suggested to contribute to further T1D complications, such as osteoporosis [145]. Intestinal changes that have been reported to precede or be caused by T1D which can influence bone health include intestinal barrier function or permeability and the intestinal microbiota [162–169].

Intestinal barrier function has been implicated in the development of T1D [162, 166, 169–172]. In rodent models of T1D, intestinal permeability or the "leakiness" of the gut has been studied by measuring the amount of disaccharides and monosaccharides in the urine following their oral administration. The spontaneously diabetic biobreeding (BB) rat model of T1D shows an increased amount of permeability in the stomach, the small intestine, and the colon [164]. The increased permeability in both the stomach and small intestine appear prior to the development of overt diabetic symptoms [164]. During the prediabetes stage, BB rats that are diabetes-prone have increased intestinal permeability, altered tight junction proteins (specifically claudin-7), increased gut infiltration by neutrophils, and decreased numbers of gut natural killer cells in comparison to BB rats which were diabetes resistant [163, 173].

Examination of intestinal permeability in human patients with T1D has been limited and has shown diverse outcomes. An initial study examining the permeability of the monosaccharide mannitol in T1D patients showed an increase in intestinal permeability [167, 174]. However, a subsequent study using pediatric T1D patients showed no difference in the permeability to lactulose or mannitol except in patients with a high-risk allele for celiac disease [167]. As of now, the role of gut permeability in T1D is not well understood, and further research is needed to understand how hypoinsulinemia affects barrier function.

The gastrointestinal system has the largest immune population in the body and creates an interface between the external environment and the host and has been linked with numerous other autoimmune diseases [175] in addition to T1D. T1D is an autoimmune condition resulting from T-cell-mediated destruction of insulin-secreting pancreatic β cells. As the gut houses the largest population of immune cells, it is not surprising that alterations in the intestine can predispose patients to the development of T1D. Furthermore, microbial communities within the intestine have also been shown to alter immune cell number and differentiation [176–181].

Development of T1D has been shown to be preceded by changes in the microbiome in both human and rodent studies. Studies examining the microbiome in T1D patients and control subjects found that T1D patients had less microbial diversity, less microbial population similarities between individual patients, as well as an increase in non-butyrate-producing bacteria when compared with non-diabetic subjects [182]. In a study examining the microbiota in children prior to the development to T1D (children were negative for anti-islet antibodies, however, they possessed the predisposing HLA genotypes), several bacterial taxa correlated with the development of anti-islet antibodies found in T1D children, indicating that alterations in the microbiome may precede the development of T1D [183]. In rodent models of T1D (non-obese diabetic (NOD) mouse and bio-breeding rat (BBR)), exposure to specific bacterial strains or metabolic products as well as vivarium hygiene can modulate T1D incidence [184–194]. Furthermore, comparison of the microbiome between NOD and the genetically related but T1D-resistant mouse (NOR) demonstrated an increase in beneficial microbe populations in NOR mice [195]. Fecal transplant of NOD stool into NOR mice increased insulin resistance; however, NOR stool transplant into NOD mice did not prevent the development of T1D [195]. As the microbiome is highly influential in the development and activity of the immune system within the gut, researchers have sought to determine the role of intestinal immune function in the development of T1D. In humans with T1D, duodenal samples had increased expression of pro-inflammatory cytokines, increased leukocytic infiltration, as well as alterations in microbial populations within the microbiome (increase in *Firmicutes*), which all contributed to a pro-inflammatory environment as compared to healthy controls [196].

As the microbiome was found to be altered in both human and rodents with T1D, studies have gone on to show that T1D prevention can be achieved by supplementation with both prebiotics and probiotics. Recently, in the non-obese diabetic (NOD) mouse model, T1D severity was shown to be inversely correlated with levels of acetate and butyrate, microbial metabolites. When these metabolites were replaced in the diet, NOD mice were protected from the development of T1D [197]. Furthermore, the authors found that acetate supplementation decreased the activation of autoreactive T cells, while butyrate supplementation increased the number and function of regulatory T cells, thereby preventing autoimmune development of T1D [197]. Several studies examining different probiotics have shown that altering the microbiome can influence both the inflammatory state of the intestine and prevent the progression of T1D [198].

In addition to intestinal changes, T1D is associated with many complications including bone loss [199–201] (Fig. 3). Our lab demonstrated that treatment with the probiotic *Lactobacillus reuteri* 6475 prevents T1D-induced bone loss in mice, suggesting a role of the gut microbiota in T1D-induced bone loss. Interestingly, the probiotic treatment prevented several of the bone pathologies of T1D including marrow adiposity, suppressed Wnt10b expression, and suppressed osteoblast activity [145]. The prevention of bone loss occurred despite metabolic dysregulation as indicated by high blood glucose levels. It remains to be seen whether T1D-induced changes in gut permeability can directly influence T1D bone loss. Even if not causal, it would be of interest to examine if reversing increases in T1D-induced intestinal permeability can also reverse T1D bone loss.

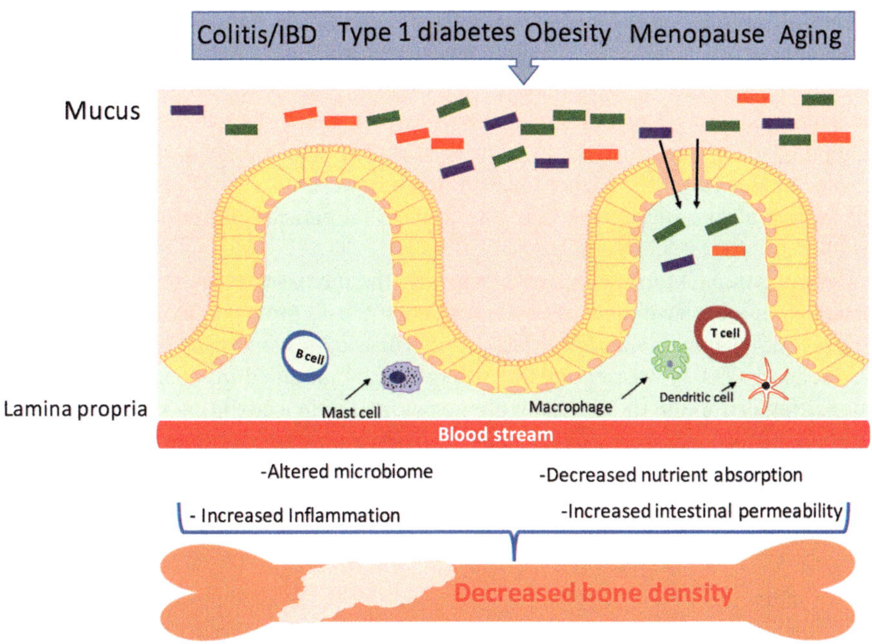

Fig. 3 Model of intestinal epithelial disruption signals that can regulate bone density. Many factors such as aging, menopause, and metabolic diseases are known to disrupt the intestinal epithelial layer. They can modulate gut microbiota composition and activity, increase intestinal permeability and inflammation, and decrease nutrient absorption. These changes can result in local and systemic responses that can affect bone density

Obesity

Obesity is typically associated with several metabolic disorders and is characterized by low-grade inflammation, the molecular origin of which remains unclear [202]. Several studies have reported that serum levels of bacterial LPS are modestly increased in a high-fat diet and that LPS is capable of inducing metabolic disease onset [203–206]. This increase in serum LPS suggests that in obesity the intestinal barrier is compromised. In vivo, animal models of obesity have demonstrated increased whole intestinal permeability via measurement of 4kD FITC-dextran transport to the serum [204, 205] (Fig. 3). Ex vivo, studies utilizing the Ussing chamber have reported increased permeability in the small intestine [206]. Interestingly, while the large intestine has the highest bacterial density and highest levels of LPS, experimental models do not support a definitive causative role for colonic gut barrier dysfunction in obesity. However, a role for increased colonic permeability in obesity cannot be conclusively ruled out as further specific studies are required [207].

The effect of obesity on intestinal permeability in humans is inconclusive. Studies using lactulose (L) and mannitol (M), two sugar probes commonly used to evaluate small intestinal permeability in humans, have shown either no change or modestly increased permeability [207]. In a study investigating obese patients with nonalcoholic steatohepatitis (NASH), the ratio of L/M excreted was similar to healthy controls suggesting no change in intestinal permeability [208]. This is supported in a study by Brignardello et al. [209] that looked at gut permeability in asymptomatic, nonsmoking obese volunteers and observed no differences compared to healthy controls. In contrast, a study by Teixeira et al. [210] reported obese females exhibited higher levels of lactulose excretion but not mannitol than the lean controls, suggesting that small intestinal paracellular permeability may be altered in obese individuals.

Investigations into the mechanisms behind the increased intestinal permeability in animal models have focused on expression of the TJ proteins. In a study by Brun et al. [206] using *ob/ob* and *db/db* mice, distribution of occludin and zonula occludens-1 (ZO-1) was reported to be profoundly modified in the small intestine, suggesting disruption of TJ links with the cytoskeleton, a condition known to compromise the sealing properties of TJs [206]. Obesity-induced changes to TJ expression are further supported in studies by Cani et al. [204, 205]. In these studies, small intestine gene expression of ZO-1 and occludin was reduced in mice fed with high-fat diet and distribution altered in *ob/ob* mice. Expression of these genes had a significant negative correlation with intestinal permeability [204, 205].

An increase in body weight due to obesity has been commonly considered to have a positive effect on bone. However, recent studies demonstrated that bone quality can be compromised in obesity [211, 212]. As mentioned before, obesity can have several effects on the gut epithelium, but their effects on bone density are not well known. It has been suggested that a high-fat diet may affect intestinal calcium absorption and therefore decrease bone formation. Free fatty acids can form unabsorbable insoluble calcium soaps and therefore decrease calcium absorption [213]. Several gut peptides whose levels are altered in obesity, such as ghrelin and incretins, may be involved in bone metabolism [214–216]. Future studies need to be performed to further understand the role of obesity and complex effects on the intestine and their subsequent impact on bone health.

Conclusions

In this chapter, we discussed the importance of the intestinal barrier in maintaining host homeostasis. We discussed different noninvasive methods such as serum, fecal, and urine biomarkers, to further understand intestinal health. We also present information on how disruption of this barrier can have detrimental effects in the host including effects on bone. Factors such as inflammation, changes in microbiota, aging, menopause, and disease have been shown to dysregulate this barrier. In addition, there is now emerging evidence that dysregulation of the intestinal barrier can affect distant organs such as the bone.

References

1. Ma TY, Anderson JM, Turner JR. Tight junctions and the intestinal barrier. Physiol Gastrointest Tract. 2012; doi:10.1016/B978-0-12-382026-6.00038-5.
2. König J, Wells J, Cani PD, García-Ródenas CL, MacDonald T, Mercenier A, Whyte J, Troost F, Brummer R-J. Human intestinal barrier function in health and disease. Clin Transl Gastroenterol. 2016;7:e196. doi:10.1038/ctg.2016.54.
3. Nagao-Kitamoto H, Kitamoto S, Kuffa P, Kamada N. Pathogenic role of the gut microbiota in gastrointestinal diseases. Intest Res. 2016;14:127–38. doi:10.5217/ir.2016.14.2.127.
4. Lee SH. Intestinal permeability regulation by tight junction: implication on inflammatory bowel diseases. Intest Res. 2015;13:11–8. doi:10.5217/ir.2015.13.1.11.
5. Soler AP, Miller RD, Laughlin KV, Carp NZ, Klurfeld DM, Mullin JM. Increased tight junctional permeability is associated with the development of colon cancer. Carcinogenesis. 1999;20:1425–31.
6. Gong J, Hu M, Huang Z, Fang K, Wang D, Chen Q, Li J, Yang D, Zou X, Xu L, Wang K, Dong H, Lu F. Berberine attenuates intestinal mucosal barrier dysfunction in type 2 diabetic rats. Front Pharmacol. 2017;8:42. doi:10.3389/fphar.2017.00042.
7. Cox AJ, Zhang P, Bowden DW, Devereaux B, Davoren PM, Cripps AW, West NP. Increased intestinal permeability as a risk factor for type 2 diabetes. Diabetes Metab. 2016;43:2–5. doi:10.1016/j.diabet.2016.09.004.
8. Li JY, Chassaing B, Tyagi AM, Vaccaro C, Luo T, Adams J, Darby TM, Weitzmann MN, Mulle JG, Gewirtz AT, Jones RM, Pacifici R. Sex steroid deficiency-associated bone loss is microbiota dependent and prevented by probiotics. J Clin Invest. 2016;126:2049–63. doi:10.1172/JCI86062.
9. Katz S, Weinerman S. Osteoporosis and gastrointestinal disease. Gastroenterol Hepatol. 2010;6:506–17.
10. Ulluwishewa D, Anderson RC, McNabb WC, Moughan PJ, Wells JM, Roy NC. Regulation of tight junction permeability by intestinal bacteria and dietary components. J Nutr. 2011;141:769–76. doi:10.3945/jn.110.135657.

11. Tulstrup MVL, Christensen EG, Carvalho V, Linninge C, Ahrné S, Højberg O, Licht TR, Bahl MI. Antibiotic treatment affects intestinal permeability and gut microbial composition in Wistar rats dependent on antibiotic class. PLoS ONE. 2015; doi:10.1371/journal.pone.0144854.
12. Al-Sadi RM, Ma TY. IL-1beta causes an increase in intestinal epithelial tight junction permeability. J Immunol. 2007;178:4641–9. 178/7/4641 [pii]
13. Ma TY, Iwamoto GK, Hoa NT, Akotia V, Pedram A, Boivin MA, Said HM. TNF-α-induced increase in intestinal epithelial tight junction permeability requires NF-κB activation. Am J Physiol Gastrointest Liver Physiol. 2004;286:G367–76. doi:10.1152/ajpgi.00173.2003.
14. Groschwitz KR, Hogan SP. Intestinal barrier function: molecular regulation and disease pathogenesis. J Allergy Clin Immunol. 2009;124:3-20–2. doi:10.1016/j.jaci.2009.05.038.
15. Zolotarevsky Y, Hecht G, Koutsouris A, Gonzalez DE, Quan C, Tom J, Mrsny RJ, Turner JR. A membrane-permeant peptide that inhibits MLC kinase restores barrier function in in vitro models of intestinal disease. Gastroenterology. 2002;123:163–72. doi:10.1053/gast.2002.34235.
16. Al-Sadi R, Guo S, Ye D, Ma TY. TNF-α modulation of intestinal epithelial tight junction barrier is regulated by ERK1/2 activation of Elk-1. Am J Pathol. 2013;183:1871–84. doi:10.1016/j.ajpath.2013.09.001.
17. Adams RB, Planchon SM, Roche JK. IFN-gamma modulation of epithelial barrier function. Time course, reversibility, and site of cytokine binding. J Immunol. 1993;150:2356–63.
18. Madsen KL, Malfair D, Gray D, Doyle JS, Jewell LD, Fedorak RN. Interleukin-10 gene-deficient mice develop a primary intestinal permeability defect in response to enteric microflora. Inflamm Bowel Dis. 1999;5:262–70. doi:10.1097/00054725-199911000-00004.
19. Murch SH, Braegger CP, Walker-Smith JA, MacDonald TT. Location of tumour necrosis factor alpha by immunohistochemistry in chronic inflammatory bowel disease. Gut. 1993;34:1705–9. doi:10.1136/gut.34.12.1705.
20. Zeissig S, Bürgel N, Günzel D, Richter J, Mankertz J, Wahnschaffe U, Kroesen AJ, Zeitz M, Fromm M, Schulzke J-D. Changes in expression and distribution of claudin 2, 5 and 8 lead to discontinuous tight junctions and barrier dysfunction in active Crohn's disease. Gut. 2007;56:61–72. doi:10.1136/gut.2006.094375.
21. Secondulfo M, de Magistris L, Fiandra R, Caserta L, Belletta M, Tartaglione MT, Riegler G, Biagi F, Corazza GR, Carratù R. Intestinal permeability in Crohn's disease patients and their first degree relatives. Dig Liver Dis. 2001;33:680–5.
22. Cornick S, Tawiah A, Chadee K. Roles and regulation of the mucus barrier in the gut. Tissue Barriers. 2015;3:e982426. doi:10.4161/21688370.2014.982426.
23. Enss ML, Cornberg M, Wagner S, Gebert A, Henrichs M, Eisenblätter R, Beil W, Kownatzki R, Hedrich HJ. Proinflammatory cytokines trigger MUC gene expression and mucin release in the intestinal cancer cell line LS180. Inflamm Res. 2000;49:162–9. doi:10.1007/s000110050576.
24. Biochemistr, Ahn D-H, Crawley SC, Hokari R, Kato S, Yang SC, Li J-D, Kim, Young S, Kim YS. TNF-alpha activates MUC2 transcription via NF-kappaB but inhibits via JNK activation. Cell Physiol Biochem. 2005;15:29–40.
25. Hasnain SZ, Tauro S, Das I, Tong H, Chen AH, Jeffery PL, McDonald V, Florin TH, McGuckin MA. IL-10 promotes production of intestinal mucus by suppressing protein misfolding and endoplasmic reticulum stress in goblet cells. Gastroenterology. 2013;144:357–368.e9. doi:10.1053/j.gastro.2012.10.043.
26. Deplancke B, Gaskins HR. Microbial modulation of innate defense: goblet cells and the intestinal mucus layer. Am J Clin Nutr. 2001;73:1131S–41S.
27. Finnie IA, Dwarakanath AD, Taylor BA, Rhodes JM. Colonic mucin synthesis is increased by sodium butyrate. Gut. 1995;36:93–9. doi:10.1136/gut.36.1.93.
28. Burger-van Paassen N, Vincent A, Puiman PJ, van der Sluis M, Bouma J, Boehm G, van Goudoever JB, van Seuningen I, Renes IB. The regulation of intestinal mucin MUC2

expression by short-chain fatty acids: implications for epithelial protection. Biochem J. 2009;420:211–9. doi:10.1042/BJ20082222.
29. Mack DR, Michail S, Wei S, McDougall L, Hollingsworth MA. Probiotics inhibit enteropathogenic E. coli adherence in vitro by inducing intestinal mucin gene expression. Am J Phys. 1999;276:941–50. doi:10.1136/gut.35.4.483.
30. Barcelo A, Claustre J, Moro F, Chayvialle JA, Cuber JC, Plaisancié P. Mucin secretion is modulated by luminal factors in the isolated vascularly perfused rat colon. Gut. 2000;46:218 24. doi:10.1136/gut.46.2.218.
31. Li JD, Feng W, Gallup M, Kim JH, Gum J, Kim Y, Basbaum C. Activation of NF-kappaB via a Src-dependent Ras-MAPK-pp90rsk pathway is required for Pseudomonas aeruginosa-induced mucin overproduction in epithelial cells. Proc Natl Acad Sci U S A. 1998;95:5718–23. doi:10.1073/pnas.95.10.5718.
32. Kim YS, Ho SB. Intestinal goblet cells and mucins in health and disease: recent insights and progress. Curr Gastroenterol Rep. 2010;12:319–30. doi:10.1007/s11894-010-0131-2.
33. Rogler G, Biedermann L. Clinical utility of biomarkers in IBD. Curr Gastroenterol Rep. 2015; doi:10.1007/s11894-015-0449-x.
34. Henriksen M, Jahnsen J, Lygren I, Stray N, Sauar J, Vatn MH, Moum B. C-reactive protein: a predictive factor and marker of inflammation in inflammatory bowel disease. Results from a prospective population-based study. Gut. 2008;57:1518–23. doi:10.1136/gut.2007.146357.
35. Shine B, Berghouse L, Jones JEL, Landon J. C-reactive protein as an aid in the differentiation of functional and inflammatory bowel disorders. Clin Chim Acta. 1985;148:105–9. doi:10.1016/0009-8981(85)90219-0.
36. Benor S, Russell GH, Silver M, Israel EJ, Yuan Q, Winter HS. Shortcomings of the inflammatory bowel disease serology 7 panel. Pediatrics. 2010;125:1230–6. doi:10.1542/peds.2009-1936.
37. Sun XQ, Fu XB, Zhang R-, Lü Y, Deng Q, Jiang XG, Sheng ZY. Relationship between plasma D(−)-lactate and intestinal damage after severe injuries in rats. World J Gastroenterol. 2001;7:555–8.
38. Vreugdenhil AC, Wolters VM, Adriaanse MP, Van den Neucker AM, van Bijnen AA, Houwen R, Buurman WA. Additional value of serum I-FABP levels for evaluating celiac disease activity in children. Scand J Gastroenterol. 2011;46:1435–41. doi:10.3109/00365521.2011.62744 7.
39. Adriaanse MPM, Tack GJ, Passos VL, Damoiseaux JGMC, Schreurs MWJ, Van Wijck K, Riedl RG, Masclee AAM, Buurman WA, Mulder CJJ, Vreugdenhil ACE. Serum I-FABP as marker for enterocyte damage in coeliac disease and its relation to villous atrophy and circulating autoantibodies. Aliment Pharmacol Ther. 2013;37:482–90. doi:10.1111/apt.12194.
40. Kanda T, Fujii H, Tani T, Murakami H, Suda T, Sakai Y, Ono T, Hatakeyama K. Intestinal fatty acid-binding protein is a useful diagnostic marker for mesenteric infarction in humans. Gastroenterology. 1996;110:339–43. doi:10.1053/gast.1996.v110.pm8566578.
41. Lewis JD. The utility of biomarkers in the diagnosis and therapy of inflammatory bowel disease. Gastroenterology. 2011;140:1817–26. doi:10.1053/j.gastro.2010.11.058.
42. Lundberg J, Hellström P. Technology insight: calprotectin, lactoferrin and nitric oxide as novel markers of inflammatory bowel disease. Nat Clin Pract .... 2005;2:96–102. doi:10.1038/ncpgasthep0094.
43. Bunn SK, Bisset WM, Main MJ, Gray ES, Olson S, Golden BE. Fecal calprotectin: validation as a noninvasive measure of bowel inflammation in childhood inflammatory bowel disease. J Pediatr Gastroenterol Nutr. 2001;33:14–22. doi:10.1097/00005176-200107000-00003.
44. Derikx JP, Luyer MD, Heineman E, Buurman WA. Non-invasive markers of gut wall integrity in health and disease. World J Gastroenterol. 2010;16:5272–9. doi:10.3748/wjg.v16.i42.5272.
45. Vojdani A. For the assessment of intestinal permeability, size matters. Altern Ther Health Med. 2013;19:12–24.

46. Fink MP. Interpreting dual-sugar absorption studies in critically ill patients: what are the implications of apparent increases in intestinal permeability to hydrophilic solutes? Intensive Care Med. 1997;23:489–92. doi:10.1007/s001340050363.
47. Kerckhoffs APM, Akkermans LMA, De Smet MBM, Besselink MGH, Hietbrink F, Bartelink IH, Busschers WB, Samsom M, Renooij W. Intestinal permeability in irritable bowel syndrome patients: effects of NSAIDs. Dig Dis Sci. 2010;55:716–23. doi:10.1007/s10620-009-0765-9.
48. Olaison G, Leandersson P, Sjödahl R, Tagesson C. Intestinal permeability to polyethyleneglycol 600 in Crohn's disease. Peroperative determination in a defined segment of the small intestine. Gut. 1988;29:196–9. doi:10.1136/gut.29.2.196.
49. Arrieta MC, Bistritz L, Meddings JB. Alterations in intestinal permeability. Gut. 2006;55:1512–20. doi:10.1136/gut.2005.085373.
50. Söderholm JD, Yates DA, Gareau MG, Yang P, MacQueen G, Perdue MH. Neonatal maternal separation predisposes adult rats to colonic barrier dysfunction in response to mild stress. Am J Physiol Gastrointest Liver Physiol. 2002;283:G1257–63. doi:10.1152/ajpgi.00314.2002.
51. Stenman LK, Holma R, Korpela R. High-fat-induced intestinal permeability dysfunction associated with altered fecal bile acids. World J Gastroenterol. 2012;18:923–9. doi:10.3748/wjg.v18.i9.923.
52. Penders J, Thijs C, Vink C, Stelma FF, Snijders B, Kummeling I, van den Brandt PA, Stobberingh EE. Factors influencing the composition of the intestinal microbiota in early infancy. Pediatrics. 2006;118:511–21. doi:10.1542/peds.2005-2824.
53. Weaver LT, Laker MF, Nelson R, Lucas A. Milk feeding and changes in intestinal permeability and morphology in the newborn. J Pediatr Gastroenterol Nutr. 1987;6:351–8. doi:10.1097/00005176-198705000-00008.
54. Bergmann KR, Liu SXL, Tian R, Kushnir A, Turner JR, Li HL, Chou PM, Weber CR, De Plaen IG. Bifidobacteria stabilize claudins at tight junctions and prevent intestinal barrier dysfunction in mouse necrotizing enterocolitis. Am J Pathol. 2013;182:1596–606. doi:10.1016/j.ajpath.2013.01.013.
55. Biol-N'garagba MC, Louisot P. Regulation of the intestinal glycoprotein glycosylation during postnatal development: role of hormonal and nutritional factors. Biochimie. 2003;85:331–52. doi:10.1016/S0300-9084(03)00039-7.
56. Arrieta M-C, Stiemsma LT, Amenyogbe N, Brown EM, Finlay B. The intestinal microbiome in early life: health and disease. Front Immunol. 2014;5:427. doi:10.3389/fimmu.2014.00427.
57. Planer JD, Peng Y, Kau AL, Blanton LV, Ndao IM, Tarr PI, Warner BB, Gordon JI. Development of the gut microbiota and mucosal IgA responses in twins and gnotobiotic mice. Nature. 2016;534:263–6. doi:10.1038/nature17940.
58. Kurl S, Heinonen K, Lansimies E, Launiala K. Determinants of bone mineral density in prematurely born children aged 6–7 years. Acta Paediatr. 1998;87:650–3. doi:10.1111/j.1651-2227.1998.tb01525.x.
59. Takada M, Shimada M, Hosono S, Tauchi M, Minato S, Takahashi M, Okuni S, Takeuchi S. Trace elements and mineral requirements for very low birth weight infants in rickets of prematurity. Early Hum Dev. 1992;29:333–8. doi:10.1016/0378-3782(92)90188-M.
60. Halpern MD, Denning PW. The role of intestinal epithelial barrier function in the development of NEC. Tissue Barriers. 2015;3:e1000707. doi:10.1080/21688370.2014.1000707.
61. Harris L, Senagore P, Young VB, McCabe LR. Inflammatory bowel disease causes reversible suppression of osteoblast and chondrocyte function in mice. Am J Physiol Gastrointest Liver Physiol. 2009;296:G1020–9. doi:10.1152/ajpgi.90696.2008.
62. Hall KE, Proctor DD, Fisher L, Rose S. American Gastroenterological Association future trends committee report: effects of aging of the population on gastroenterology practice, education, and research. Gastroenterology. 2005;129:1305–38. doi:10.1053/j.gastro.2005.06.013.
63. Hollander D, Tarnawski H. Aging-associated increase in intestinal absorption of macromolecules. Gerontology. 1985;31:133–7. doi:10.1159/000212694.
64. Ma TY, Hollander D, Dadufalza V, Krugliak P. Effect of aging and caloric restriction on intestinal permeability. Exp Gerontol. 1992;27:321–33. doi:10.1016/0531-5565(92)90059-9.

65. Greenwood-Van Meerveld B, Johnson AC, Grundy D. Gastrointestinal physiology and function. Handb Exp Pharmacol. 2017; doi:10.1007/164_2016_118.
66. Mitchell EL, Davis AT, Brass K, Dendinger M, Barner R, Gharaibeh R, Fodor AA, Kavanagh K. Reduced intestinal motility, mucosal barrier function, and inflammation in aged monkeys. J Nutr Health Aging. 2017;21:354–61. doi:10.1007/s12603-016-0725-y.
67. Valentini L, Ramminger S, Haas V, Postrach E, Werich M, Fischer A, Koller M, Swidsinski A, Bereswill S, Lochs H, Schulzke J D. Small intestinal permeability in older adults. Physiol Rep. 2014;2:e00281. doi:10.14814/phy2.281.
68. Man AL, Bertelli E, Rentini S, Regoli M, Briars G, Marini M, Watson AJM, Nicoletti C. Age-associated modifications of intestinal permeability and innate immunity in human small intestine. Clin Sci. 2015;129:515–27. doi:10.1042/CS20150046.
69. Weng N p. Aging of the immune system: how much can the adaptive immune system adapt? Immunity. 2006;24:495–9. doi:10.1016/j.immuni.2006.05.001.
70. Gomez CR, Gomez CR, Boehmer ED, Boehmer ED, Kovacs EJ, Kovacs EJ. The aging innate immune system. Curr Opin Immunol. 2005;17:457–62. doi:10.1016/j.coi.2005.07.013.
71. Ahmadi-Abhari S, Luben RN, Wareham NJ, Khaw KT. Distribution and determinants of C-reactive protein in the older adult population: European prospective investigation into cancer-norfolk study. Eur J Clin Investig. 2013;43:899–911. doi:10.1111/eci.12116.
72. Nicoletti C. Age-associated changes of the intestinal epithelial barrier: local and systemic implications. Expert Rev Gastroenterol Hepatol. 2015;9:1467–9. doi:10.1586/17474124.2015.1092872.
73. Man AL, Gicheva N, Nicoletti C. The impact of ageing on the intestinal epithelial barrier and immune system. Cell Immunol. 2014;289:112–8. doi:10.1016/j.cellimm.2014.04.001.
74. Tran L, Greenwood-Van Meerveld B. Age-associated remodeling of the intestinal epithelial barrier. J Gerontol Ser A Biol Sci Med Sci. 2013;68:1045–56. doi:10.1093/gerona/glt106.
75. Kovacs EJ, Gomez CR, Karavitis J, Palmer JL, Faunce DE, Ramirez L, Nomellini V. Interleukin-6 contributes to age-related alteration of cytokine production by macrophages. Mediat Inflamm. 2010; doi:10.1155/2010/475139.
76. Turnbaugh PJ, Ley RE, Mahowald MA, Magrini V, Mardis ER, Gordon JI. An obesity-associated gut microbiome with increased capacity for energy harvest. Nature. 2006;444:1027–131. doi:10.1038/nature05414.
77. Peterson DA, Frank DN, Pace NR, Gordon JI. Metagenomic approaches for defining the pathogenesis of inflammatory bowel diseases. Cell Host Microbe. 2008;3:417–27. doi:10.1016/j.chom.2008.05.001.
78. Neish AS. Microbes in gastrointestinal health and disease. Gastroenterology. 2009;136:65–80. doi:10.1053/j.gastro.2008.10.080.
79. Biagi E, Nylund L, Candela M, Ostan R, Bucci L, Pini E, Nikkïla J, Monti D, Satokari R, Franceschi C, Brigidi P, de Vos W. Through ageing, and beyond: gut microbiota and inflammatory status in seniors and centenarians. PLoS ONE. 2010; doi:10.1371/journal.pone.0010667.
80. Hopkins MJ, Macfarlane GT. Changes in predominant bacterial populations in human faeces with age and with Clostridium difficile infection. J Med Microbiol. 2002;51:448–54.
81. Woodmansey EJ, McMurdo MET, Macfarlane GT, Macfarlane S. Comparison of compositions and metabolic activities of fecal microbiotas in young adults and in antibiotic-treated and non-antibiotic-treated elderly subjects. Appl Environ Microbiol. 2004;70:6113–22. doi:10.1128/AEM.70.10.6113-6122.2004.
82. Mariat D, Firmesse O, Levenez F, Guimarães V, Sokol H, Doré J, Corthier G, Furet J-P. The Firmicutes/Bacteroidetes ratio of the human microbiota changes with age. BMC Microbiol. 2009;9:123. doi:10.1186/1471-2180-9-123.
83. Irwin R, Lee T, Young VB, Parameswaran N, McCabe LR. Colitis-induced bone loss is gender dependent and associated with increased inflammation. Inflamm Bowel Dis. 2013;19:1586–97. doi:10.1097/MIB.0b013e318289e17b.
84. Irwin R, Raehtz S, Parameswaran N, McCabe LR. Intestinal inflammation without weight loss decreases bone density and growth. Am J Phys Regul Integr Comp Phys. 2016;311:R1149–57. doi:10.1152/ajpregu.00051.2016.

85. Dahlman-Wright K, Cavailles V, Fuqua SA, Jordan VC, Katzenellenbogen JA, Korach KS, Maggi A, Muramatsu M, Parker MG, Gustafsson J-A. International Union of Pharmacology. LXIV. Estrogen receptors. Pharmacol Rev. 2006;58:773–81. doi:10.1124/pr.58.4.8.
86. Campbell-Thompson M, Lynch IJ, Bhardwaj B. Expression of estrogen receptor (ER) subtypes and ERbeta isoforms in colon cancer. Cancer Res. 2001;61:632–40.
87. Konstantinopoulos PA, Kominea A, Vandoros G, Sykiotis GP, Andricopoulos P, Varakis I, Sotiropoulou-Bonikou G, Papavassiliou AG. Oestrogen receptor beta (ERbeta) is abundantly expressed in normal colonic mucosa, but declines in colon adenocarcinoma paralleling the tumour's dedifferentiation. Eur J Cancer. 2003;39:1251–8.
88. Braniste V, Leveque M, Buisson-Brenac C, Bueno L, Fioramonti J, Houdeau E. Oestradiol decreases colonic permeability through oestrogen receptor beta-mediated up-regulation of occludin and junctional adhesion molecule-A in epithelial cells. J Physiol. 2009;587:3317–28. doi:10.1113/jphysiol.2009.169300.
89. Heijmans J, Wielenga MCB, Rosekrans SL, van Lidth de Jeude JF, Roelofs J, Groothuis P, Ederveen A, de Jonge-Muller ESM, Biemond I, Hardwick JCH, D'Haens G, Hommes DW, Muncan V, van den Brink GR. Oestrogens promote tumorigenesis in a mouse model for colitis-associated cancer. Gut. 2014;63:310–6. doi:10.1136/gutjnl-2012-304216.
90. Braniste V, Jouault A, Gaultier E, Polizzi A, Buisson-Brenac C, Leveque M, Martin PG, Theodorou V, Fioramonti J, Houdeau E. Impact of oral bisphenol A at reference doses on intestinal barrier function and sex differences after perinatal exposure in rats. Proc Natl Acad Sci U S A. 2010;107:448–53. doi:10.1073/pnas.0907697107.
91. Looijer-van Langen M, Hotte N, Dieleman LA, Albert E, Mulder C, Madsen KL, Langen ML, Kl M. Estrogen receptor-β signaling modulates epithelial barrier function. Am J Physiol Gastrointest Liver Physiol. 2011;300:G621–6. doi:10.1152/ajpgi.00274.2010.
92. Wada-Hiraike O, Imamov O, Hiraike H, Hultenby K, Schwend T, Omoto Y, Warner M, Gustafsson J-A. Role of estrogen receptor beta in colonic epithelium. Proc Natl Acad Sci U S A. 2006;103:1605–8. doi:10.1073/pnas.0511271103.
93. Collins FL, Rios-Arce ND, Atkinson S, Bierhalter H, Schoenherr D, Bazil JN, McCabe LR, Parameswaran N. Temporal and regional intestinal changes in permeability, tight junction, and cytokine gene expression following ovariectomy-induced estrogen deficiency. Physiol Rep. 2017;5:e13263. doi:10.14814/phy2.13263.
94. Dong XL, Zhang Y, Wong MS. Estrogen deficiency-induced Ca balance impairment is associated with decrease in expression of epithelial Ca transport proteins in aged female rats. Life Sci. 2014;96:26–32. doi:10.1016/j.lfs.2013.12.025.
95. Hoenderop JGJ, Nilius B, Bindels RJM. Calcium absorption across epithelia. Physiol Rev. 2005;85:373–422. doi:10.1152/physrev.00003.2004.
96. Prince RL, Smith M, Dick IM, Price RI, Webb PG, Henderson NK, Harris MM. Prevention of postmenopausal osteoporosis. A comparative study of exercise, calcium supplementation, and hormone-replacement therapy. N Engl J Med. 1991;325:1189–95. doi:10.1056/NEJM199110243251701.
97. Heshmati HM, Khosla S, Burritt MF, O'Fallon WM, Riggs BL. A defect in renal calcium conservation may contribute to the pathogenesis of postmenopausal osteoporosis. J Clin Endocrinol Metab. 1998;83:1916–20. doi:10.1210/jcem.83.6.4854.
98. O'Loughlin PD, Morris HA. Oestrogen deficiency impairs intestinal calcium absorption in the rat. J Physiol. 1998;511(Pt 1):313–22. doi:10.1111/j.1469-7793.1998.313bi.x.
99. Schwartz B, Smirnoff P, Shany S, Liel Y. Estrogen controls expression and bioresponse of 1,25-dihydroxyvitamin D receptors in the rat colon. Mol Cell Biochem. 2000;203:87–93. doi:10.1023/A:1007015027268.
100. Gilad LA, Bresler T, Gnainsky J, Smirnoff P, Schwartz B. Regulation of vitamin D receptor expression via estrogen-induced activation of the ERK 1/2 signaling pathway in colon and breast cancer cells. J Endocrinol. 2005;185:577–92. doi:10.1677/joe.1.05770.
101. Gabriel SE, Tosteson AN, Leibson CL, Crowson CS, Pond GR, Hammond CS, Melton LJ. Direct medical costs attributable to osteoporotic fractures. Osteoporos Int. 2002;13:323–30. doi:10.1007/s001980200033.

102. Britton RA, Irwin R, Quach D, Schaefer L, Zhang J, Lee T, Parameswaran N, McCabe LR. Probiotic L. reuteri treatment prevents bone loss in a menopausal ovariectomized mouse model. J Cell Physiol. 2014;229:1822–30. doi:10.1002/jcp.24636.
103. O'Hara AM, Shanahan F. The gut flora as a forgotten organ. EMBO Rep. 2006;7:688–93. doi:10.1038/sj.embor.7400731.
104. Sun J, Chang EB. Exploring gut microbes in human health and disease: pushing the envelope. Genes Dis. 2014,1.132–9. doi:10.1016/j.gendis.2014.08.001.
105. Frey A, Giannasca KT, Weltzin R, Giannasca PJ, Reggio H, Lencer WI, Neutra MR. Role of the glycocalyx in regulating access of microparticles to apical plasma membranes of intestinal epithelial cells: implications for microbial attachment and oral vaccine targeting. J Exp Med. 1996;184:1045–59.
106. McAuley JL, Linden SK, Png CW, King RM, Pennington HL, Gendler SJ, Florin TH, Hill GR, Korolik V, McGuckin MA. MUC1 cell surface mucin is a critical element of the mucosal barrier to infection. J Clin Invest. 2007;117:2313–24. doi:10.1172/JCI26705.
107. Salzman NH, Ghosh D, Huttner KM, Paterson Y, Bevins CL. Protection against enteric salmonellosis in transgenic mice expressing a human intestinal defensin. Nature. 2003;422:522–6. doi:10.1038/nature01520.
108. Ganz T. Defensins: antimicrobial peptides of innate immunity. Nat Rev Immunol. 2003;3:710–20. doi:10.1038/nri1180.
109. Shen L, Turner JR. Role of epithelial cells in initiation and propagation of intestinal inflammation. Eliminating the static: tight junction dynamics exposed. Am J Physiol Gastrointest Liver Physiol. 2006:290, G577–G282.
110. Jakobsson HE, Jernberg C, Andersson AF, Sjölund-Karlsson M, Jansson JK, Engstrand L. Short-term antibiotic treatment has differing long-term impacts on the human throat and gut microbiome. PLoS ONE. 2010;5:e9836. doi:10.1371/journal.pone.0009836.
111. Dethlefsen L, Relman DA. Incomplete recovery and individualized responses of the human distal gut microbiota to repeated antibiotic perturbation. Proc Natl Acad Sci. 2011;108:4554–61. doi:10.1073/pnas.1000087107.
112. Cryan JF, Dinan TG. Mind-altering microorganisms: the impact of the gut microbiota on brain and behaviour. Nat Rev Neurosci. 2012;13:701–12. doi:10.1038/nrn3346.
113. Kang SS, Jeraldo PR, Kurti A, Miller ME, Cook MD, Whitlock K, Goldenfeld N, Woods JA, White BA, Chia N, Fryer JD. Diet and exercise orthogonally alter the gut microbiome and reveal independent associations with anxiety and cognition. Mol Neurodegener. 2014;9:36. doi:10.1186/1750-1326-9-36.
114. Touchefeu Y, Montassier E, Nieman K, Gastinne T, Potel G, Bruley des Varannes S, Le Vacon F, de La Cochetière MF. Systematic review: the role of the gut microbiota in chemotherapy- or radiation-induced gastrointestinal mucositis – current evidence and potential clinical applications. Aliment Pharmacol Ther. 2014;40:409–421. n/a-n/a. doi:10.1111/apt.12878.
115. Myers SP. The causes of intestinal dysbiosis. Altern Med Rev Altern Med Rev. 2004;99:180–97.
116. Takaishi H, Matsuki T, Nakazawa A, Takada T, Kado S, Asahara T, Kamada N, Sakuraba A, Yajima T, Higuchi H, Inoue N, Ogata H, Iwao Y, Nomoto K, Tanaka R, Hibi T. Imbalance in intestinal microflora constitution could be involved in the pathogenesis of inflammatory bowel disease. Int J Med Microbiol. 2008;298:463–72. doi:10.1016/j.ijmm.2007.07.016.
117. Spiller RC. Role of infection in irritable bowel syndrome. J Gastroenterol. 2007;42(Suppl 1):41–7. doi:10.1007/s00535-006-1925-8.
118. Evans CC, LePard KJ, Kwak JW, Stancukas MC, Laskowski S, Dougherty J, Moulton L, Glawe A, Wang Y, Leone V, Antonopoulos DA, Smith D, Chang EB, Ciancio MJ. Exercise prevents weight gain and alters the gut microbiota in a mouse model of high fat diet-induced obesity. PLoS ONE. 2014;9:e92193. doi:10.1371/journal.pone.0092193.
119. Scher JU, Abramson SB. The microbiome and rheumatoid arthritis. Nat Rev Rheumatol. 2011;7:569. doi:10.1038/nrrheum.2011.121.
120. Nenci A, Becker C, Wullaert A, Gareus R, van Loo G, Danese S, Huth M, Nikolaev A, Neufert C, Madison B, Gumucio D, Neurath MF, Pasparakis M. Epithelial NEMO links innate immunity to chronic intestinal inflammation. Nature. 2007;446:557–61. doi:10.1038/nature05698.

121. Capaldo CT, Nusrat A. Cytokine regulation of tight junctions. Biochim Biophys Acta. 2009;1788:864–71. doi:10.1016/j.bbamem.2008.08.027.
122. Ozaki H, Ishii K, Horiuchi H, Arai H, Kawamoto T, Okawa K, Iwamatsu A, Kita T. Cutting edge: combined treatment of TNF-alpha and IFN-gamma causes redistribution of junctional adhesion molecule in human endothelial cells. J Immunol. 1999;163:553–7.
123. Youakim A, Ahdieh M. Interferon-gamma decreases barrier function in T84 cells by reducing ZO-1 levels and disrupting apical actin. Am J Phys. 1999;276:G1279–88.
124. Wrzosek L, Miquel S, Noordine M-L, Bouet S, Chevalier-Curt M, Robert V, Philippe C, Bridonneau C, Cherbuy C, Robbe-Masselot C, Langella P, Thomas M. Bacteroides thetaiotaomicron and Faecalibacterium prausnitzii influence the production of mucus glycans and the development of goblet cells in the colonic epithelium of a gnotobiotic model rodent. BMC Biol. 2013;11:61. doi:10.1186/1741-7007-11-61.
125. Png CW, Lindén SK, Gilshenan KS, Zoetendal EG, McSweeney CS, Sly LI, McGuckin MA, Florin THJ. Mucolytic bacteria with increased prevalence in IBD mucosa augment in vitro utilization of mucin by other bacteria. Am J Gastroenterol. 2010;105:2420–8. doi:10.1038/ajg.2010.281.
126. van der Post S, Subramani DB, Backstrom M, Johansson MEV, Vester-Christensen MB, Mandel U, Bennett EP, Clausen H, Dahlen G, Sroka A, Potempa J, Hansson GC. Site-specific O-glycosylation on the MUC2 mucin protein inhibits cleavage by the Porphyromonas gingivalis secreted cysteine protease (RgpB). J Biol Chem. 2013;288:14636–46. doi:10.1074/jbc.M113.459479.
127. Sun J, Shen X, Li Y, Guo Z, Zhu W, Zuo L, Zhao J, Gu L, Gong J, Li J. Therapeutic potential to modify the mucus barrier in inflammatory bowel disease. Forum Nutr. 2016; doi:10.3390/nu8010044.
128. Yan J, Herzog JW, Tsang K, Brennan CA, Bower MA, Garrett WS, Sartor BR, Aliprantis AO, Charles JF. Gut microbiota induce IGF-1 and promote bone formation and growth. Proc Natl Acad Sci. 2016;113:E7554–63. doi:10.1073/pnas.1607235113.
129. Villa CR, Ward WE, Comelli EM. Gut microbiota-bone axis. Crit Rev Food Sci Nutr. 2015; doi:10.1080/10408398.2015.1010034.
130. Sjögren K, Engdahl C, Henning P, Lerner UH, Tremaroli V, Lagerquist MK, Bäckhed F, Ohlsson C. The gut microbiota regulates bone mass in mice. J Bone Miner Res. 2012;27:1357–67. doi:10.1002/jbmr.1588.
131. Zhang K, Hornef MW, Dupont A. The intestinal epithelium as guardian of gut barrier integrity. Cell Microbiol. 2015;17:1561–1569. n/a-n/a. doi:10.1111/cmi.12501.
132. McCabe LR, Irwin R, Schaefer L, Britton RA. Probiotic use decreases intestinal inflammation and increases bone density in healthy male but not female mice. J Cell Physiol. 2013;228:1793–8. doi:10.1002/jcp.24340.
133. Collins FL, Irwin R, Bierhalter H, Schepper J, Britton RA, Parameswaran N, McCabe LR. Lactobacillus reuteri 6475 increases bone density in intact females only under an inflammatory setting. PLoS ONE. 2016;11:e0153180. doi:10.1371/journal.pone.0153180.
134. Yan J, Charles JF. Gut microbiome and bone: to build, destroy, or both? Curr Osteoporos Rep. 2017; doi:10.1007/s11914-017-0382-z.
135. Kobayashi K, Takahashi N, Jimi E, Udagawa N, Takami M, Kotake S, Nakagawa N, Kinosaki M, Yamaguchi K, Shima N, Yasuda H, Morinaga T, Higashio K, Martin TJ, Suda T. Tumor necrosis factor alpha stimulates osteoclast differentiation by a mechanism independent of the ODF/RANKL-RANK interaction. J Exp Med. 2000;191:275–86.
136. Kudo O, Fujikawa Y, Itonaga I, Sabokbar A, Torisu T, Athanasou NA. Proinflammatory cytokine (TNF?/IL-1?) induction of human osteoclast formation. J Pathol. 2002;198:220–7. doi:10.1002/path.1190.
137. Weitzmann MN, Pacifici R. T cells: unexpected players in the bone loss induced by estrogen deficiency and in basal bone homeostasis. Ann N Y Acad Sci. 2007;1116:360–75. doi:10.1196/annals.1402.068.
138. Roberfroid M, Gibson GR, Hoyles L, McCartney AL, Rastall R, Rowland I, Wolvers D, Watzl B, Szajewska H, Stahl B, Guarner F, Respondek F, Whelan K, Coxam V, Davicco

M-J, Léotoing L, Wittrant Y, Delzenne NM, Cani PD, Neyrinck AM, Meheust A. Prebiotic effects: metabolic and health benefits. Br J Nutr. 2010;104(Suppl):S1–63. doi:10.1017/S0007114510003363.

139. Wu RY, Abdullah M, Määttänen P, Pilar AVC, Scruten E, Johnson-Henry KC, Napper S, O'Brien C, Jones NL, Sherman PM. Protein kinase C δ signaling is required for dietary prebiotic-induced strengthening of intestinal epithelial barrier function. Sci Rep. 2017;7:40820. doi:10.1038/srep40820.
140. García-Vieyra MI, Del Real A, López MG. Agave fructans: their effect on mineral absorption and bone mineral content. J Med Food. 2014;17:1247–55. doi:10.1089/jmf.2013.0137.
141. Takahara S, Morohashi T, Sano T, Ohta A, Yamada S, Sasa R. Fructooligosaccharide consumption enhances femoral bone volume and mineral concentrations in rats. J Nutr. 2000;130:1792–5.
142. Chonan O, Matsumoto K, Watanuki M. Effect of galactooligosaccharides on calcium absorption and preventing bone loss in ovariectomized rats. Biosci Biotechnol Biochem. 1995;59:236–9. doi:10.1271/bbb.59.236.
143. Ohta A, Uehara M, Sakai K, Takasaki M, Adlercreutz H, Morohashi T, Ishimi Y. A combination of dietary fructooligosaccharides and isoflavone conjugates increases femoral bone mineral density and equol production in ovariectomized mice. J Nutr. 2002;132:2048–54.
144. Liu H-Y, Roos S, Jonsson H, Ahl D, Dicksved J, Lindberg JE, Lundh T. Effects of Lactobacillus johnsonii and Lactobacillus reuteri on gut barrier function and heat shock proteins in intestinal porcine epithelial cells. Physiol Rep. 2015;3:e12355. doi:10.14814/phy2.12355.
145. Zhang J, Motyl KJ, Irwin R, MacDougald OA, Britton RA, McCabe LR. Loss of bone and Wnt10b expression in male type 1 diabetic mice is blocked by the probiotic Lactobacillus reuteri. Endocrinology. 2015;156:3169–82. doi:10.1210/EN.2015-1308.
146. Bamias G, Okazawa A, Rivera-Nieves J, Arseneau KO, De La Rue SA, Pizarro TT, Cominelli F. Commensal bacteria exacerbate intestinal inflammation but are not essential for the development of murine ileitis. J Immunol. 2007;178:1809–18. 178/3/1809 [pii]
147. Suenaert P, Maerten P, Van Assche G, Van Driessche W, Geboes K, Bulteel V, Simaels J, Augustijns P, Ceuppens JL, Rutgeerts P, Perrier C. Effects of T cell-induced colonic inflammation on epithelial barrier function. Inflamm Bowel Dis. 2010;16:1322–31. doi: 10.1002/ibd.21211.
148. Utech M, Mennigen R, Bruewer M. Endocytosis and recycling of tight junction proteins in inflammation. J Biomed Biotechnol. 2010;2010:484987. doi:10.1155/2010/484987.
149. Suzuki T, Yoshinaga N, Tanabe S. Interleukin-6 (IL-6) regulates claudin-2 expression and tight junction permeability in intestinal epithelium. J Biol Chem. 2011;286:31263–71. doi:10.1074/jbc.M111.238147.
150. Amasheh M, Fromm A, Krug SM, Amasheh S, Andres S, Zeitz M, Fromm M, Schulzke J-D. TNFalpha-induced and berberine-antagonized tight junction barrier impairment via tyrosine kinase, Akt and NFkappaB signaling. J Cell Sci. 2010;123:4145–55. doi:10.1242/jcs.070896.
151. Ye D, Guo S, Al-Sadi R, Ma TY. MicroRNA regulation of intestinal epithelial tight junction permeability. Gastroenterology. 2011;141:1323–33. doi:10.1053/j.gastro.2011.07.005.
152. Vetrano S, Rescigno M, Cera MR, Correale C, Rumio C, Doni A, Fantini M, Sturm A, Borroni E, Repici A, Locati M, Malesci A, Dejana E, Danese S. Unique role of junctional adhesion molecule-a in maintaining mucosal homeostasis in inflammatory bowel disease. Gastroenterology. 2008;135:173–84. doi:10.1053/j.gastro.2008.04.002.
153. Blair SA, Kane SV, Clayburgh DR, Turner JR. Epithelial myosin light chain kinase expression and activity are upregulated in inflammatory bowel disease. Lab Investig. 2006;863700373:191–201. doi:10.1038/labinvest.3700373.
154. Schneider MR, Dahlhoff M, Horst D, Hirschi B, Trülzsch K, Müller-Höcker J, Vogelmann R, Allgäuer M, Gerhard M, Steininger S, Wolf E, Kolligs FT. A key role for E-cadherin in intestinal homeostasis and Paneth cell maturation. PLoS ONE. 2010;5:e14325. doi:10.1371/journal.pone.0014325.

155. Ghishan FK, Kiela PR. Advances in the understanding of mineral and bone metabolism in inflammatory bowel diseases. Am J Physiol Gastrointest Liver Physiol. 2010;300:G191–201. doi:10.1152/ajpgi.00496.2010.
156. Bianchi ML. Inflammatory bowel diseases, celiac disease, and bone. Arch Biochem Biophys. 2010;503:54–65. doi:10.1016/j.abb.2010.06.026.
157. Compston JE, Judd D, Crawley EO, Evans WD, Evans C, Church HA, Reid EM, Rhodes J. Osteoporosis in patients with inflammatory bowel disease. Gut. 1987;28:410–5. doi:10.1136/gut.28.4.410.
158. Nielsen OH, Vainer B, Madsen SM, Seidelin JB, Heegaard NHH. Established and emerging biological activity markers of inflammatory bowel disease. Am J Gastroenterol. 2000;95:359–67. doi:10.1016/S0002-9270(99)00849-7.
159. Daft JG, Lorenz RG. Role of the gastrointestinal ecosystem in the development of type 1 diabetes. Pediatr Diabetes. 2015;16:407–18. doi:10.1111/pedi.12282.
160. Paun A, Yau C, Danska JS. Immune recognition and response to the intestinal microbiome in type 1 diabetes. J Autoimmun. 2016;71:10–8. doi:10.1016/j.jaut.2016.02.004.
161. Gülden E, Wong FS, Wen L. The gut microbiota and type 1 diabetes. Clin Immunol. 2015;159:143–53. doi:10.1016/j.clim.2015.05.013.
162. Vaarala O, Atkinson M A, Neu J. The "perfect storm" for type 1 diabetes: the complex interplay between intestinal microbiota, gut permeability, and mucosal immunity. Diabetes. 2008;57:2555–62. doi:10.2337/db08-0331.
163. Neu J, Reverte CM, Mackey AD, Liboni K, Tuhacek-Tenace LM, Hatch M, Li N, Caicedo RA, Schatz DA, Atkinson M. Changes in intestinal morphology and permeability in the bio-breeding rat before the onset of type 1 diabetes. J Pediatr Gastroenterol Nutr. 2005;40:589–95.
164. Meddings JB, Jarand J, Urbanski SJ, Hardin J, Gall DG. Increased gastrointestinal permeability is an early lesion in the spontaneously diabetic BB rat. Am J Phys. 1999;276:G951–7.
165. Graham S, Courtois P, Malaisse WJ, Rozing J, Scott FW, Mowat AMI. Enteropathy precedes type 1 diabetes in the BB rat. Gut. 2004;53:1437–44. doi:10.1136/gut.2004.042481.
166. Bosi E, Molteni L, Radaelli MG, Folini L, Fermo I, Bazzigaluppi E, Piemonti L, Pastore MR, Paroni R. Increased intestinal permeability precedes clinical onset of type 1 diabetes. Diabetologia. 2006;49:2824–7. doi:10.1007/s00125-006-0465-3.
167. Kuitunen M, Saukkonen T, Ilonen J, Akerblom HK, Savilahti E. Intestinal permeability to mannitol and lactulose in children with type 1 diabetes with the HLA-DQB1*02 allele. Autoimmunity. 2002;35:365–8.
168. Sapone A, de Magistris L, Pietzak M, Clemente MG, Tripathi A, Cucca F, Lampis R, Kryszak D, Cartenì M, Generoso M, Iafusco D, Prisco F, Laghi F, Riegler G, Carratu R, Counts D, Fasano A. Zonulin upregulation is associated with increased gut permeability in subjects with type 1 diabetes and their relatives. Diabetes. 2006;55:1443–9.
169. Secondulfo M, Iafusco D, Carratù R, DeMagistris L, Sapone A, Generoso M, Mezzogiorno A, Sasso FC, Cartenì M, De Rosa R, Prisco F, Esposito V. Ultrastructural mucosal alterations and increased intestinal permeability in non-celiac, type I diabetic patients. Dig Liver Dis. 2004;36:35–45. doi:10.1016/j.dld.2003.09.016.
170. Vaarala O. Is the origin of type 1 diabetes in the gut? Immunol Cell Biol. 2012;90:271–6. doi:10.1038/icb.2011.115.
171. Lee AS, Gibson DL, Zhang Y, Sham HP, Vallance BA, Dutz JP. Gut barrier disruption by an enteric bacterial pathogen accelerates insulitis in NOD mice. Diabetologia. 2010;53:741–8. doi:10.1007/s00125-009-1626-y.
172. Hadjiyanni I, Li KK, Drucker DJ. Glucagon-like peptide-2 reduces intestinal permeability but does not modify the onset of type 1 diabetes in the nonobese diabetic mouse. Endocrinology. 2009;150:592–9. doi:10.1210/en.2008-1228.
173. Todd DJ, Forsberg EM, Greiner DL, Mordes JP, Rossini AA, Bortell R. Deficiencies in gut NK cell number and function precede diabetes onset in BB rats. J Immunol. 2004;172:5356–62.
174. Carratù R, Secondulfo M, de Magistris L, Iafusco D, Urio A, Carbone MG, Pontoni G, Cartenì M, Prisco F. Altered intestinal permeability to mannitol in diabetes mellitus type I. J Pediatr Gastroenterol Nutr. 1999;28:264–9.

175. Fasano A. Leaky gut and autoimmune diseases. Clin Rev Allergy Immunol. 2012;42:71–8. doi:10.1007/s12016-011-8291-x.
176. Ivanov II, Atarashi K, Manel N, Brodie EL, Shima T, Karaoz U, Wei D, Goldfarb KC, Santee CA, Lynch SV, Tanoue T, Imaoka A, Itoh K, Takeda K, Umesaki Y, Honda K, Littman DR. Induction of intestinal Th17 cells by segmented filamentous bacteria. Cell. 2009;139:485–98. doi:10.1016/j.cell.2009.09.033.
177. Gaboriau-Routhiau V, Rakotobe S, Lécuyer E, Mulder I, Lan A, Bridonneau C, Rochet V, Pisi A, De Paepe M, Brandi G, Eberl G, Snel J, Kelly D, Cerf Bensussan N. The key role of segmented filamentous bacteria in the coordinated maturation of gut helper T cell responses. Immunity. 2009;31:677–89. doi:10.1016/j.immuni.2009.08.020.
178. Atarashi K, Nishimura J, Shima T, Umesaki Y, Yamamoto M, Onoue M, Yagita H, Ishii N, Evans R, Honda K, Takeda K. ATP drives lamina propria TH17 cell differentiation. Nature. 2008;455:808–12. doi:10.1038/nature07240.
179. Atarashi K, Tanoue T, Shima T, Imaoka A, Kuwahara T, Momose Y, Cheng G, Yamasaki S, Saito T, Ohba Y, Taniguchi T, Takeda K, Hori S, Ivanov II, Umesaki Y, Itoh K, Honda K. Induction of colonic regulatory T cells by indigenous clostridium species. Science. (80-. 2011;331:337–41.
180. Arpaia N, Campbell C, Fan X, Dikiy S, van der Veeken J, deRoos P, Liu H, Cross JR, Pfeffer K, Coffer PJ, Rudensky AY. Metabolites produced by commensal bacteria promote peripheral regulatory T-cell generation. Nature. 2013;504:451–5. doi:10.1038/nature12726.
181. Smith PM, Howitt MR, Panikov N, Michaud M, Gallini CA, Bohlooly-Y M, Glickman JN, Garrett WS. The microbial metabolites, short-chain fatty acids, regulate colonic Treg cell homeostasis. Science. (80–.). 2013;341:569–73.
182. Brown CT, Davis-Richardson AG, Giongo A, Gano KA, Crabb DB, Mukherjee N, Casella G, Drew JC, Ilonen J, Knip M, Hyöty H, Veijola R, Simell T, Simell O, Neu J, Wasserfall CH, Schatz D, Atkinson MA, Triplett EW. Gut microbiome metagenomics analysis suggests a functional model for the development of autoimmunity for type 1 diabetes. PLoS ONE. 2011;6:e25792. doi:10.1371/journal.pone.0025792.
183. de Goffau MC, Luopajarvi K, Knip M, Ilonen J, Ruohtula T, Harkonen T, Orivuori L, Hakala S, Welling GW, Harmsen HJ, Vaarala O. Fecal microbiota composition differs between children with -cell autoimmunity and those without. Diabetes. 2013;62:1238–44. doi:10.2337/db12-0526.
184. Bach J-F. The effect of infections on susceptibility to autoimmune and allergic diseases. N Engl J Med. 2002;347:911–20. doi:10.1056/NEJMra020100.
185. Pozzilli P, Signore A, Williams AJK, Beales PE. NOD mouse colonies around the world- recent facts and figures. Immunol Today. 1993;14:193–6. doi:10.1016/0167-5699(93)90160-M.
186. Cooke A, Tonks P, Jones FM, O'shea H, Hutchings P, Fulford AJC, Dunne. Infection with Schistosoma mansoni prevents insulin dependent diabetes mellitus in non-obese diabetic mice. Parasite Immunol. 1999;21:169–76. doi:10.1046/j.1365-3024.1999.00213.x.
187. Shaper AG, Phillips AN, Pocock SJ, Walker M, Macfarlane PW. Risk factors for stroke in middle aged British men. BMJ. 1991;302:1111–5.
188. Liu Q, Sundar K, Mishra PK, Mousavi G, Liu Z, Gaydo A, Alem F, Lagunoff D, Bleich D, Gause WC. Helminth infection can reduce insulitis and type 1 diabetes through CD25- and IL-10-independent mechanisms. Infect Immun. 2009;77:5347–58. doi:10.1128/IAI.01170-08.
189. Raine T, Zaccone P, Mastroeni P, Cooke A. Salmonella typhimurium infection in nonobese diabetic mice generates immunomodulatory dendritic cells able to prevent type 1 diabetes. J Immunol. 2006;177
190. King C, Sarvetnick N, Volchkov P, Stranges P, Avanesyan L. The incidence of type-1 diabetes in NOD mice is modulated by restricted flora not germ-free conditions. PLoS ONE. 2011;6:e17049. doi:10.1371/journal.pone.0017049.
191. Kriegel MA, Sefik E, Hill JA, Wu H-J, Benoist C, Mathis D. Naturally transmitted segmented filamentous bacteria segregate with diabetes protection in nonobese diabetic mice. Proc Natl Acad Sci U S A. 2011;108:11548–53. doi:10.1073/pnas.1108924108.

192. Rossini AA, Williams RM, Mordes JP, Appel MC, Like AA. Spontaneous diabetes in the Gnotobiotic BB/W rat. Diabetes. 1979;28
193. Brugman S, Klatter FA, Visser JTJ, Wildeboer-Veloo ACM, Harmsen HJM, Rozing J, Bos NA. Antibiotic treatment partially protects against type 1 diabetes in the bio-breeding diabetes-prone rat. Is the gut flora involved in the development of type 1 diabetes? Diabetologia. 2006;49:2105–8. doi:10.1007/s00125-006-0334-0.
194. Roesch LF, Lorca GL, Casella G, Giongo A, Naranjo A, Pionzio AM, Li N, Mai V, Wasserfall CH, Schatz D, Atkinson MA, Neu J, Triplett EW. Culture-independent identification of gut bacteria correlated with the onset of diabetes in a rat model. ISME J. 2009;3:536–48. doi:10.1038/ismej.2009.5.
195. Brown K, Godovannyi A, Ma C, Zhang Y, Ahmadi-Vand Z, Dai C, Gorzelak MA, Chan Y, Chan JM, Lochner A, Dutz JP, Vallance BA, Gibson DL. Prolonged antibiotic treatment induces a diabetogenic intestinal microbiome that accelerates diabetes in NOD mice. ISME J. 2016;10:321–32. doi:10.1038/ismej.2015.114.
196. Pellegrini S, Sordi V, Bolla AM, Saita D, Ferrarese R, Canducci F, Clementi M, Invernizzi F, Mariani A, Bonfanti R, Barera G, Testoni PA, Doglioni C, Bosi E, Piemonti L. Duodenal mucosa of patients with type 1 diabetes shows distinctive inflammatory profile and microbiota. J Clin Endocrinol Metab. 2017;102:1468–77. doi:10.1210/jc.2016-3222.
197. Marino E, Richards JL, McLeod KH, Stanley D, Yap YA, Knight J, McKenzie C, Kranich J, Oliveira AC, Rossello FJ, Krishnamurthy B, Nefzger CM, Macia L, Thorburn A, Baxter AG, Morahan G, Wong LH, Polo JM, Moore RJ, Lockett TJ, Clarke JM, Topping DL, Harrison LC, Mackay CR. Gut microbial metabolites limit the frequency of autoimmune T cells and protect against type 1 diabetes. Nat Immunol. 2017;18:552–62. doi:10.1038/ni.3713.
198. Dolpady J, Sorini C, Di Pietro C, Cosorich I, Ferrarese R, Saita D, Clementi M, Canducci F, Falcone M. Oral probiotic VSL#3 prevents autoimmune diabetes by modulating microbiota and promoting indoleamine 2,3-dioxygenase-enriched tolerogenic intestinal environment. J Diabetes Res. 2016;2016:1–12. doi:10.1155/2016/7569431.
199. Motyl K, McCabe LR. Streptozotocin, type 1 diabetes severity and bone. Biol Proced Online. 2009;11:296–315. doi:10.1007/s12575-009-9000-5.
200. Motyl KJ, Botolin S, Irwin R, Appledorn DM, Kadakia T, Amalfitano A, Schwartz RC, McCabe LR. Bone inflammation and altered gene expression with type I diabetes early onset. J Cell Physiol. 2009;218:575–83. doi:10.1002/jcp.21626.
201. Botolin S, McCabe LR. Bone loss and increased bone adiposity in spontaneous and pharmacologically induced diabetic mice. Endocrinology. 2007;148:198–205. doi:10.1210/en.2006-1006.
202. Alberti KGMM, Zimmet P, Shaw J. The metabolic syndrome – a new worldwide definition. Lancet. 2005;366:1059–62. doi:10.1016/S0140-6736(05)67402-8.
203. Cani PD, Amar J, Iglesias MA, Poggi M, Knauf C, Bastelica D, Neyrinck AM, Fava F, Tuohy KM, Chabo C, Waget A, Delmée E, Cousin B, Sulpice T, Chamontin B, Ferrières J, Tanti JF, Gibson GR, Casteilla L, Delzenne NM, Alessi MC, Burcelin R. Metabolic endotoxemia initiates obesity and insulin resistance. Diabetes. 2007;56:1761–72. doi:10.2337/db06-1491.
204. Cani PD, Bibiloni R, Knauf C, Neyrinck AM, Delzenne NM. Changes in gut microbiota control metabolic diet–induced obesity and diabetes in mice. Diabetes. 2008;57:1470–81. doi:10.2337/db07-1403.Additional.
205. Cani PD, Possemiers S, Van de Wiele T, Guiot Y, Everard A, Rottier O, Geurts L, Naslain D, Neyrinck A, Lambert DM, Muccioli GG, Delzenne NM. Changes in gut microbiota control inflammation in obese mice through a mechanism involving GLP-2-driven improvement of gut permeability. Gut. 2009;58:1091–103. doi:10.1136/gut.2008.165886.
206. Brun P, Castagliuolo I, di Leo V, Buda A, Pinzani M, Palu G, Martines D. Increased intestinal permeability in obese mice: new evidence in the pathogenesis of nonalcoholic steatohepatitis. Am J Physiol Gastrointest Liver Physiol. 2007;292:518–25. doi:10.1152/ajpgi.00024.2006.
207. Teixeira TFS, Collado MC, Ferreira CLLF, Bressan J, Peluzio Mdo CG. Potential mechanisms for the emerging link between obesity and increased intestinal permeability. Nutr Res. 2012;32:637–47. doi:10.1016/j.nutres.2012.07.003.

208. Farhadi A, Gundlapalli S, Shaikh M, Frantzides C, Harrell L, Kwasny MM, Keshavarzian A. Susceptibility to gut leakiness: a possible mechanism for endotoxaemia in non-alcoholic steatohepatitis. Liver Int. 2008;28:1026–33. doi:10.1111/j.1478-3231.2008.01723.x.
209. Brignardello J, Morales P, Diaz E, Romero J, Brunser O, Gotteland M. Pilot study: alterations of intestinal microbiota in obese humans are not associated with colonic inflammation or disturbances of barrier function. Aliment Pharmacol Ther. 2010;32:1307–14. doi:10.1111/j.1365-2036.2010.04475.x.
210. Teixeira TFS, Souza NCS, Chiarello PG, Franceschini SCC, Bressan J, Ferreira CLLF, Peluzio Mdo CG. Intestinal permeability parameters in obese patients are correlated with metabolic syndrome risk factors. Clin Nutr. 2012;31:735–40. doi:10.1016/j.clnu.2012.02.009.
211. Shapses SA, Sukumar D. Bone metabolism in obesity and weight loss. Annu Rev Nutr. 2012;32:287–309. doi:10.1146/annurev.nutr.012809.104655.
212. Cao JJ. Effects of obesity on bone metabolism. J Orthop Surg Res. 2011;6:30. doi:10.1186/1749-799X-6-30.
213. Carnielli VP, Luijendijk IH, Van Goudoever JB, Sulkers EJ, Boerlage AA, Degenhart HJ, Sauer PJ. Structural position and amount of palmitic acid in infant formulas: effects on fat, fatty acid, and mineral balance. J Pediatr Gastroenterol Nutr. 1996;23:553–60. doi:10.1097/00005176-199612000-00007.
214. Klok MD, Jakobsdottir S, Drent ML. The role of leptin and ghrelin in the regulation of food intake and body weight in humans: a review. Obes Rev. 2007;8:21–34. doi:10.1111/j.1467-789X.2006.00270.x.
215. Fukushima N, Hanada R, Teranishi H, Fukue Y, Tachibana T, Ishikawa H, Takeda S, Takeuchi Y, Fukumoto S, Kangawa K, Nagata K, Kojima M. Ghrelin directly regulates bone formation. J Bone Miner Res. 2004;20:790–8. doi:10.1359/JBMR.041237.
216. Dicembrini I, Mannucci E, Rotella CM. Bone: incretin hormones perceiver or receiver? Exp Diabetes Res. 2012;2012:519784. doi:10.1155/2012/519784.

Impact of Enteric Health and Mucosal Permeability on Skeletal Health and Lameness in Poultry

L.R. Bielke, B.M. Hargis, and J.D. Latorre

GIT Health, Microbiome, and Pioneer Colonizers

In rodent medical models, intestinal inflammation and skeletal health are known to implicate the importance of gut health in the ontogenesis of normal bone development [1–4]. Intestinal barrier leakage and/or altered gut microbial composition has been shown to markedly impact both osteoblast and osteoclast activities, systemically through circulation of gut immune cells and cytokines and locally by causing inflammation of extraintestinal organs (such as the liver and bone marrow). Even mild cases of heightened intestinal inflammation can cause bone loss in male mice in the absence of any overt nutritional deficiencies or weight loss [4, 5]. Similarly, using broiler chicken studies, *Salmonella* Enteritidis challenge did not affect growth or feed utilization, but did decrease bone volume fraction and bone mineral density (Fig. 1). In chickens, paratyphoid salmonellae are documented to cause mild enteric inflammation, most notably increases in IL-6, IL-8, TNF-α, and IFN-γ, and decreased expression of IL-2, a cytokine that regulates tolerance of self [6, 7]. Conversely, selected probiotics are known to promote immune quiescence in the GIT [7, 8]. A commercially used poultry probiotic that was selected for anti-*Salmonella* activity was analyzed for its immune pathway effects by transcriptional microarray analysis for regulation of multiple genes related to innate immune function. Particularly interesting were changes in expression of genes related to LPS response and NF-κB and apoptosis pathways [9]. Similarly, Carey and Kostrzynska [10] reported decreased IL-8 chemokine associated with lactic acid bacteria prophylactic treatment of HT-29 cells exposed to *Salmonella* Typhimurium.

L.R. Bielke (✉)
The Ohio State University, Animal Sciences, 2029 Fyffe Road, Columbus, OH, USA
e-mail: Bielke.1@osu.edu

B.M. Hargis • J.D. Latorre
University of Arkansas, Department of Poultry Science, Fayetteville, AR, USA

L.R. McCabe, N. Parameswaran (eds.), *Understanding the Gut-Bone Signaling Axis*, Advances in Experimental Medicine and Biology 1033,
DOI 10.1007/978-3-319-66653-2_9

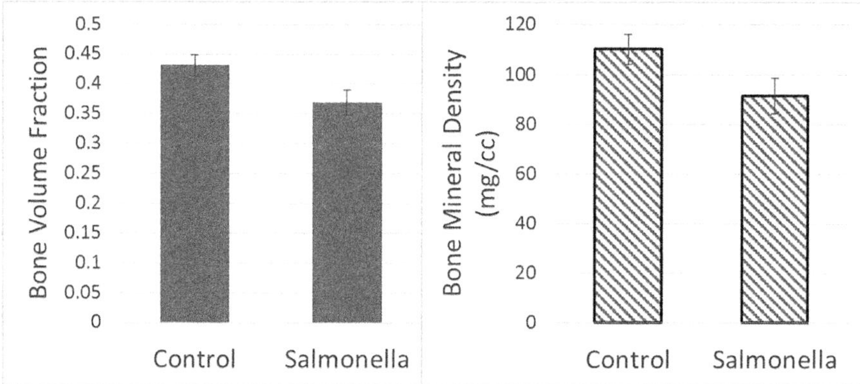

Fig. 1 Changes in (**a**) bone volume fraction (expressed as a percent of total region volume) and (**b**) bone mineral density in broiler chickens after *Salmonella* Enteritidis challenge. Chicks were challenged on day of hatch by oral gavage, femur bone samples were collected on d21 and measured by μCT analysis. *indicates significantly different from control ($P \leq 0.05$)

On the other hand, ingredients selected for feed formulation have also a significant impact on gut health, intestinal microbiota, bone quality, and performance parameters. Feedstuffs included in livestock diets are often contaminated with anti-nutritional factors such as phytates, trypsin inhibitors, tannins, lectins, and non-starch polysaccharides (NSP; [11, 12]). Diets containing wheat, barley, or rye, ingredients with elevated concentrations of highly branched arabinoxylans and β-glucans, are poorly digested by poultry and other monogastric animals because they do not produce endogenous enzymes capable of hydrolyzing NSP [13]. The presence of these soluble NSP in the intestinal lumen can increase digesta viscosity, which has a negative impact on nutrient absorption and interferes with access of digestive enzymes to the endosperm of cereal grains [14]. Furthermore, an increment in the feed passage rate through the GIT in elevated NSP diets, together with abundance of unabsorbed nutrients in the intestinal lumen, promotes bacterial overgrowth, generating dysbacteriosis and occasionally the presentation of enteric diseases such as necrotic enteritis, a multifactorial disease caused by *Clostridium perfringens* in chickens [15]. Consumption of diets with a high content of soluble NSP can affect bone quality parameters by reducing the amount of conjugated bile acids in the intestine, therefore diminishing the absorption of fat-soluble vitamins such as vitamin D and minerals like calcium and phosphorus [16]. Recent enteric inflammation studies have shown that high NSP-containing diets have effects on intestinal viscosity, bone mineral content, and breaking strength, along with increased fluorescein isothiocyanate-dextran (FITC-d) leakage, an indicator of GIT mucosal permeability and enteric inflammation. FITC-d is a chemical molecule (3–5 kDa) that usually does not pass the intestinal epithelial barrier. However, when any condition affects tight junction stability between enterocytes, the dextran-labeled molecule can enter systemic circulation [17]. Mucosal integrity was also measured by translocation of recovered Gram-negative enteric bacteria to the liver, showing that incorporation of feed ingredients containing elevated soluble NSP levels has a detrimental effect on gut barrier permeability, resulting from intestinal dysbacteriosis and inflammation. Decreased mucosal integrity was associated with a reduction in all bone quality parameters evaluated including tibia strength, tibia diameter, and tibia mineral content, among other observations resulting in poor performance parameters (Table 1; [18]).

Microflora within the intestinal tract provide valuable and well-known functions that impact not only gut health, but the overall health of the host. The interaction between the immune system of the gut and commensal microflora in animals begins immediately after hatch or birth and leads to a low level of inflammation characterized by increased IL-8 expression [19]. This results in the infiltration of heterophils and lymphocytes into the lamina propria and normalization of the intestinal immune system [20–22]. Establishment of appropriate commensal bacteria in the intestines as soon as possible after hatch is fundamental for the immune system to develop properly and, in turn, to promote the effectiveness of some vaccines. However, inflammatory intestinal responses may also directly and profoundly alter the gut microbiota [23], leading to a cycle of increased enteritis and diseases such as necrotic enteritis and even lameness.

Effect of Probiotics on Enteric Health

Recently, much focus in enteric microbial research has turned toward immune modulatory effects of probiotics, and much of this interaction takes place at the mucosal interface of the GIT. Generally, the modes of action by which probiotics affect

Table 1 Evaluation of bone and mucosal leakage parameters in chickens fed with corn-based or rye-based (high NSP) diets

	Tibia strength load at yield (kg/mm^2)	Tibia diameter (mm)	Total ash from tibia (%)	Calcium (% of ash)	Phosphorous (% of ash)
Experiment 1					
Corn	5.04 ± 0.011[a]	3.34 ± 0.17[a]	55.01 ± 0.41[a]	29.48 ± 0.27[a]	18.15 ± 0.12[a]
Rye	1.58 ± 0.009[b]	1.61 ± 0.28[b]	34.87 ± 0.35[b]	18.48 ± 0.27[b]	13.15 ± 0.12[b]
Experiment 2					
Corn	6.14 ± 0.01[a]	4.55 ± 0.32[a]	65.61 ± 0.81[a]	37.65 ± 0.07[a]	21.35 ± 0.52[a]
Rye	2.58 ± 0.03[b]	1.82 ± 0.78[b]	30.87 ± 0.75[b]	21.32 ± 0.46[b]	15.67 ± 0.29[b]
	Body weight (g)	Intestinal viscosity (cP Log_{10})	Serum FITC-d (μ g/mL)	Bacterial translocation (CFU Log_{10})	
Experiment 1					
Corn	283.21 ± 10.57[a]	0.30 ± 0.04[b]	0.20 ± 0.01[b]	0.00 ± 0.00[b]	
Rye	110.69 ± 5.21[b]	2.84 ± 0.57[a]	0.42 ± 0.05[a]	1.35 ± 0.45[a]	
Experiment 2					
Corn	301.46 ± 10.57[a]	0.23 ± 0.35[b]	0.31 ± 0.03[b]	0.00 ± 0.00[b]	
Rye	140.89 ± 5.21[b]	2.90 ± 0.83[a]	0.52 ± 0.07[a]	2.40 ± 0.73[a]	

[a]Within columns indicate significant difference $P < 0.05$ (Adapted from Tellez et al. [18])
[b]Within columns indicate significant difference $P < 0.05$ (Adapted from Tellez et al. [18])

inflammation fall into three categories, as reviewed by Lebeer et al. [24, 25]. These include competitive exclusion, enhanced mucosal barrier function, and modulation of local and systemic immune responses. While the former effect is dependent on direct microbe-microbe interaction, the latter two are dependent on microbe-host relationships, especially related to M-cells and monocyte lineage cells such as dendritic cells [24, 26–28]. Microorganism-associated molecular patterns (MAMPs) interact with pattern recognition receptors (PRRs) on dendritic cells and macrophages to induce signaling cascades that mount molecular responses against microorganisms [29]. Flagella, fimbrial structure proteins, and lipopolysaccharides are among the most common MAMPs on Gram-negative bacteria, whereas lipoteichoic acid, peptidoglycans, and cell wall-associated polysaccharides are common MAMPs on Gram-positive bacteria [25, 27]. Though many of these MAMPs are shared between pathogens and probiotics alike, the host immune system clearly responds differently to each, perhaps through different PRRs such as toll-like receptors and lectin binding sites [30].

Some probiotics and prebiotics, as potential antibiotic growth promoter (AGP) alternatives, have been shown to increase performance and have anti-inflammatory effects, through a variety of mechanisms including increased production of volatile fatty acids (VFA), known to have profound anti-inflammatory activity [31]. For example, some work has indicated that increased VFA production was associated with lactose supplementation [32], and improved production, especially in combination with specific lactic acid bacterial cultures [33–35]. These studies have shown that application of these cultures reduced *Salmonella* colonization and improved performance of poultry under laboratory and field trial conditions. Reduced inflammation has been associated with stabilization of beneficial commensal bacterial populations known to reduce expression of inflammatory cytokines and reduce goblet cell size and mucus secretion, supporting the premise that inflammation is a precursor to opportunistic diseases [36–39]. This may explain why multiple classes of antibiotics and effective probiotics enhance performance and indicate that effective AGP alternatives, through anti-inflammatory activity, could replace AGP for improved performance and animal well-being.

Role of Mucosal Permeability in Development of Lameness Disorders

Heavy bird breeds, especially commercial broilers, grow at an exceptionally fast rate, with a 100-fold increase in body weight form day of hatch to 8 weeks of age, when they reach processing weight. While selection for growth has primarily focused on meat yield, rate of bone growth has also increased, but has not necessarily kept up with lean tissue gains. Increases in incidence of lameness, especially compared to that of lighter egg-laying breeds, suggest that there is a disproportionate growth between bone and body mass [40]. Avian bone growth is rapid compared to that of mammals, with growth plate turnover estimated at just 21 h. This is attributed to increased thickness of growth plates and unevenly aligned chondrocyte columns [41]. Wideman and Prisby [41] suggested that chronic mechanical stress, such as long periods of inactivity in the resting posture of broiler, creates osteochondrotic clefts (OC) between and within cartilage layers, which can be recognized as osteochondrosis dissecans (OCD) or physeal osteochondrosis (POC). Osteochondrosis dissecans was also recently described as a major predisposing factor in the development of spondylitis in broilers, discussed in detail below [42]. The OC cause poor blood flow that results in focal ischemia and necrosis, encouraged by compression of cartilage layers. Colonization of these OC by bacteria, distributed through blood, is a major contributor to the pathogenesis of lameness pathologies such as bacterial chondronecrosis with osteomyelitis (BCO; reference), turkey osteomyelitis complex (TOC; [43]), and enterococcal spondylitis (ES; 43). A primary source of hematogenous bacteria is the GIT, especially in avian species that lack lacteals, emphasizing the importance of controlling leaky gut in animals in order to reduce lameness [45]. Bacteria transmitted to chicks from breeder parents, contaminated eggshells, or hatchery sources, or that enter circulation via translocation through the integument, respiratory system, or GIT, can spread hematogenously to colonize OC. Translocated opportunistic pathogenic bacteria form bacterial foci and sequestrate within the growth plate and adjacent metaphysis, where they are notoriously inaccessible to antibiotics and cellular components of the immune system. The ability of bacteria to bind cartilage and exposed bone collagen has been implicated as a major pathogenicity factor for development of these diseases [46–48]. Femoral and tibial BCO has been studied and reviewed extensively by Wideman and Prisby [41] and will not be discussed in detail in this chapter.

Enterococcal spondylitis is a disease of broilers characterized by the formation of an abscess in the free thoracic vertebrae (FTV), which leads to lameness. The free thoracic vertebrae in birds are the freely movable thoracic vertebrae, T4, which separates fused vertebrae of the notarium and synsacrum and is subjected to repeated torsional and mechanical stress [49]. Birds affected by ES develop leg paralysis, due to compression of the thoracolumbar spinal cord in FTV, at about 5–8 weeks of age with 5–15% mortality in flocks due to inability of affected birds to eat and drink [50]. Due to the late age of onset of lameness, this particular disease is incredibly costly to producers, as affected chickens are near processing stage and a majority of the cost of raising has been invested.

Compression of the spinal cord is due to growth of an abscess related to infection by *E. cecorum* in the FTV. The FTV may be especially susceptible to infection due to the fact that it is the only vertebra of chickens that has weight-bearing articulations as the connection between the notarium and synsacrum. Borst and coworkers [44] recently documented OCD, a condition in which cartilage in a joint dies due to lack of blood flow and separates from its associated bone, in production flocks as young as 7 days of age. Incidence of OCD in case flocks and control flocks was similar, but genotype of intestinal *E. cecorum* varied, as well as age of initial isolation. This suggests, as in BCO cases, that skeletal deformities do play a role in development of disease, but ultimately, the presence of bacteria at the lesion site is necessary.

The presence of enteric bacteria in FTV, as a key component of the etiology of kinky back, suggests that mucosal permeability may play a necessary role in the pathogenesis of this disease. Thus, a means of prevention of kinky back could be control of enteric inflammation that leads to mucosal permeability and systemic circulation of bacteria. Some probiotics and prebiotics, as potential AGP alternatives, have been shown to increase performance and have anti-inflammatory effects, through a variety of mechanisms including increased production of volatile fatty acids (VFA), known to have profound anti-inflammatory activity [31]. Interestingly, inflammation of the gut mucosal epithelium has been shown as a key mechanism for mucosal colonization by several pathogens and is supported by emerging data, primarily from rodent studies, indicating that inflammation in the gut directly results in dysbiosis where the overall diversity and abundance of bacteria are reduced [23]. Studies investigating the role of mucosal permeability in translocation of bacteria to liver and *Enterococcus cecorum* to FTV after induction of mucosal permeability models in broilers showed that increased levels of hematogenous bacteria, measured as liver CFU, matched the increased levels of *E. cecorum* recovery in FTV. Additionally, elevated FITC-d in serum was noted in several inflammation treatments (Table 2). With increased inflammation, regardless of cause, it is likely that mucosal permeability is also increased, which would allow bacterial translocation of potential pathogens into systemic circulation, a key component of ES.

Table 2 Effect of enteric inflammation treatment on mucosal permeability and recovery of *E. cecorum* from FTV. All groups, except control, were inoculated with *E. cecorum* on d11. Control and EC groups were reared under normal conditions. Dexamethasone in feed, 0.56 mg/kg feed (DEX, d4–11), rye-based diet (RYE, d7–11), and 15% dried distillers grain with solubles (DDGS). Liver and FTV samples were aseptically collected on d15 and cultured on selective agar for *E. cecorum*

	Serum FITC-d (μg/mL) d11	Liver aerobic bacteria (Log_{10} CFU/g) d11	FTV *E. cecorum* d15
Experiment 1			Log10 CFU/g
EC	0.29 ± 0.02^{a}	1.15 ± 0.44^{abc}	0.92 ± 0.33^{a}
DEX	0.48 ± 0.03^{b}	2.46 ± 0.34^{b}	3.18 ± 0.33^{b}
RYE	0.38 ± 0.01^{a}	2.53 ± 0.43^{b}	3.57 ± 0.22^{b}
DDGS	0.33 ± 0.03^{a}	0.88 ± 0.39^{a}	1.47 ± 0.43^{a}
Experiment 2			(% incidence)
Control	0.06 ± 0.02^{a}	0.96 ± 0.43^{a}	$10\%^{d}$
EC	0.10 ± 0.01^{a}	1.29 ± 0.43^{ab}	$15\%^{d}$
DEX	0.23 ± 0.02^{b}	1.97 ± 0.49^{ab}	$75\%^{b}$
RYE	0.19 ± 0.01^{b}	2.51 ± 0.43^{b}	$50\%^{a}$
DDGS	0.20 ± 0.02^{b}	2.08 ± 0.47^{ab}	$55\%^{ab}$

[a]Letters with different superscripts, within experiment and column, significantly different ($P < 0.05$)

[b]Letters with different superscripts, within experiment and column, significantly different ($P < 0.05$)

[c]EC significantly different from DEX and RYE when tested independently

[d]Letters with different superscripts, within experiment and column, significantly different ($P < 0.05$)

Effect of Probiotics on Growth Performance and Bone Quality

An appropriate balance of microflora is suggested to play an important role in growth and development of agriculture animals, but poultry-rearing practices prevent exposure of hatching chicks to parental microflora and may have a negative impact on the development of beneficial GIT microbial communities. To mitigate the effect of dysbiosis in the GIT, diets have historically been supplemented with subtherapeutic levels of antibiotics that effectively decrease the incidence of digestive disorders and increase growth performance [51]. However, consumer demand and pending government regulations challenge the poultry industry to find economically viable strategies to the conventional use of subtherapeutic antibiotics in poultry diets without affecting production parameters [52]. Continuous and extensive research of suitable alternatives include feed additives, such as probiotics and direct-fed microbials (DFM; 43, [53]), organic acids, and essential oils from plant extracts [54], or bacteriophage therapy [55]. In the case of probiotics, a common commercial type is lactic acid bacteria (LAB) that include the genus *Lactobacillus* and *Pediococcus*, which are normally part of the microflora of many animal species [56]. However, LAB probiotics are not feed stable and must be microencapsulated, refrigerated, and/or lyophilized to prolong storage shelf life and usually are administered in the drinking water. In this regard, among the large number of probiotic products in use today, some are bacterial spore formers, mostly of the genus *Bacillus*. Used primarily in their spore form, some *Bacillus* direct-fed microbials (DFM) have been shown to prevent selected gastrointestinal disorders with an astonishing diversity of species and applications [57]. While not all *Bacillus* spores are highly heat tolerant, some isolates are the toughest life form known on earth [58] and can be used under extreme heat and pressure conditions (pelletization). Moreover, it has been previously investigated that selected *Bacillus* strains can produce antimicrobial compounds against Gram-negative enteropathogens that may promote enteric inflammation and mucosal permeability, such as *Salmonella* spp., *Escherichia coli*, and *Campylobacter* spp. [56]. Additionally, evidence suggests that *Bacillus* spores can germinate in the GIT into metabolically active vegetative cells, which colonize and can be considered as part of the microflora rather than just transiently present in the gut [59]. *Bacillus* spp. are the most widely studied Gram-positive genera, and are a model organism for research, and together with other bacteria and fungal species, have been extensively used as a source of industrial enzymes and antibiotics by biotechnology companies [60]. Not all *Bacilli* synthesize the same type of enzymes and require selection and characterization of adequate isolates according to specific target substrates in the diet.

It has been demonstrated that inclusion of an exogenous enzyme producing *Bacillus* DFM in diets containing ingredients with high levels of soluble NSP significantly reduced both viscosity and *C. perfringens* proliferation in an in vitro digestive model [61]. These results were supported by later in vivo studies in which chickens and turkeys fed with a rye-based diet. Added *Bacillus*-DFM candidate was included in the experimental rye-based diet of 10-day-old turkey poults, and signifi-

Table 3 Evaluation of body weight, digesta viscosity, and bacterial translocation to the liver in neonatal turkey poults fed with a rye-soybean-based diet with *Bacillus* direct-fed microbial (DFM) supplementation

Measurement	Control[a]	*Bacillus* DFM[b]
Experiment 1		
Body weight (g)[c]	65.91 ± 3.6[d]	82.85 ± 4.2[e]
Digesta viscosity (cP Log_{10})[f]	2.03 ± 0.3[e]	1.54 ± 0.2[d]
Bacterial translocation (cfu Log_{10})[g]	3.03 ± 0.5[e]	1.24 ± 0.5[d]
Experiment 2		
Body weight (g)[c]	74.47 ± 1.6[d]	95.60 ± 2.2[e]
Digesta viscosity (cP Log_{10})[f]	2.80 ± 0.4[e]	1.62 ± 0.5[d]
Bacterial translocation (cfu Log_{10})[g]	2.13 ± 0.7[e]	0.35 ± 0.4[d]

Adapted from Latorre et al. [45]
[a]Control rye-based diet without DFM
[b]Control rye-based diet with candidate DFM (10^6 spores/g of feed)
[c]Body weight $n = 25$; Data is express as Mean ± SE.
[d]Superscripts within rows indicate significant difference at $P < 0.05$
[e]Superscripts within rows indicate significant difference at $P < 0.05$
[f]Digesta viscosity is expressed in Log_{10} (in centipoise, cP = 1/100 dyne s/cm^2), $n = 12$
[g]Liver bacterial translocation (expressed in cfu Log_{10}/g of tissue), $n = 12$

Table 4 Evaluation of bone strength and bone composition in neonatal turkey poults fed with a rye-soybean-based diet without or with *Bacillus* direct-fed microbial (DFM) supplementation

Measurement	Control[a]	*Bacillus* DFM[b]
Tibia strength load at yield (kg)	1.14 ± 0.2[c]	2.55 ± 0.1[d]
Tibia diameter (mm)	4.45 ± 0.3[c]	5.82 ± 0.8[d]
Total ash from tibia (%)	35.61 ± 0.8[c]	50.87 ± 0.7[d]
Calcium (% of ash)	27.35 ± 0.1[c]	40.31 ± 0.5[d]
Phosphorus (% of ash)	16.35 ± 0.5[c]	22.67 ± 0.3[d]

Adapted from Latorre et al. [45]
Tibias were collected to evaluate bone quality ($n = 12$). Data is expressed as mean ± SE
[a]Control rye-based diet without DFM
[b]Control rye-based diet with candidate DFM (106 spores/g of feed)
[c]Superscripts within rows indicate significant difference at $P < 0.05$
[d]Superscripts within rows indicate significant difference at $P < 0.05$

cant improvements in intestinal viscosity, performance parameters, bacterial translocation, and bone quality were observed in supplemented animals (Tables 3 and 4), suggesting that the consumption of a selected *Bacillus* DFM producing a variable set of enzymes could enhance nutrient digestibility and promote healthy intestinal integrity [45].

Additionally, the effects of dietary inclusion of a *Bacillus* DFM on bone quality in chickens fed with a rye-based diet resulted in significant improvement in all bone quality measurements of 10-day-old broilers (Table 5). Similar results were also obtained in experiment 3, where bone quality variables were measured in 28 days of age, showing an increase in bone strength, percentage of ash, calcium, and phosphorus when a *Bacillus* DFM was included in the diet ($P < 0.05$). The study presented in Table 4 resulted in increased tibia diameter in DFM-fed birds, but the latter chicken experiments (Table 5) did not result. Both studies demonstrated that inclusion of exogenous enzyme producing DFM in rye-based diets not only enhanced bone quality parameters but also significantly improved growth performance, gut health, and intestinal microbiota balance in comparison to animals consuming non-supplemented diets.

Table 5 Evaluation of bone breaking strength and bone composition in broiler chickens consuming a rye-based diet with or without dietary inclusion of a selected *Bacillus* direct-fed microbial candidate

Item	Rye diet	Rye diet + DFM
Experiment 1[a]		
Load at break (kg)	1.67 ± 0.01[b]	2.68 ± 0.01[c]
Tibia diameter (mm)	2.61 ± 0.28[c]	2.85 ± 0.28[c]
Breaking strength (kg/mm^2)	0.64 ± 0.02[b]	0.94 ± 0.01[c]
Total ash (%)	34.87 ± 0.35[b]	54.68 ± 0.39[c]
Calcium (%)	18.48 ± 0.27[b]	36.48 ± 0.87[c]
Phosphorus (%)	13.12 ± 0.12[b]	26.11 ± 0.82[c]
Experiment 2[a]		
Load at break (kg)	1.75 ± 0.03[b]	2.81 ± 0.09[c]
Tibia diameter (mm)	2.92 ± 0.78[c]	2.90 ± 0.28[c]
Breaking strength (kg/mm^2)	0.60 ± 0.03[b]	0.97 ± 0.09[c]
Total ash (%)	30.87 ± 0.75[b]	56.57 ± 0.44[c]
Calcium (%)	21.32 ± 0.46[b]	40.28 ± 0.21[c]
Phosphorus (%)	15.67 ± 0.29[b]	29.75 ± 0.10[c]
Experiment 3[d]		
Load at break (kg)	22.15 ± 0.93[b]	26.51 ± 1.68[c]
Tibia diameter (mm)	5.47 ± 0.08[c]	5.58 ± 0.20[c]
Breaking strength (kg/mm^2)	4.05 ± 0.23[b]	4.75 ± 0.18[c]
Total ash (%)	44.87 ± 0.95[b]	55.01 ± 0.61[c]
Calcium (%)	17.47 ± 0.26[b]	29.48 ± 0.27[c]
Phosphorus (%)	9.15 ± 0.11[b]	15.15 ± 0.13[c]

Adapted from Latorre et al. [62]

[a]Bone measurements evaluated from 10-day-old broilers, $n = 12$/group

[b]Superscripts within rows indicate significant difference at $P < 0.05$ within each experiment. Data are expressed as mean ± SE

[c]Superscripts within rows indicate significant difference at $P < 0.05$ within each experiment. Data are expressed as mean ± SE

[d]Bone measurements evaluated from 28-day-old broilers, $n = 16$/group

Utilization of cereal grains with an elevated content of NSP in poultry diets has been related to malabsorption of lipids, deterioration of bone mineralization, and reduced leg soundness [63]. This negative effect on bone quality could be related to an elevated digesta viscosity, therefore enhancing the deconjugation of bile acids by the overgrowth intestinal microflora, resulting in a reduction of micelle formation, affecting fat solubilization and absorption of fat-soluble vitamins and minerals [64]. Since monogastric animals do not have endogenous enzymes capable of hydrolyzing the β-linkages present in soluble NSP, dietary inclusion of selected *Bacillus* spp. that produce carbohydrases among other enzymes (xylanase, β-glucanase, β-mannanase, α-galactosidase, and pectinase) could be an alternative feed additive in poultry diets in attempt to reduce the adverse impact of these and other anti-nutritional factors on performance, intestinal integrity, and bone quality.

References

1. Sylvester FA, Wyzga N, Hyams JS, Davis PM, Lerer T, Vance K, et al. Natural history of bone metabolism and bone mineral density in children with inflammatory bowel disease. Inflamm Bowel Dis. 2007;13(1):42–50.
2. Harris L, Senagore P, Young VB, McCabe LR. Inflammatory bowel disease causes reversible suppression of osteoblast and chondrocyte function in mice. Am J Physiol Gastrointest Liver Physiol. 2009;296(5):G1020–9.
3. Agrawal M, Arora S, Li J, Rahmani R, Sun L, Steinlauf AF, et al. Bone, inflammation, and inflammatory bowel disease. Curr Osteoporos Rep. 2011;9(4):251–7.
4. McCabe LR, Irwin R, Schaefer L, Britton RA. Probiotic use decreases intestinal inflammation and increases bone density in healthy male but not female mice. J Cell Physiol. 2013;228(8):1793–8.
5. Irwin R, Lee T, Young VB, Parameswaran N, McCabe LR. Colitis-induced bone loss is gender dependent and associated with increased inflammation. Inflamm Bowel Dis. 2013;19(8):1586–97.
6. Kaiser P, Rothwell L, Galyov EE, Barrow PA, Burnside J, Wigley P. Differential cytokine expression in avian cells in response to invasion by Salmonella typhimurium, Salmonella enteritidis and Salmonella gallinarum. Microbiol Read Engl. 2000;146(Pt 12):3217–26.
7. Haghighi HR, Abdul-Careem MF, Dara RA, Chambers JR, Sharif S. Cytokine gene expression in chicken cecal tonsils following treatment with probiotics and Salmonella infection. Vet Microbiol. 2008;126(1–3):225–33.
8. Neish AS. Microbes in gastrointestinal health and disease. Gastroenterology. 2009;136(1):65–80.
9. Higgins SE, Wolfenden AD, Tellez G, Hargis BM, Porter TE. Transcriptional profiling of cecal gene expression in probiotic- and Salmonella-challenged neonatal chicks. Poult Sci. 2011;90(4):901–13.
10. Carey CM, Kostrzynska M. Lactic acid bacteria and bifidobacteria attenuate the proinflammatory response in intestinal epithelial cells induced by Salmonella enterica serovar Typhimurium. Can J Microbiol. 2013;59(1):9–17.
11. Choct M, Annison G. Anti-nutritive effect of wheat pentosans in broiler chickens: roles of viscosity and gut microflora. Br Poult Sci. 1992;33(4):821–34.
12. Erdaw MM, Bhuiyan MM, Iji PA. Enhancing the nutritional value of soybeans for poultry through supplementation with new-generation feed enzymes. Worlds Poult Sci J. 2016;72(2):307–22.

13. Knudsen KEB. Fiber and nonstarch polysaccharide content and variation in common crops used in broiler diets. Poult Sci. 2014;93(9):2380–93.
14. Choct M, Hughes RJ, Bedford MR. Effects of a xylanase on individual bird variation, starch digestion throughout the intestine, and ileal and caecal volatile fatty acid production in chickens fed wheat. Br Poult Sci. 1999;40(3):419–22.
15. Annett CB, Viste JR, Chirino-Trejo M, Classen HL, Middleton DM, Simko E. Necrotic enteritis: effect of barley, wheat and corn diets on proliferation of Clostridium perfringens type A. Avian Pathol. 2002;31(6):598–601.
16. Langhout DJ, Schutte JB, Geerse C, Kies AK, Jong JD, Verstegen MWA. Effects on chick performance and nutrient digestibility of an endo-xylanase added to a wheat- and rye-based diet in relation to fat source. Br Poult Sci. 1997;38(5):557–63.
17. Kuttappan VA, Vicuña EA, Latorre JD, Wolfenden AD, Téllez GI, Hargis BM, et al. Evaluation of gastrointestinal leakage in multiple enteric inflammation models in chickens. Front Vet Sci [Internet]. 2015;2:66. [cited 2016 Mar 8] Available from: http://www.ncbi.nlm.nih.gov/pmc/articles/PMC4677096/
18. Tellez G, Latorre JD, Kuttappan VA, Kogut MH, Wolfenden A, Hernandez-Velasco X, et al. Utilization of rye as energy source affects bacterial translocation, intestinal viscosity, microbiota composition, and bone mineralization in broiler chickens. Front Genet. 2014;25:5.
19. Bar-Shira E, Friedman A. Development and adaptations of innate immunity in the gastrointestinal tract of the newly hatched chick. Dev Comp Immunol. 2006;30(10):930–41.
20. Bar-Shira E, Sklan D, Friedman A. Establishment of immune competence in the avian GALT during the immediate post-hatch period. Dev Comp Immunol. 2003;27(2):147–57.
21. Van Immerseel F, De Buck J, De Smet I, Mast J, Haesebrouck F, Ducatelle R. Dynamics of immune cell infiltration in the caecal lamina propria of chickens after neonatal infection with a Salmonella Enteritidis strain. Dev Comp Immunol. 2002;26(4):355–64.
22. Lowenthal JW, York JJ, O'neil TE, Rhodes S, Prowse SJ, Strom ADG, et al. In vivo effects of chicken interferon-γ during infection with Eimeria. J Interf Cytokine Res. 1997;17(9):551–8.
23. Craven M, Egan CE, Dowd SE, McDonough SP, Dogan B, Denkers EY, et al. Inflammation drives dysbiosis and bacterial invasion in murine models of Ileal Crohn's disease. PLoS One. 2012;7(7):e41594.
24. Lebeer S, Vanderleyden J, Keersmaecker SCJD. Genes and molecules of lactobacilli supporting probiotic action. Microbiol Mol Biol Rev. 2008;72(4):728–64.
25. Lebeer S, Vanderleyden J, De Keersmaecker SCJ. Host interactions of probiotic bacterial surface molecules: comparison with commensals and pathogens. Nat Rev Microbiol. 2010;8(3):171–84.
26. McCracken VJ, Gaskins HR. Probiotics and the immune system. In: Probiotics: a critical review. Food and Agriculture Organization of the United Nations. 1999. p. 85–111.
27. Sánchez B, Bressollier P, Urdaci MC. Exported proteins in probiotic bacteria: adhesion to intestinal surfaces, host immunomodulation and molecular cross-talking with the host. FEMS Immunol Med Microbiol. 2008;54(1):1–17.
28. Gareau MG, Sherman PM, Walker WA. Probiotics and the gut microbiota in intestinal health and disease. Nat Rev Gastroenterol Hepatol. 2010;7(9):503–14.
29. Medzhitov R. Recognition of microorganisms and activation of the immune response. Nature. 2007;449(7164):819–26.
30. García CEV, Petrova M, Claes IJJ, Boeck ID, Verhoeven TLA, Dilissen E, et al. Piliation of Lactobacillus rhamnosus GG promotes adhesion, phagocytosis and cytokine modulation in macrophages. Appl Environ Microbiol. 2015;81:2050. AEM.03949–14
31. Liu Z-H, Huang M-J, Zhang X-W, Wang L, Huang N-Q, Peng H, et al. The effects of perioperative probiotic treatment on serum zonulin concentration and subsequent postoperative infectious complications after colorectal cancer surgery: a double-center and double-blind randomized clinical trial. Am J Clin Nutr. 2013;97(1):117–26.
32. Tellez G, Dean CE, Corrier DE, Deloach JR, Jaeger L, Hargis BM. Effect of dietary lactose on cecal morphology, pH, organic acids, and Salmonella enteritidis organ invasion in Leghorn chicks. Poult Sci. 1993;72(4):636–42.

33. Torres-Rodriguez A, Donoghue AM, Donoghue DJ, Barton JT, Tellez G, Hargis BM. Performance and condemnation rate analysis of commercial turkey flocks treated with a Lactobacillus spp.-based probiotic. Poult Sci. 2007;86(3):444–6.
34. Torres-Rodriguez A, Higgins SE, Vicente JLS, Wolfenden AD, Gaona-Ramirez G, Barton JT, et al. Effect of lactose as a prebiotic on Turkey body weight under commercial conditions. J Appl Poult Res. 2007;16(4):635–41.
35. Vicente J, Wolfenden A, Torres-Rodriguez A, Higgins S, Tellez G, Hargis B. Effect of a lactobacillus species-based probiotic and dietary lactose prebiotic on Turkey poult performance with or without Salmonella Enteritidis challenge. J Appl Poult Res. 2007;16(3):361–4.
36. Fraune S, Bosch TCG. Why bacteria matter in animal development and evolution. BioEssays News Rev Mol Cell Dev Biol. 2010;32(7):571–80.
37. Maslowski KM, Mackay CR. Diet, gut microbiota and immune responses. Nat Immunol. 2011;12(1):5–9.
38. Salzman NH. Microbiota-immune system interaction: an uneasy alliance. Curr Opin Microbiol. 2011;14(1):99–105.
39. Kau AL, Ahern PP, Griffin NW, Goodman AL, Gordon JI. Human nutrition, the gut microbiome and the immune system. Nature. 2011;474(7351):327–36.
40. Wideman RF. Bacterial chondronecrosis with osteomyelitis and lameness in broilers: a review. Poult Sci. 2016;95:325–44.
41. Wideman RF, Prisby RD. Bone circulatory disturbances in the development of spontaneous bacterial chondronecrosis with osteomyelitis: a translational model for the pathogenesis of femoral head necrosis. Front Endocrinol. 2013;3:183. doi:10.3389/fendo.2012.00183.
42. Borst LB, Suyemoto MM, Sarsour AH, Harris MC, Martin MP, Strickland JD, et al. Pathogenesis of Enterococcal Spondylitis caused by Enterococcus cecorum in broiler chickens. Vet Pathol. 2017;54(1):61–73.
43. Huff GR, Huff WE, Balog JM, Rath NC. The effects of dexamethasone immunosuppression on turkey osteomyelitis complex in an experimental Escherichia coli respiratory infection. Poult Sci. 1998;77(5):654–61.
44. Borst LB, Suyemoto MM, Sarsour AH, Harris MC, Martin MP, Strickland JD, et al. Pathogenesis of Enterococcal Spondylitis caused by Enterococcus cecorum in broiler chickens. Vet Pathol. 2016;10:0300985816658098.
45. Latorre JD, Hernandez-Velasco X, Kogut MH, Vicente JL, Wolfenden R, Wolfenden A, et al. Role of a bacillus subtilis direct-fed microbial on digesta viscosity, bacterial translocation, and bone mineralization in turkey poults fed with a rye-based diet. Front Vet Sci [Internet]. 2014;1:26. [cited 2017 Jan 25]. Available from: http://journal.frontiersin.org/article/10.3389/fvets.2014.00026/full
46. Borst LB, Suyemoto MM, Robbins KM, Lyman RL, Martin MP, Barnes HJ. Molecular epidemiology of Enterococcus cecorum isolates recovered from enterococcal spondylitis outbreaks in the southeastern United States. Avian Pathol. 2012;41(5):479–85.
47. Kibenge FSB, Wilcox GE, Pass DA. Pathogenicity of four strains of Staphylococcus aureus ioslated from chickens with clinical tenosynovitis. Avian Pathol. 1983;12:213–20.
48. Smeltzer MS, Gillaspy AF. Molecular pathogenesis of staphylococcal osteomyelitis. Poult Sci. 2000;79:1042–9.
49. Whittow GC. Front matter. In: Sturkie's avian physiology. 5th ed. [Internet]. San Diego: Academic; 2000:iii [Cited 25 Jan 2017]. Available from: //www.sciencedirect.com/science/article/pii/B9780127476056500298
50. Martin LT, Martin MP, Barnes HJ. Experimental reproduction of Enterococcal Spondylitis in male broiler breeder chickens. Avian Dis. 2011;55(2):273–8.
51. Parker DS, Armstrong DG. Antibiotic feed additives and livestock production. Proc Nutr Soc. 1987;46(3):415–21.
52. Boyle EC, Bishop JL, Grassl GA, Finlay BB. Salmonella: from pathogenesis to therapeutics. J Bacteriol. 2007;189(5):1489–95.
53. Hong HA, Duc LH, Cutting SM. The use of bacterial spore formers as probiotics. FEMS Microbiol Rev. 2005;29(4):813–35.

54. Ricke SC. Perspectives on the use of organic acids and short chain fatty acids as antimicrobials. Poult Sci. 2003;82(4):632–9.
55. Filho RLA, Higgins JP, Higgins SE, Gaona G, Wolfenden AD, Tellez G, et al. Ability of bacteriophages isolated from different sources to reduce Salmonella enterica Serovar Enteritidis in vitro and in vivo. Poult Sci. 2007;86(9):1904–9.
56. Tellez G, Pixley C, Wolfenden RE, Layton SL, Hargis BM. Probiotics/direct fed microbials for Salmonella control in poultry. Food Res Int. 2012;45(2):628–33.
57. Hong HA, Huang J M, Khaneja R, Hiep LV, Urdaci MC, Cutting SM. The safety of Bacillus subtilis and Bacillus indicus as food probiotics. J Appl Microbiol. 2008;105(2):510–20.
58. Vreeland RH, Rosenzweig WD, Powers DW. Isolation of a 250 million-year-old halotolerant bacterium from a primary salt crystal. Nature. 2000;407(6806):897–900.
59. Hoa TT, Duc LH, Isticato R, Baccigalupi L, Ricca E, Van PH, et al. Fate and dissemination of Bacillus subtilis spores in a murine model. Appl Environ Microbiol. 2001;67(9):3819–23.
60. Kunst F, Ogasawara N, Moszer I, Albertini AM, Alloni G, Azevedo V, et al. The complete genome sequence of the gram-positive bacterium Bacillus subtilis. Nature. 1997;390(6657):249–56.
61. Latorre JD, Hernandez-Velasco X, Kuttappan VA, Wolfenden RE, Vicente JL, Wolfenden AD, et al. Selection of Bacillus spp. for cellulase and xylanase production as direct-fed microbials to reduce digesta viscosity and Clostridium perfringens proliferation using an in vitro digestive model in different poultry diets. Front Vet Sci [Internet]. 2015;2:25. Cited 25 Jan 2017. Available from: http://www.ncbi.nlm.nih.gov/pmc/articles/PMC4672186/
62. Latorre JD, Hernandez-Velasco X, Bielke LR, Vicente JL, Wolfenden R, Menconi A, et al. Evaluation of a Bacillus direct-fed microbial candidate on digesta viscosity, bacterial translocation, microbiota composition and bone mineralisation in broiler chickens fed on a rye-based diet. Br Poult Sci. 2015;56(6):723–32.
63. MacAuliffe T, McGinnis J. Effect of antibiotic supplements to diets containing rye on chick growth. Poult Sci. 1971;50(4):1130–4.
64. Grammer JC, McGinnis J, Pubols MH. The effects of a pectic enzyme on the growth-depressing and rachitogenic properties of rye for chicks. Poult Sci. 1982;61(9):1891–6.

Part III
Modulation of Gut Microbiota to Support Bone Health

Prebiotics and Bone

Corrie M. Whisner and Connie M. Weaver

Introduction

Gut Microbiome

Until recently, little attention had been given to the influence of microbial cells found in and on the human body and their effects on overall health. These microbes as a collective unit are now referred to as the microbiota, while the genes they encode have been coined the microbiome. Collectively, the human body harbors an estimated 100 trillion microbial cells, a number that surpasses the sum of human cells ten to one. Microbial communities vary by body site with the intestinal tract containing nearly 1,000 microbial species with metabolic activity that rivals only that of the human liver [1]. The gut microbiota play an important role in physiological processes which include energy metabolism, nutrient supply, and immune and inflammatory responses [2, 3]. Additionally, research has linked the intestinal microbiota to more than 25 human diseases and conditions [1]. The majority of evidence supports links with obesity, allergies, behavioral disturbances, cardiovascular diseases, and some cancers [4], but emerging evidence also suggests the importance of gut microbiota for mineral absorption and osteoporosis risk.

The gut microbiota contribute to a dynamic symbiotic relationship with humans allowing for efficient energy harvesting from carbohydrate-rich diets [5, 6]. Dietary intake of the host is integral to the gut microbiota as it provides a source of energy

C.M. Whisner
Arizona State University, School of Nutrition and Health Promotion, Arizona Biomedical Collaborative, 425 N 5th St, Phoenix, AZ 85004, USA

C.M. Weaver (✉)
Purdue University, Nutrition Science, Stone Hall, 700 W State St, West Lafayette, IN 47907, USA
e-mail: weavercm@purdue.edu

L.R. McCabe, N. Parameswaran (eds.), *Understanding the Gut-Bone Signaling Axis*, Advances in Experimental Medicine and Biology 1033,
DOI 10.1007/978-3-319-66653-2_10

for microbes residing in the lower small intestine and colon [4]. Furthermore, gut microbiota can impact energy extraction from the diet by providing additional metabolic capacity and by regulating genes integral to carbohydrate and lipid metabolism [7, 8]. Dietary carbohydrates, specifically dietary fibers, have been linked to the species composition, quantity, and fermentation potential of microbes in the intestine [4].

Prebiotics

The rise of functional foods has brought about new approaches for the treatment and prevention of adverse health outcomes. Dietary prebiotics have functional effects (Fig. 1) and have been defined as "selectively fermented ingredients that result in specific changes, thus conferring benefit(s) upon host health" [7]. While any nutrient that enters the large intestine may have prebiotic effects, the majority of known prebiotics are carbohydrates. Specifically, oligosaccharides such as inulin-type fructans and galactooligosaccharides are well supported in the literature for their prebiotic effects on microbial composition primarily through increased proportions of bifidobacteria and lactobacillus [8]. It is important to note that although prebiotic oligosaccharides can be classified as soluble fiber, this does not mean that all dietary fibers are capable of eliciting prebiotic effects. Both prebiotics and dietary fibers resist digestion in the small intestine and, in the case of soluble fibers, are fermented in the lower gut and colon. However, what sets prebiotics apart is their ability to stimulate the growth of a selective set of microbial species in the complex microbial environment of the human and/or animal gut.

With recent advances in molecular biology techniques, identification of the entire microbiota community, including strict anaerobes, is now possible. These new

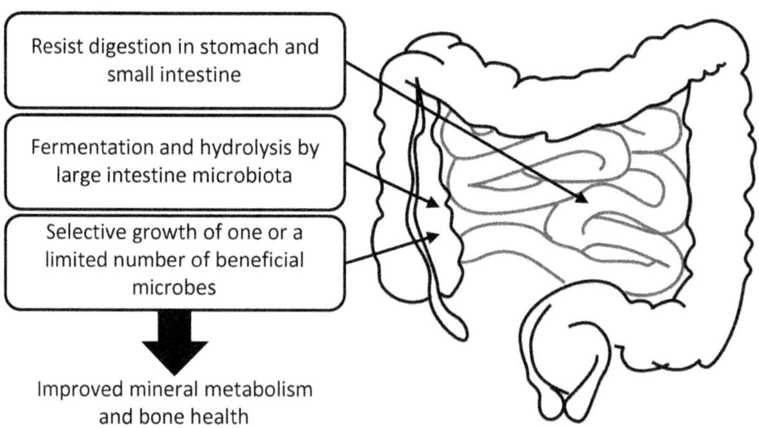

Fig. 1 Prebiotic food ingredients are metabolized by intestinal microbes which results in beneficial health effects for the host

methods have allowed for more rapid progress in the scientific literature, specifically in relation to studying real-time shifts of the microbiome in response to prebiotic interventions. Additionally, careful implementation of clinical trials with prebiotics has resulted in a better understanding of the mechanisms by which microbiota impact overall health and reduce the risk of disease. These health benefits extend beyond that of the intestines, impacting tissues such as the vasculature, skin, and even bone.

Both human and animal models provide clear evidence that shifts in microbiota following prebiotic consumption are linked to skeletal health, specifically through modifications in calcium metabolism. Animal models provide evidence that microbiota changes positively correlate with measures of improved bone density and strength while data from human studies suggest that gut microbiota shifts are associated with increased calcium absorption in the lower gut. Despite these findings, the exact mechanisms by which gut microbiota enhance bone health via prebiotic intervention remain poorly understood. Currently, the most popular mechanism is thought to be microbial fermentation and short-chain fatty acid (SCFA) production in the colon which lowers the pH of the intestinal lumen and improves the absorption of calcium and other divalent minerals. The overarching aim of this chapter is to provide an overview of the current evidence on prebiotic supplementation for maximizing calcium absorption, improving measures of bone density and strength, and preventing bone loss. Common prebiotics and dosing will be discussed as well as data to support potential microbial mechanisms and public health strategies.

Microbiome-Bone Interactions

The microbiome is an incredibly rich community of bacteria, fungi, viruses, and archaea. Together these organisms have profound metabolic activity which has led to the human microbiome being coined a "forgotten organ" [9]. Collectively, more than three million unique microbial genes are present in the gut microbiome which extends the functional capacity of human genes twofold [10, 11]. The functional capacity of these microbial genes moves beyond the intestinal environment in which these organisms live and accounts for more than 200 and 220 metabolites excreted in urine and feces, respectively [12]. These metabolites represent the efforts of both host and microbial metabolism, the study of which provides great insight into the larger metagenome and its link to organ systems, metabolic pathways and states of health and disease. The microbiome therefore, has large impacts on the metabolic phenotypes of their host which have been shown to affect the immune system, alter the digestion and absorption of dietary components, and prevent infection by pathogens through displacement and/or improved intestinal barrier function. An emerging area of microbiome-host interaction research includes the exploration of direct microbial effects on the skeletal tissue which have been categorized below into three overarching categories (Fig. 2).

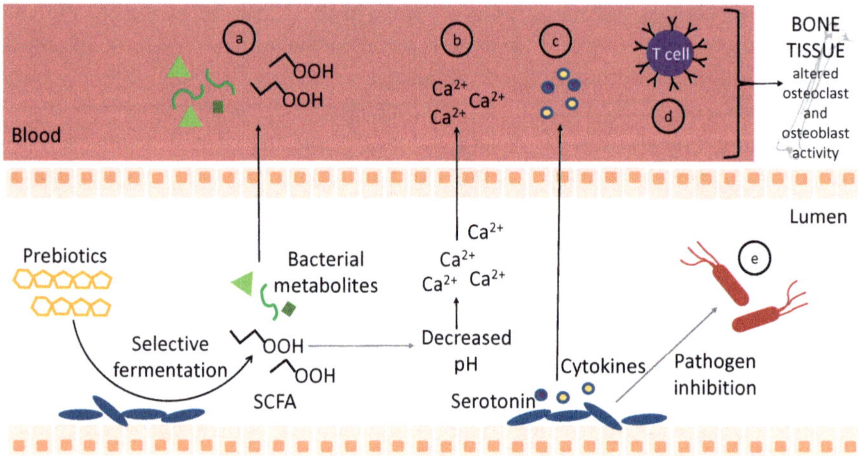

Fig. 2 Gut microbes alter the intestinal microbial environment which fosters the production of various signaling molecules, immune cells, and metabolites that may benefit bone. Microbial interactions with prebiotic ingredients influence the larger intestinal environment through (*a*) the production of short-chain fatty acids and other metabolites, (*b*) decreased pH and greater mineral bioavailability, (*c*) microbial production of serotonin and cytokines, (*d*) immune system modifications, and (*e*) pathogen displacement

Nutrient Bioavailability and Absorption

The intestinal tract is a unique organ because it acts as a conduit by which environmental components including nutrients, drugs, and toxins are able to interact with human cells. It is well accepted that gut microbes are involved in the production of B and K vitamins as well as short-chain fatty acids (SCFAs), an important fuel for intestinal cells. SCFAs are also thought to play a role in the absorption of minerals like calcium, magnesium, iron, and zinc [8]. Although many studies have evaluated the effects of prebiotic consumption on mineral absorption, our understanding of how gut microbes communicate with bone remains ill defined. Germ-free (raised in a sterile environment without intestinal microbial colonization) mice are a useful tool for studying relationships between the microbiome and bone. Sjögren et al. [13] compared germ-free mice to those inoculated with gut microbiota from conventionally raised mice. Findings suggested that germ-free mice experienced significant increases in trabecular bone mineral density, trabecular bone volume, trabecular number, and cortical bone area compared to conventionally raised mice. Additionally, per area of bone osteoclast numbers were lower in germ-free mice compared to conventionally raised counterparts. These data suggest that microbiome actions may be involved in bone catabolism.

Immune Pathways

Systemic inflammation and immune signaling have been associated with bone health. Gut microbiota are also thought to communicate with the immune system; primary mechanisms include protection against pathogenic microbes and immune suppression against symbiotic or beneficial microbes. The absence of gut microbiota (germ-free mice) results in an immature mucosal immune system [14] and reduced immune signaling [13]. The current mechanism by which these immune compounds impact bone is through activated T cells increasing TNFα expression in the bone marrow [15]. TNFα promotes the development and maturation of osteoclasts which may interrupt the delicate balance between bone formation and resorption thereby resulting in bone loss [16, 17]. It has been speculated that the gut microbiota may be responsible for activating T cells through the production of stimulating antigens [18]. This makes sense given that T cells and osteoclasts are both the result of hematopoietic stem cells produced in the skeleton. Bone loss has been linked to colitis-induced inflammation of the gut which is also thought to impact immune responses in the bone marrow [19]. Ovariectomy has also been shown to increase inflammation which may impact bone health in the same way [18, 20].

Signaling Molecules/Hormones

SCFAs influence signaling pathways which may impact intestinal cell metabolism. The gut microbiota is thought to play a large role in the production of peripheral serotonin which may impact bone health [21, 22]. Selective serotonin reuptake inhibitors, drugs commonly prescribed for anxiety and depression, have been shown to reduce bone formation in animals [23, 24] and significantly increase risk for fracture in humans [25]. Early in vitro studies using osteoblast cultures suggest that the serotonin 6 G-protein-coupled receptor (5-HT_6R) is highly expressed during bone remodeling and osteoblast differentiation, and upon stimulation with serotonin, alkaline phosphatase activity and bone mineralization are inhibited [26]. Other serotonin receptor types, 5-HT1 and 5-HT2, have resulted in similar effects [23]. Interestingly, inhibition of gut-derived serotonin production resulted in greater bone formation and improved bone mass in ovariectomized animals. This suggests that modulation of gut-derived serotonin production may have implications for osteoporosis treatment [22, 27].

Deficiencies in sex steroids have also been associated with chronic inflammation which contributes to osteoporosis [28]. The loss of both androgens and estrogen decreases the body's ability to combat oxidative stress from reactive oxygen species (ROS). ROS may have negative effects on Wnt signaling which ultimately leads to decreased osteoblastogenesis and increased bone resorption as a result of androgen loss [28, 29]. The gut microbiome is believed to contribute to peripheral androgen concentrations and activity, but currently it remains difficult to study the effects of microbial and human-derived androgens separately. Additionally, gut microbes respond to androgens such that the microbiome diverges at puberty when sex hormones drastically change between males and females [30]. In relation to bone health, it remains unclear how microbe-androgen interactions influence bone physiology in males versus females, but sex differences have been observed for immune signaling in relation to type 1 diabetes development [30]. In young weaning mice receiving penicillin, male mice experienced reductions in bone mineral content and bone area, while female animals experienced improved bone mineral content and density suggesting that sex-specific effects may be impacted by the microbiome.

Prebiotic Effects on Bone

Products with Established Prebiotic Effects

Poorly digested carbohydrates classified as nondigestible oligosaccharides (NDOs) are currently regarded as the most promising prebiotics for bone health (Fig. 3). These compounds include galactooligosaccharides (GOS), fructooligosaccharides (FOS), oligofructose, and inulin. Oligosaccharides generally vary in chain length, with a degree of polymerization (DP; number of sugar monomers included in each chain) between 4 and 10, but other short-chain disaccharides and longer-chain polysaccharides (DP 10–60) also exist [8]. Prebiotic disaccharides include milk sugar derivatives, lactulose, and lactitol, while polysaccharides with prebiotic effects include long-chain fructooligosaccharides and high molecular weight inulin.

Stimulated growth and proliferation of bifidobacteria in the colon, a microbe associated with beneficial health effects, has been observed following FOS [31–33]

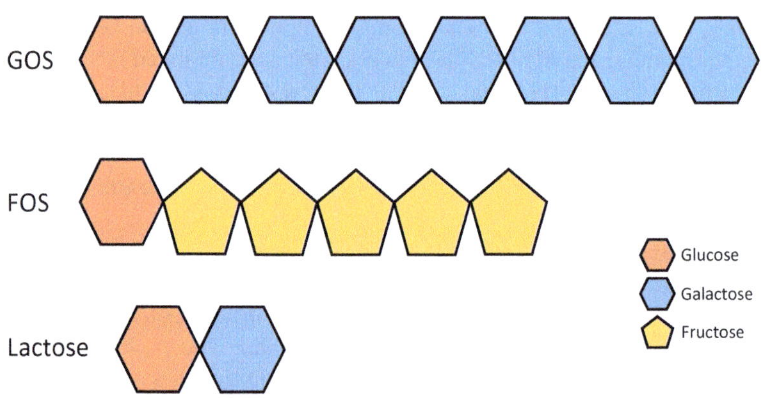

Fig. 3 Prebiotics involve differential arrangements of the monosaccharides glucose, galactose, and fructose

and GOS [34–37] consumption. A larger variety of compounds have been implicated as functional fibers or prebiotics for various health outcomes including reduced inflammation, improved immune function, weight loss/maintenance, greater insulin sensitivity, and gastrointestinal health/regularity [8, 38]. With regard to bone health, the majority of evidence supports FOS and GOS as a prebiotic agent, but new data are emerging for compounds such as soluble corn fiber, gums, and synbiotics.

Characteristics and Functions of Common Prebiotics

Inulin and Fructooligosaccharides

FOS are NDOs that can be synthetically made but also occur naturally in plant-based foods including chicory root, artichoke, wheat, onion, asparagus, and banana [39]. They are comprised of fructose units with varying types of bonds between monomers and have a DP ranging between 2 and 60. FOS with a DP of three to six units are referred to as short-chain fructooligosaccharides with a more specific type called oligofructose which has a mean DP of 4. Inulin from chicory root typically has an average DP of 12. Longer NDOs in this category include long-chain FOS (lcFOS) and high molecular weight inulin (DP 25) [8]. Fructose units in FOS chains are connected by β(2-1) fructosyl-fructose linkages and commonly carry a glucose unit on one end [39]. FOS are not digested by human enzymes allowing these polymers to be hydrolyzed and fermented by bacteria in the colon. A comprehensive review of prebiotics, including FOS, suggested their ability to increase calcium absorption, improve BMD in growing rats, and decrease the loss of bone mineral in postmenopausal rat models by improving mineral solubility and increasing the surface area available for absorption in the large intestine [8]. Additionally, significant and selective growth of beneficial gut microbes has been reported in humans; bifidobacteria increased in humans after 1 week of FOS (10 g) consumption [40].

Galactooligosaccharides

GOS occur naturally in human breast milk and account for a large portion of its nutrients (5–10 g/l) behind both lactose and lipids. The presence of GOS in breast milk makes this unique oligosaccharide an important contributor to immunity and gut health in nursing infants by fostering the growth of protective and beneficial gut microbes such as bifidobacteria and lactobacillus [41–44]. GOS can also be prepared through the enzymatic conversion of lactose with beta-D-galactosidase [45]. This results in lactose units bound to chains of galactose that are 2–8 monomers in length and connected via glycosidic β(1-2), β(1-3), β(1-4), and β(1-6) linkages [46, 47]. While humans lack digestive enzymes to break down these different chain linkages, recent data suggest that the specific combination of bonds between galactose units and galactose-glucose units may affect their ability to promote the growth of beneficial gut microbes such as bifidobacteria [46]. Previous studies have observed significant increases in skeletal calcium content in postmenopausal rat models [48] and increased calcium absorption in postmenopausal women [49] following GOS consumption.

Lactose Derivatives

Lactose, a naturally occurring disaccharide in milk, is composed of the sugar monomers glucose and galactose. As a reducing sugar, it can easily be transformed into prebiotic disaccharides such as lactulose and lactitol. Consumption of lactose, in combination with calcium, has been shown to improve bone mineral content and strength in rats [50]. With age humans experience diminished lactase activity resulting in greater quantities of lactose reaching the colon. Data suggest that this allows for prebiotic-induced bone effects such that lactase-deficient individuals absorbed more calcium from lactose-containing milk than did individuals with normal lactase activity [51]. The increased absorption among lactose-intolerant individuals may be the result of upregulated calcium absorption in response to the lower intakes common in lactose-intolerant individuals; however, this effect may also be explained by colonic fermentation and β-galactosidase activity of colonic microflora. However, other studies report no benefit of lactose on calcium absorption [52, 53]. Lactulose, a product of heat treated lactose, has been utilized in the medical and pharmaceutical industry for decades for its effects on gut microbes and laxation [54, 55]. It is frequently commercially produced by alkaline isomerization of lactose which results in β(1-4) glycosidic bonds between galactose and fructose [54]. Lactulose has been shown to increase calcium absorption in both rats [56] and postmenopausal women [57].

Soluble Corn Fiber

Soluble corn fiber (SCF), a maize-based soluble fiber, is formed by the enzymatic hydrolysis of starch. This process results in glucose chains containing a mixture of α(1-2), α(1-3), α(1-4), and α(1-6) glucosidic bonds which resist digestion in the small intestine thereby allowing for microbial fermentation in the lower gut. SCF is well tolerated with data suggesting that total daily intakes up to 65 g were better tolerated than inulin at lower doses [58]. Further, SCF has a low viscosity and is resistant to processing and manufacturing techniques with heat and variable pH [59]. Consumption of SCF has been associated with increased proportions of bifidobacteria in feces following consumption of 8–21 g/day [60]. With regard to bone health, this fiber has been associated with improved calcium absorption in rats [61] and adolescent boys and girls [62, 63], improved calcium retention in postmenopausal women, [64] as well as improved bone strength in rats [61].

Synbiotics and Other Prebiotics

Synbiotics are mixtures of more than one kind of prebiotic or a combination of prebiotics with other bioactive ingredients ranging from probiotics (live microorganisms) to polyphenolic compounds. The benefit of combining prebiotics is the ability to increase the DP of the supplement which may have prolonged impact in the gut as the varying chain lengths can be fermented and hydrolyzed along the entire length of the lower intestine and thereby maximize their benefits on host health [65]. Shorter-chain prebiotics such as oligofructose are believed to be metabolized in the proximal colon, while longer-chain compounds like inulin are metabolized by gut microbes in the distal colon. Inulin-type fructan mixtures (ITF-mix) are the most cited synbiotic with positive effects on calcium absorption and bone health outcomes [66–71]. Mixtures of GOS and FOS have also shown benefits in growing rats, resulting in greater trabecular bone mineral density, bone volume, osteoblast surface area, and measures of stiffness and elasticity [72].

Soy isoflavones in conjunction with prebiotics have had mixed results. Two studies evaluating the effects of FOS and isoflavones together in rodents resulted in greater femoral BMD when compared to each compound individually [73, 74], while other studies did not observe a synergistic effect on BMD [75–77]. However, one study reported improvements in trabecular microarchitectural properties of the tibia despite no change in BMD following treatment with FOS and isoflavones [77]. Other plant polyphenols combined with prebiotics have shown synergistic effects on bone [78]. FOS combined with dried plum fractions (purees, skins, juice, extract), whole raisins, dates, and figs, and β-hydroxy-β-methylbutyrate in ovariectomized rats suggested that FOS with dried plum resulted in greater femur and lumbar BMD.

The combination of prebiotics and probiotics, especially *Bifidobacterium* species, has also proven to be an effective way of improving bone health. FOS from yacon flour combined with *Bifidobacterium longum* increased the calcium, magnesium, and phosphorus content of bone as well as bone breaking force in rats [79]. The work by Perez-Conesa et al. reported that combining *Bifidobacterium bifidum* and *longum* species with GOS acutely improved calcium, magnesium, and phosphorus bioavailability and absorption in one study [80] while improving mineral content of the femur and tibia in a second study [81]. Similar work with *Bifidobacterium* and lactulose improved bone strength while also increasing the number of bifidobacteria and SCFAs in the cecum [82].

Evidence from Animal Models

Animal models have proven to be especially helpful in elucidating the mechanisms by which prebiotics influence health outcomes including bone health. Currently, the primary mechanism reported in rats has been a decrease in pH following microbial fermentation in the cecum and colon. In addition to improved mineral absorption, supplementation with prebiotics in rats has been associated with improved measures of bone density and strength in both growing and postmenopausal animal models [15, 83–85].

Calcium Absorption

Prebiotic supplementation in animals increases intestinal bioavailability and absorption of calcium [86–91] as well as other divalent minerals [87, 89–97]. Varying types of prebiotics including GOS [48], FOS/inulin [98, 99], and polydextrose [92] have been associated with improvements in calcium absorption. Animal studies have found prebiotics to have a dose-dependent effect on calcium absorption. Inulin-type fructans (up to 20% of the diet) [100] and lactulose (5% and 10%) [56] resulted in greater absorption as the dose increased. Despite encouraging results regarding dose, results can vary depending on a variety of factors including animal age, experimental conditions, duration of treatment, and selected outcome measures.

Mineral absorption has been found to decrease with age. In rats consuming inulin, both calcium and magnesium absorption were lower in 10- and 20-month-old rats compared to rats aged 2 and 5 months [101]. However, the dose and type of prebiotic may also influence these responses. Adult male rats consuming different fructan prebiotic combinations that contained a range of chain lengths and branching found that only an oligofructose-inulin combination resulted in significantly increased calcium absorption [102]. The combination of short- and long-chain fructans in this study may have resulted in synergistic effects by allowing for prolonged fermentation and absorption of calcium throughout the large intestine. Prebiotics have also been shown to have positive effects during and after menopause, a time when estrogen deficiency negatively impacts calcium absorption. GOS supplementation for 20 days resulted in greater calcium absorption in ovariectomized rats [48], while inulin in combination with FOS for 21 days resulted in improved calcium balance [103].

Experimental conditions impact the ability of prebiotics to increase calcium absorption. Difructose anhydride III (DFAIII), a nondigestible disaccharide, increased calcium absorption in vitamin D-deficient ovariectomized rats [104], while oligofructose was more effective that other prebiotics at increasing calcium absorption among rats consuming high calcium diets [105]. Treatment duration with prebiotics also has an influence on calcium absorption in animal models with effects occurring in 1–3 days with doses ranging from 5 g/100 kg to 50 g/kg of body weight [56, 87, 106]. In young, growing rats, treatment with inulin for 40 days suggested that the calcium absorption response was dependent on calcium dose (0.25%, 0.50%, and 0.75%) [107]. By day 13, calcium absorption had increased with all calcium intakes but after 40 days inulin consumption only improved calcium absorption on the lowest calcium diet. Despite these findings, studies with longer durations beyond 40 days suggest that prebiotic effects on calcium absorption and retention may persist long-term where 3-month consumption of oligofructose resulted in greater calcium absorption compared to control rats [108]. This study attributed these effects to morphological changes and greater expression of calbindin-D9K, an intracellular calcium transport protein, in the intestine. Interestingly, the same effects have been observed in gastrectomized animals consuming FOS [86] and DFAIII [109].

BMD and Bone Strength

Despite data that supports increased calcium absorption following prebiotic consumption, improvements in bone health require that the benefits of prebiotic consumption translate to increased bone mass and improvements in bone architecture. Ultimately, the goal of these treatments should be a reduction in the risk for fracture. Rats and mice are beneficial models for assessing these effects as their shorter lifespan allows long-term effectiveness studies in relation to functional bone health outcomes.

Supplementation with prebiotics has resulted in improved tibial and femoral calcium content [91, 110, 111] but conflicting results exist among growing animals [76, 102]. Beyond improved measures of calcium content, beneficial changes in bone microarchitecture have been reported, and the data suggest that trabecular-rich bone may be more responsive to prebiotic-induced effects. Treatment of growing rats with GOS for 4 weeks increased the breaking strength of the tibia, trabecular volumetric bone mineral density (vBMD) of the distal femur, as well as the area and vBMD of the proximal tibia [91].

Animal models of menopause suggest similar findings with oligofructose at varying doses (25, 50, and 100 g/kg) resulting in reduced bone loss following ovariectomy [105]. Interestingly, bone microarchitecture was differentially affected by calcium intake and oligofructose dose. The lowest dose of oligofructose resulted in greater trabecular thickness only at adequate calcium intakes, while the two larger oligofructose doses at the same calcium intake resulted in greater trabecular circumference. These effects disappeared when rats were fed with high calcium diets. Overall, data from this study suggested that weight-bearing skeletal sites benefit most from prebiotic consumption as lumbar spine calcium content responded positively to only the high calcium and 100 g/kg oligofructose treatment. A mixture of inulin and FOS reduced bone resorption in ovariectomized rats such that femoral calcium content, BMD, and bone balance were significantly increased after 21 days [103]. Similar synbiotic studies administering FOS and isoflavones to ovariectomized mice resulted in improved bone mass of the femur [73], while reports of FOS alone resulted in nonsignificant improvements in bone mineral calcium content [92].

Beyond changes in bone architecture, improved bone strength has also been reported [89, 110, 112, 113], even in the absence of BMD improvements [89, 112]. Soluble corn fiber and soluble fiber dextrin had the greatest effects on structural bone properties in young rats which included improvements in the peak breaking force of the distal femur when compared to six other fibers and cellulose [61]. Similar effects were observed for the femur and tibia of rats consuming GOS for 8 weeks [91].

Evidence from Humans

The effects of prebiotics on mineral absorption and bone health have resulted in contradictory findings in humans. The lack of robust findings may be the result of large variation in intervention parameters and length of treatment, prebiotic dose, baseline mineral status, prebiotic vehicle composition and structure, and age of study cohorts. Further, longer gastrointestinal transit time in humans has informed research on prebiotic consumption and mineral absorption. Study designs using urinary excretion of stable calcium isotopes require urine collection beyond 24 h in order to observe the effects of prebiotics in the large intestine [114, 115].

Calcium Absorption

Increases in calcium absorption have been observed in the majority of studies involving young, adult, and elderly individuals consuming a range (8–20 g) of prebiotic doses [49, 57, 62, 63, 66, 69, 70, 114, 116]; however, a few studies have observed no prebiotic effect on calcium absorption [117–119]. Calcium intake may, at least in part, explain the null effects observed in these studies. In one of the studies, calcium intakes (1,500 mg/day) exceeded the recommended intake of 1,300 mg/day suggesting that at high intakes, the prebiotic effect may be overpowered by increased luminal calcium concentrations. Age and prebiotic doses have varied widely in the current calcium absorption literature which may influence functional outcomes. Small doses (0.75–1.25 g/day) of short-chain inulin in 6–12-month-old infants had no influence on calcium absorption but resulted in improved iron and magnesium retention [121]. These findings may be explained by limited SCFA production which is thought to influence calcium absorption [121] while magnesium absorption responds to decreases in luminal pH which did occur in this infant study [122]. Larger doses (5 g/day) of GOS in weaning infants have been shown to improve calcium absorption [123]. This study also reported increases in the number of intestinal bifidobacteria which may have contributed to the observed improvement in mineral absorption.

The ability to increase calcium absorption at critical times in the life cycle can be beneficial for improved skeletal health. During the pubertal growth spurt when habitual calcium intakes are often inadequate, prebiotics have been shown to improve calcium absorption which may increase peak bone mass and prevent fractures later in life. Work in growing children and adolescents has shown that FOS, GOS, and SCF have been effective at improving calcium absorption [62, 63, 66, 69, 70, 114, 116]. While the majority of studies have reported 6–12% increases in calcium absorption relative to control treatments [62, 63, 66, 70, 114, 116], oligofructose consumption for 3 weeks in young girls near menarche resulted in a 30 % increase in calcium absorption [69]. Most studies to date have evaluated calcium absorption following acute periods (9 days to 4 weeks) of prebiotic consumption, but the long-term effects require further research. One group studied the effects of ITF-mix over 1 year and found that calcium absorption increased by 8 weeks, and this effect persisted across the entire year [66].

Postmenopausal women are another group with greater risk of adverse skeletal health outcomes. In this population, lactulose, FOS, inulin, and GOS have had beneficial effects on calcium absorption [49, 57, 71, 124]. Lactose at both 5 and 10 g/day had dose-dependent effects on calcium absorption with only the 10 g supplement resulting in significant improvements [57]. Consumption of other milk-based prebiotics (20 g/day of GOS) resulted in a 16% increase in calcium absorption relative to placebo in postmenopausal women consuming products for 9 days [49]. FOS and inulin products have had mixed effects with 8 g/day as chicory fructans increasing calcium absorption by 42% over a 3-month supplementation period [124] and 10 g/day of ITF-mix consumption for 6 weeks increasing absorption by 7% [71]. Conversely, a similar dose of short chain FOS had no effect on calcium absorption [119].

BMD and/or Bone Biomarkers

Few studies have evaluated the effects of prebiotics on bone mineral due to the need to follow humans for long periods of time before changes are observable in bone by dual-energy x-ray absorptiometry, the gold standard for measuring bone mineral density and content. Among adolescents, 1 year of ITF-mix consumption resulted in significant increases in whole-body bone mineral content and density by 35 g and 0.015 g/cm^2, respectively [66]. Postmenopausal women who are at risk for increased bone loss have also experienced skeletal benefits from prebiotic consumption. In a dose-response study with 0, 10, and 20 g/day as soluble corn fiber, skeletal calcium retention was increased in a dose-dependent manner with 10 g and 20 g increasing retention by 5% and 7%, respectively [64]. This study was able to measure calcium retention using a novel technique with rare long-lived (half-life of 10^5 year) calcium radioisotope ^{41}Ca which can be used to label bone and study bone loss over long periods of time.

Biochemical markers of bone metabolism have proven helpful in shorter studies to evaluate bone turnover in response to prebiotic intake. Among postmenopausal women, urinary excretion of deoxypyridinoline, a systemic bone resorption marker, decreased following chicory fructan fiber consumption [124]. In the ^{41}Ca study above, markers of bone turnover N-terminal telopeptide and osteocalcin were unchanged, while formation marker, bone-specific alkaline phosphatase, increased significantly by 8% [64]. Data from 300 non-osteoporotic postmenopausal women consuming a combination of calcium and short-chain FOS for 12 months suggested that prebiotic consumption may impact bone turnover and modeling in the absence of BMD effects with declines in both osteocalcin and C-telopeptides of type I collagen over the duration of the study [125]. These data suggest the need for further study in order to evaluate whether these changes in bone turnover markers translate to reduced fracture risk later in life.

Mechanisms

In many of the studies of prebiotic effects on calcium absorption and bone health outcomes, gut microbiome profiles were not evaluated. However, with recent advances in high-throughput sequencing techniques, investigations of prebiotic effects on bone have begun to include these important measures. Despite these new studies, our current knowledge of the microbiota's role in prebiotic-bone mechanisms remain limited. In an evaluation of growing male rats receiving 0%, 2%, 4%, 6%, or 8% of GOS, regression modeling was used to study intermediate mechanistic outcomes, including gut microbiome measures, which mediate prebiotic effects on bone mineral and strength measures [91]. GOS treatment resulted in dose-dependent differences in gut microbial communities, and quantitative PCR measures suggested that GOS increased the prevalence of bifidobacteria. In conjunction with these measures, decreases in pH and greater cecal content and wall weights were observed which may contribute to improvements in observed bone strength measures. Human studies of GOS and SCF have both indicated that the gut microbiota change with prebiotic consumption. Among healthy preadolescent girls, consumption of 0, 5, and 10 g/day GOS in combination with a calcium-containing smoothie drink resulted in greater bifidobacteria numbers as measured by quantitative PCR [116]. In this study, calcium absorption was increased by 10% following GOS consumption. Using more advanced techniques to measure microbial diversity and specific community composition, high-throughput sequencing of the 16S ribosomal RNA gene, changes in the microbiome were correlated with improvements in calcium absorption following consumption of SCF among adolescent girls and boys [62]. Many of the microbes that increased with SCF treatment and were correlated with fractional calcium absorption measures following dual-stable calcium isotope administration were from the genera *Bacteroides*, *Butyricicoccus*, *Oscillibacter*, and *Dialister* and also known to ferment fiber. In a follow-up study evaluating the dose-response effects of SCF in free-living, healthy adolescent girls, significant increases in fecal microbial community diversity occurred after consuming SCF. The proportion of the community that was comprised of the genus *Parabacteroides* significantly increased with SCF dose and increased calcium absorption was positively correlated with increases in *Clostridium* and unclassified Clostridiaceae. Overall, these data suggest the importance of microbial community shifts in contributing to improved mineral absorption and for functional benefits to bone, but further research is needed to fully understand these mechanisms. The following sections provide a summary of data to support current mechanistic theories mediated by the microbiome.

Short-Chain Fatty Acids (SCFA)

Currently, the most prominent theory to explain prebiotic effects on bone is that prebiotic fibers resist digestion in the small intestine and upon reaching the colon are fermented into SCFAs (Fig. 4). This conversion to SCFA results in reduced pH which is thought to prevent calcium from complexing with other compounds such as phosphates and oxalates. As a result, more calcium is available for absorption and bone mineralization. Despite wide support, data exist to refute this theory. Using chamber experiments comparing calcium absorption following exposure to SCFA or hydrochloric acid suggest that only the SCFA treatment increased calcium transport across colonic cells [121]. Further, neither total nor individual (acetic, propionic, butyric, and valeric acid) concentrations of SCFA were correlated with calcium absorption or skeletal measures [61]. Alternatively, SCFA may elicit a response on calcium absorption by stimulating signaling pathways which increase cellular metabolism and proliferation. Butyrate, as the preferred energy source for colonocytes [126], may also be involved in regulating gene expression through nucleosome and histone modifications [127, 128].

Morphologic Changes and Calcium Transport Proteins

Trophic changes have been observed in the colon which increase the area available for mineral absorption [129]. SCFA have been associated with increased proliferation of intestinal mucosal cells [130] which results in greater epithelial cell density, greater crypt depth, and improved cecal blood flow in animals [81]. Further, these morphological changes have been associated with increased calcium absorption following consumption of prebiotics [129].

Additional changes at the cellular level following prebiotic consumption may include increases in calcium transport proteins (Fig. 4). Animal models have shown increased expression of calbindin-D9K in the cecum and colon following FOS supplementation [86]. Later research to further elucidate this mechanism suggested that increased calbindin-D9k expression was mediated through transcription of the vitamin D receptor [131].

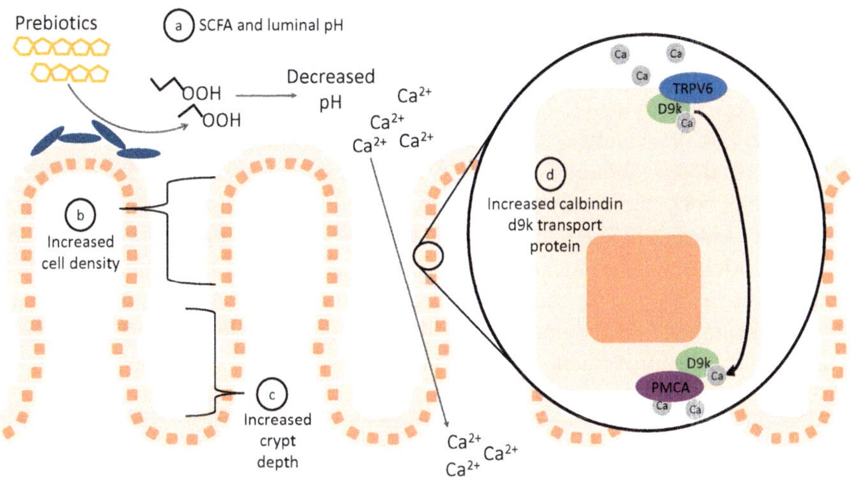

Fig. 4 Prebiotic mechanisms for improved calcium absorption include (*a*) fermentation by saccharolytic microbes in the large intestine to form SCFAs such as butyrate and acetate which decrease the luminal pH, ionize calcium, and allow for greater absorption, (*b*) increased cell proliferation and density of the mucosal lining, (*c*) increased depth of intestinal crypts to maximize surface area for absorption, and (*d*) greater expression of the intracellular calcium transport protein calbindin-D9K. *SCFAs* short-chain fatty acids, *D9k* calbindin-D9k

Immune Signaling

The microbiota may also impact mineral absorption through interactions with the immune system (Fig. 1). Germ-free mouse studies have suggested that the absence of gut microbes is associated with reduced TNFα and T cell expression as well as greater bone mass as a result of reduced osteoclast numbers [13]. Conversely, conventionally raised mice had increased osteoclast numbers and lower cortical bone and trabecular bone volume compared to germ-free mice. TNFα expression has been shown to stimulate the differentiation of osteoclasts which may increase bone resorption [16, 17]. Despite these findings, further research is needed to elucidate how prebiotic supplementation affects these responses in both germ-free and conventionally raised animals.

Public Health Relevance and Strategies for Bone Health Across Life

Prebiotics and Microbiota Help Achieve Peak Bone Mass

Strategies to reduce risk of fracture include building peak bone mass as high as possible within one's genetic potential and reducing the rate of loss later in life. Nutrition and other lifestyle choices, although important throughout life, have the greatest impact during growth. Peak bone mass is achieved soon after adolescence which demarks the end of being able to build additional mass. Optimizing bone accrual during growth has tremendous fracture prevention potential. For every standard deviation decrease in size-adjusted bone mass, there is an 89% increase in fracture risk in childhood [132]. A 10–15% increase in peak bone mass is estimated to decrease risk of fracture by 25–50% later in life [133].

Out of 18 lifestyle factors evaluated by a systematic review to influence development of peak bone mass, only dietary calcium and physical activity received grade A level of evidence [133]. Calcium is a shortfall nutrient intake according to the Dietary Guidelines or Americans [134]. Prebiotics that increase calcium absorption as described in this chapter can improve calcium nutrition, especially for individuals not consuming the recommended intakes of calcium. However, prebiotics provide a modest improvement and cannot correct a large deficiency in dietary calcium.

Prebiotics and Microbiota Help Prevent Osteoporosis and Fracture

At the other end of the lifespan, bone mass is being lost. When sufficient bone is lost to weaken bones, fracture may result leading to a diagnosis of osteoporosis. Lifestyle factors including prebiotic fiber also influence rate of loss of bone. It is not known whether prebiotic fibers have equal protection during the early rapid loss of bone with menopause-induced estrogen deficiency or in the more stable period. The study of soluble corn fiber [64], which showed a benefit to bone calcium retention using a calcium tracer, was in women stable to menopause.

Future Areas for Research

Longer-term studies which measure the dose menopause effects of various prebiotic fibers on bone density, bone strength, and fracture will be more convincing for making public health recommendations. Future research that focuses on understanding the mechanisms of action of prebiotic fibers promises to be an exciting area. Some work has been done on microbial community shifts, and their relationship to increasing calcium absorption has been done as described in this chapter. Metagenomics are yet to be performed to explore pathways that are perturbed which could explain the increases in mineral absorption. Involving carbohydrate chemists to design optimal substrates for preferred microbiota would enhance prebiotic food and supplements available to the consumer. Understanding individual differences in fiber fermentation could help predict responders. Understanding the effects of metabolites such as SCFA on epigenetics and metabalomics and their effect on bone health are future areas of research. The future is bright for interdisciplinary research in this area.

References

1. de Vos WM, de Vos EAJ. Role of the intestinal microbiome in health and disease: from correlation to causation. Nutr Rev. 2012;70(Suppl 1):S45–56. doi:10.1111/j.1753--4887.2012.00505.x. ADDIN Mendeley Bibliography CSL_BIBLIOGRAPHY
2. Ly NP, Litonjua A, Gold DR, Celedón JC. Gut microbiota, probiotics, and vitamin D: interrelated exposures influencing allergy, asthma, and obesity? J Allergy Clin Immunol. 2011;127:1087–1094.–6. doi:10.1016/j.jaci.2011.02.015.
3. Sanz Y. Gut microbiota and probiotics in maternal and infant health. Am J Clin Nutr. 2011;94:2000S–5S. doi:10.3945/ajcn.110.001172.
4. Flint HJ. The impact of nutrition on the human microbiome. Nutr Rev. 2012;70(Suppl 1):S10–3. doi:10.1111/j.1753-4887.2012.00499.x.
5. Clemente JC, Ursell LK, Parfrey LW, Knight R. The impact of the gut microbiota on human health: an integrative view. Cell. 2012;148:1258–70. doi:10.1016/j.cell.2012.01.035.
6. Hooper LV, Littman DR, Macpherson AJ. Interactions between the microbiota and the immune system. Science. 2012;336:1268–73. doi:10.1126/science.1223490.
7. Gibson GR, Scott KP, Rastall RA, Tuohy KM, Hotchkiss A, Dubert-Ferrandon A, et al. Dietary prebiotics: current status and new definition. Food Sci Technol Bull Funct Foods. n.d. 2010;7:1–19. doi:10.1616/1476-2137.
8. Roberfroid M, Gibson GR, Hoyles L, McCartney AL, Rastall R, Rowland I, et al. Prebiotic effects: metabolic and health benefits. Br J Nutr. 2010;104(Suppl):S1–63. doi:10.1017/S0007114510003363.

9. O'Hara AM, Shanahan F. The gut flora as a forgotten organ. EMBO Rep. 2006;7:688–93. doi:10.1038/sj.embor.7400731.
10. Qin J, Li R, Raes J, Arumugam M, Burgdorf KS, Manichanh C, et al. A human gut microbial gene catalogue established by metagenomic sequencing. Nature. 2010;464:59–65. doi:10.1038/nature08821.
11. Ley RE, Peterson DA, Gordon JI. Ecological and evolutionary forces shaping microbial diversity in the human intestine. Cell. 2006;124:837–48. doi:10.1016/j.cell.2006.02.017.
12. Zheng X, Xie G, Zhao A, Zhao L, Yao C, Chiu NHL, et al. The footprints of gut microbial-mammalian co-metabolism. J Proteome Res. 2011;10:5512–22. doi:10.1021/pr2007945.
13. Sjögren K, Engdahl C, Henning P, Lerner UH, Tremaroli V, Lagerquist MK, et al. The gut microbiota regulates bone mass in mice. J Bone Miner Res. 2012;27:1357–67. doi:10.1002/jbmr.1588.
14. Macpherson AJ, Harris NL. Interactions between commensal intestinal bacteria and the immune system. Nat Rev Immunol. 2004;4:478–85. doi:10.1038/nri1373.
15. McCabe L, Britton RA, Parameswaran N. Prebiotic and probiotic regulation of bone health: role of the intestine and its microbiome. Curr Osteoporos Rep. 2015;13:363–71. doi:10.1007/s11914-015-0292-x.
16. Kobayashi K, Takahashi N, Jimi E, Udagawa N, Takami M, Kotake S, et al. Tumor necrosis factor alpha stimulates osteoclast differentiation by a mechanism independent of the ODF/RANKL-RANK interaction. J Exp Med. 2000;191:275–86.
17. Kudo O, Fujikawa Y, Itonaga I, Sabokbar A, Torisu T, Athanasou NA. Proinflammatory cytokine (TNFalpha/IL-1alpha) induction of human osteoclast formation. J Pathol. 2002;198:220–7. doi:10.1002/path.1190.
18. Weitzmann MN, Pacifici R. T cells: unexpected players in the bone loss induced by estrogen deficiency and in basal bone homeostasis. Ann N Y Acad Sci. 2007;1116:360–75. doi:10.1196/annals.1402.068.
19. Irwin R, Lee T, Young VB, Parameswaran N, McCabe LR. Colitis-induced bone loss is gender dependent and associated with increased inflammation. Inflamm Bowel Dis. 2013;19:1586–97. doi:10.1097/MIB.0b013e318289e17b.
20. Gao Y, Grassi F, Ryan MR, Terauchi M, Page K, Yang X, et al. IFN-gamma stimulates osteoclast formation and bone loss in vivo via antigen-driven T cell activation. J Clin Invest. 2007;117:122–32. doi:10.1172/JCI30074.
21. Rosen CJ. Breaking into bone biology: serotonin's secrets. Nat Med. 2009;15:145–6. doi:10.1038/nm0209-145.
22. Yadav VK, Ryu J-H, Suda N, Tanaka KF, Gingrich JA, Schütz G, et al. Lrp5 controls bone formation by inhibiting serotonin synthesis in the duodenum. Cell. 2008;135:825–37. doi:10.1016/j.cell.2008.09.059.
23. Nam S-S, Lee JC, Kim HJ, Park J-W, Lee J-M, Suh J-Y, et al. Serotonin inhibits osteoblast differentiation and bone regeneration in rats. J Periodontol. 2016;87:461–9. doi:10.1902/jop.2015.150302.
24. Warden SJ, Robling AG, Sanders MS, Bliziotes MM, Turner CH. Inhibition of the serotonin (5-hydroxytryptamine) transporter reduces bone accrual during growth. Endocrinology. 2005;146:685–93. doi:10.1210/en.2004-1259.
25. Eom C-S, Lee H-K, Ye S, Park SM, Cho K-H. Use of selective serotonin reuptake inhibitors and risk of fracture: a systematic review and meta-analysis. J Bone Miner Res. 2012;27:1186–95. doi:10.1002/jbmr.1554.
26. Yun H-M, Park K-R, Hong JT, Kim E-C, Mann JJ, Cote F, et al. Peripheral serotonin-mediated system suppresses bone development and regeneration via serotonin 6 G-protein-coupled receptor. Sci Rep. 2016;6:30985. doi:10.1038/srep30985.
27. Yadav VK, Balaji S, Suresh PS, Liu XS, Lu X, Li Z, et al. Pharmacological inhibition of gut-derived serotonin synthesis is a potential bone anabolic treatment for osteoporosis. Nat Med. 2010;16:308–12. doi:10.1038/nm.2098.
28. Manolagas SC. From estrogen-centric to aging and oxidative stress: a revised perspective of the pathogenesis of osteoporosis. Endocr Rev. 2010;31:266–300. doi:10.1210/er.2009-0024.

29. Weitzmann MN, Pacifici R, Bouxsein ML, Komm BS, Tomkinson A, Gevers EF, et al. Estrogen deficiency and bone loss: an inflammatory tale. J Clin Invest. 2006;116:1186–94. doi:10.1172/JCI28550.
30. Yurkovetskiy L, Burrows M, Khan AA, Graham L, Volchkov P, Becker L, et al. Gender bias in autoimmunity is influenced by microbiota. Immunity. 2013;39:400–12. doi:10.1016/j.immuni.2013.08.013.
31. Gibson GR, Beatty ER, Wang X, Cummings JH. Selective stimulation of bifidobacteria in the human colon by oligofructose and inulin. Gastroenterology. 1995;108:975–82.
32. Yazawa K, Imai K, Tamura Z. Oligosaccharides and polysaccharides specifically utilizable by bifidobacteria. Chem Pharm Bull (Tokyo). 1978;26:3306–11.
33. Mao B, Li D, Zhao J, Liu X, Gu Z, Chen YQ, et al. Metagenomic insights into the effects of fructo-oligosaccharides (FOS) on the composition of fecal microbiota in mice. J Agric Food Chem. 2015;63:856–63. doi:10.1021/jf505156h.
34. Minami Y, Yazawa K, Nakamura K, Tamura Z. Selectivity and efficiency of utilization of galactosyl-oligosaccharides by bifidobacteria. Chem Pharm Bull (Tokyo). 1985;33:710–4.
35. Barile D, Rastall RA. Human milk and related oligosaccharides as prebiotics. Curr Opin Biotechnol. 2013;24:214–9. doi:10.1016/j.copbio.2013.01.008.
36. Vulevic J, Juric A, Tzortzis G, Gibson GR. A mixture of trans-galactooligosaccharides reduces markers of metabolic syndrome and modulates the fecal microbiota and immune function of overweight adults. J Nutr. 2013;143:324–31. doi:10.3945/jn.112.166132.
37. Vulevic J, Juric A, Walton GE, Claus SP, Tzortzis G, Toward RE, et al. Influence of galacto-oligosaccharide mixture (B-GOS) on gut microbiota, immune parameters and metabonomics in elderly persons. Br J Nutr. 2015;114:586–95. doi:10.1017/S0007114515001889.
38. Slavin J. Fiber and prebiotics: mechanisms and health benefits. Forum Nutr. 2013;5:1417–35. doi:10.3390/nu5041417.
39. Sabater-Molina M, Larqué E, Torrella F, Zamora S. Dietary fructooligosaccharides and potential benefits on health. J Physiol Biochem. 2009;65:315–28. doi:10.1007/BF03180584.
40. Bouhnik Y, Raskine L, Simoneau G, Vicaut E, Neut C, Flourié B, et al. The capacity of non-digestible carbohydrates to stimulate fecal bifidobacteria in healthy humans: a double-blind, randomized, placebo-controlled, parallel-group, dose-response relation study. Am J Clin Nutr. 2004;80:1658–64.
41. Bode L. Recent advances on structure, metabolism, and function of human milk oligosaccharides. J Nutr. 2006;136:2127–30.
42. Boehm G, Moro G. Structural and functional aspects of prebiotics used in infant nutrition. J Nutr. 2008;138:1818S–28S.
43. Coppa GV, Zampini L, Galeazzi T, Gabrielli O. Prebiotics in human milk: a review. Dig Liver Dis. 2006;38(Suppl 2):S291–4. doi:10.1016/S1590-8658(07)60013-9.
44. Fanaro S, Boehm G, Garssen J, Knol J, Mosca F, Stahl B, et al. Galacto-oligosaccharides and long-chain fructo-oligosaccharides as prebiotics in infant formulas: a review. Acta Paediatr. 2007;94:22–6. doi:10.1111/j.1651-2227.2005.tb02150.x.
45. Torres DPM, Gonçalves M, do Pilar F, Teixeira JA, Rodrigues LR. Galacto-oligosaccharides: production, properties, applications, and significance as prebiotics. Compr Rev Food Sci Food Saf. 2010;9:438–54. doi:10.1111/j.1541-4337.2010.00119.x.
46. Hernandez-Hernandez O, Marin-Manzano MC, Rubio LA, Moreno FJ, Sanz ML, Clemente A. Monomer and linkage type of galacto-oligosaccharides affect their resistance to ileal digestion and prebiotic properties in rats. J Nutr. 2012;142:1232–9. doi:10.3945/jn.111.155762.
47. van Leeuwen SS, Kuipers BJH, Dijkhuizen L, Kamerling JP. Comparative structural characterization of 7 commercial galacto-oligosaccharide (GOS) products. Carbohydr Res. 2016;425:48–58. doi:10.1016/j.carres.2016.03.006.
48. Chonan O, Matsumoto K, Watanuki M. Effect of galactooligosaccharides on calcium absorption and preventing bone loss in ovariectomized rats. Biosci Biotechnol Biochem. 1995;59:236–9. doi:10.1271/bbb.59.236.
49. van den Heuvel E, Schoterman M, Muijs T. Trans-galactooligosaccharides stimulate calcium absorption in postmenopausal women. J Nutr. 2000;130:2938–42.

50. Schaafsma G, Visser WJ, Dekker PR, Van Schaik M. Effect of dietary calcium supplementation with lactose on bone in vitamin D-deficient rats. Bone. 1987;8:357–62. doi:10.1016/8756-3282(87)90067-6.
51. Griessen M, Cochet B, Infante F, Jung A, Bartholdi P, Donath A, et al. Calcium absorption from milk in lactase-deficient subjects. Am J Clin Nutr. 1989;49:377–84.
52. Cochet B, Jung A, Griessen M, Bartholdi P, Schaller P, Alfred D, et al. Effects of lactose on intestinal calcium absorption in normal and lactase-deficient subjects. Gastroenterology. 1960;84:935–40. doi:10.5555/URI:PII:0016508583901944.
53. Tremaine WJ, Newcomer AD, Riggs BL, McGill DB. Calcium absorption from milk in lactase-deficient and lactase-sufficient adults. Dig Dis Sci. 1986;31:376–8.
54. Mizota T, Tamura Y, Tomita M, Okonogi S. Lactulose as a sugar with physiological significance (activity as a Bifidus Factor). Bull – Fed Int Lait (Belgium) Int Dairy Fed No 212 1987. n.d..
55. Seki N, Hamano H, Iiyama Y, Asano Y, Kokubo S, Yamauchi K, et al. Effect of lactulose on calcium and magnesium absorption: a study using stable isotopes in adult men. J Nutr Sci Vitaminol (Tokyo). 2007;53:5–12.
56. Brommage R, Binacua C, Antille S, Carrié AL. Intestinal calcium absorption in rats is stimulated by dietary lactulose and other resistant sugars. J Nutr. 1993;123:2186–94.
57. van den Heuvel EGHM, Muijs T, van Dokkum W, Schaafsma G. Lactulose stimulates calcium absorption in postmenopausal women. J Bone Miner Res. 1999;14:1211–6. doi:10.1359/jbmr.1999.14.7.1211.
58. Housez B, Cazaubiel M, Vergara C, Bard J-M, Adam A, Einerhand A, et al. Evaluation of digestive tolerance of a soluble corn fibre. J Hum Nutr Diet. 2012;25:488–96. doi:10.1111/j.1365-277X.2012.01252.x.
59. Tate & Lyle. Soluble corn fiber: health benefits and product applications 2014. http://www.foodnutritionknowledge.info/. Accessed 30 Aug 2016.
60. Costabile A, Deaville ER, Morales AM, Gibson GR, Huttenhower C, Ding T, et al. Prebiotic potential of a maize-based soluble fibre and impact of dose on the human gut microbiota. PLoS One. 2016;11:e0144457. doi:10.1371/journal.pone.0144457.
61. Weaver CM, Martin BR, Story JA, Hutchinson I, Sanders L. Novel fibers increase bone calcium content and strength beyond efficiency of large intestine fermentation. J Agric Food Chem. 2010;58:8952–7. doi:10.1021/jf904086d.
62. Whisner CM, Martin BR, Nakatsu CH, McCabe GP, McCabe LD, Peacock M, et al. Soluble maize fibre affects short-term calcium absorption in adolescent boys and girls: a randomised controlled trial using dual stable isotopic tracers. Br J Nutr. 2014;112:446–56. doi:10.1017/S0007114514000981.
63. Whisner CM, Martin BR, Nakatsu CH, Story JA, MacDonald-Clarke CJ, McCabe LD, et al. Soluble corn fiber increases calcium absorption associated with shifts in the gut microbiome: a randomized dose-response trial in free-living pubertal females. J Nutr. 2016;146:1298–306. doi:10.3945/jn.115.227256.
64. Jakeman SA, Henry CN, Martin BR, McCabe GP, McCabe LD, Jackson GS, et al. Soluble corn fiber increases bone calcium retention in postmenopausal women in a dose-dependent manner: a randomized crossover trial. Am J Clin Nutr. 2016; doi:10.3945/ajcn.116.132761.
65. Coxam V. Inulin-type fructans and bone health: state of the art and perspectives in the management of osteoporosis. Br J Nutr. 2005;93(Suppl 1):S111–23.
66. Abrams SA, Griffin IJ, Hawthorne KM, Liang L, Gunn SK, Darlington G, et al. A combination of prebiotic short- and long-chain inulin-type fructans enhances calcium absorption and bone mineralization in young adolescents. Am J Clin Nutr. 2005;82:471–6.
67. Abrams SA, Griffin IJ, Hawthorne KM. Young adolescents who respond to an inulin-type fructan substantially increase total absorbed calcium and daily calcium accretion to the skeleton. J Nutr. 2007;137:2524S–6S.
68. Abrams SA, Hawthorne KM, Aliu O, Hicks PD, Chen Z, Griffin IJ. An inulin-type fructan enhances calcium absorption primarily via an effect on colonic absorption in humans. J Nutr. 2007;137:2208–12.

69. Griffin IJ, Davila PM, Abrams SA. Non-digestible oligosaccharides and calcium absorption in girls with adequate calcium intakes. Br J Nutr. 2002;87(Suppl 2):S187–91. doi:10.1079/BJNBJN/2002536.
70. Griffin IJ, Hicks PM, Heaney RP, Abrams SA, Roberfroid MB, Griffin IJ, et al. Enriched chicory inulin increases calcium absorption mainly in girls with lower calcium absorption. Nutr Res. 2003;23:901–9. doi:10.1016/S0271-5317(03)00085-X.
71. Holloway L, Moynihan S, Abrams SA, Kent K, Hsu AR, Friedlander AL. Effects of oligofructose-enriched inulin on intestinal absorption of calcium and magnesium and bone turnover markers in postmenopausal women. Br J Nutr. 2007;97:365–72. doi:10.1017/S000711450733674X.
72. Bryk G, Coronel MZ, Pellegrini G, Mandalunis P, Rio ME, de Portela MLPM, et al. Effect of a combination GOS/FOS® prebiotic mixture and interaction with calcium intake on mineral absorption and bone parameters in growing rats. Eur J Nutr. 2015;54:913–23. doi:10.1007/s00394-014-0768-y.
73. Ohta A, Uehara M, Sakai K, Takasaki M, Adlercreutz H, Morohashi T, et al. A combination of dietary fructooligosaccharides and isoflavone conjugates increases femoral bone mineral density and equol production in ovariectomized mice. J Nutr. 2002;132:2048–54.
74. Hooshmand S, Juma S, Arjmandi BH. Combination of genistin and fructooligosaccharides prevents bone loss in ovarian hormone deficiency. J Med Food. 2010;13:320–5. doi:10.1089/jmf.2009.0059.
75. Legette LL, Lee W-H, Martin BR, Story JA, Arabshahi A, Barnes S, et al. Genistein, a phytoestrogen, improves total cholesterol, and synergy, a prebiotic, improves calcium utilization, but there were no synergistic effects. Menopause. 2011;18:923–31. doi:10.1097/gme.0b013e3182116e81.
76. Zafar TA, Weaver CM, Jones K, Moore DR, Barnes S. Inulin effects on bioavailability of soy isoflavones and their calcium absorption enhancing ability. J Agric Food Chem. 2004;52:2827–31. doi:10.1021/jf035080f.
77. Devareddy L, Khalil DA, Korlagunta K, Hooshmand S, Bellmer DD, Arjmandi BH. The effects of fructo-oligosaccharides in combination with soy protein on bone in osteopenic ovariectomized rats. Menopause. 2006;13:692–9. doi:10.1097/01.gme.0000195372.74944.71.
78. Arjmandi BH, Johnson CD, Campbell SC, Hooshmand S, Chai SC, Akhter MP. Combining fructooligosaccharide and dried plum has the greatest effect on restoring bone mineral density among select functional foods and bioactive compounds. J Med Food. 2010;13:312–9. doi:10.1089/jmf.2009.0068.
79. Rodrigues FC, Castro ASB, Rodrigues VC, Fernandes SA, Fontes EAF, de Oliveira TT, et al. Yacon flour and Bifidobacterium longum modulate bone health in rats. J Med Food. 2012;15:664–70. doi:10.1089/jmf.2011.0296.
80. Pérez-Conesa D, López G, Abellán P, Ros G. Bioavailability of calcium, magnesium and phosphorus in rats fed probiotic, prebiotic and synbiotic powder follow-up infant formulas and their effect on physiological and nutritional parameters. J Sci Food Agric. 2006;86:2327–36. doi:10.1002/jsfa.2618.
81. Pérez-Conesa D, López G, Ros G. Effects of probiotic, prebiotic and synbiotic follow-up infant formulas on large intestine morphology and bone mineralisation in rats. J Sci Food Agric. 2007;87:1059–68. doi:10.1002/jsfa.2812.
82. Igarashi M, Iiyama Y, Kato R, Tomita M, Omi N, Ezawa I. Effect of Bifidobacterium longum and lactulose on the strength of bone in ovariectomized osteoporosis model rats. Bifidus Flores Fruct Semin. 1994;7:139–47.
83. Weaver CM. Diet, gut microbiome, and bone health. Curr Osteoporos Rep. 2015;13:125–30. doi:10.1007/s11914-015-0257-0.
84. Park C, Weaver C. Calcium and bone health: influence of prebiotics. Funct Food Rev. 2011;3:62–72.
85. Whisner C, Weaver C. Interactions of probiotics and prebiotics with minerals. In: Ötleş S, editor. Probiotics prebiotics food. Nutr. Boca Raton, FL Heal. CRC Press; 2013. p. 200–31.

86. Ohta A, Motohashi Y, Sakai K, Hirayama M, Adachi T, Sakuma K. Dietary fructooligosaccharides increase calcium absorption and levels of mucosal calbindin-D9k in the large intestine of gastrectomized rats. Scand J Gastroenterol. 1998;33:1062–8.
87. Ohta A, Ohtsuki M, Baba S, Adachi T, Sakata T, Sakaguchi E. Calcium and magnesium absorption from the colon and rectum are increased in rats fed fructooligosaccharides. J Nutr. 1995;125:2417–24.
88. Chonan O, Watanuki M. The effect of 6′-galactooligosaccharides on bone mineralization of rats adapted to different levels of dietary calcium. Int J Vitam Nutr Res Int Z Vitam Ernährungsforschung J Int Vitaminol Nutr. 1996;66:244–9.
89. Lobo AR, Colli C, Filisetti TMCC. Fructooligosaccharides improve bone mass and biomechanical properties in rats. Nutr Res. 2006;26:413–20. doi:10.1016/j.nutres.2006.06.019.
90. Wang Y, Zeng T, Wang S, Wang W, Wang Q, Yu H-X. Fructo-oligosaccharides enhance the mineral absorption and counteract the adverse effects of phytic acid in mice. Nutrition. 2010;26:305–11. doi:10.1016/j.nut.2009.04.014.
91. Weaver CM, Martin BR, Nakatsu CH, Armstrong AP, Clavijo A, McCabe LD, et al. Galactooligosaccharides improve mineral absorption and bone properties in growing rats through gut fermentation. J Agric Food Chem. 2011;59:6501–10. doi:10.1021/jf2009777.
92. Legette LL, Lee W, Martin BR, Story JA, Campbell JK, Weaver CM. Prebiotics enhance magnesium absorption and inulin-based fibers exert chronic effects on calcium utilization in a postmenopausal rodent model. J Food Sci. 2012;77:H88–94. doi:10.1111/j.1750-3841.2011.02612.x.
93. Ohta A, Ohtuki M, Takizawa T, Inaba H, Adachi T, Kimura S. Effects of fructooligosaccharides on the absorption of magnesium and calcium by cecectomized rats. Int J Vitam Nutr Res Int Z Vitam Ernährungsforschung J Int Vitaminol Nutr. 1994;64:316–23.
94. Delzenne N, Aertssens J, Verplaetse H, Roccaro M, Roberfroid M. Effect of fermentable fructo-oligosaccharides on mineral, nitrogen and energy digestive balance in the rat. Life Sci. 1995;57:1579–87.
95. Lopez HW, Coudray C, Levrat-Verny MA, Feillet-Coudray C, Demigné C, Rémésy C. Fructooligosaccharides enhance mineral apparent absorption and counteract the deleterious effects of phytic acid on mineral homeostasis in rats. J Nutr Biochem. 2000;11:500–8.
96. Rondón LJ, Rayssiguier Y, Mazur A. Dietary inulin in mice stimulates Mg2+ absorption and modulates TRPM6 and TRPM7 expression in large intestine and kidney. Magnes Res. 2008;21:224–31.
97. Coudray C, Feillet-Coudray C, Gueux E, Mazur A, Rayssiguier Y. Dietary inulin intake and age can affect intestinal absorption of zinc and copper in rats. J Nutr. 2006;136:117–22.
98. Roberfroid MB, Cumps J, Devogelaer JP. Dietary chicory inulin increases whole-body bone mineral density in growing male rats. J Nutr. 2002;132:3599–602.
99. García-Vieyra MI, Del Real A, López MG. Agave fructans: their effect on mineral absorption and bone mineral content. J Med Food. 2014;17:1247–55. doi:10.1089/jmf.2013.0137.
100. Levrat MA, Rémésy C, Demigné C. High propionic acid fermentations and mineral accumulation in the cecum of rats adapted to different levels of inulin. J Nutr. 1991;121:1730–7.
101. Coudray C, Rambeau M, Feillet-Coudray C, Tressol JC, Demigne C, Gueux E, et al. Dietary inulin intake and age can significantly affect intestinal absorption of calcium and magnesium in rats: a stable isotope approach. Nutr J. 2005;4:29. doi:10.1186/1475-2891-4-29.
102. Coudray C, Tressol JC, Gueux E, Rayssiguier Y. Effects of inulin-type fructans of different chain length and type of branching on intestinal absorption and balance of calcium and magnesium in rats. Eur J Nutr. 2003;42:91–8. doi:10.1007/s00394-003-0390-x.
103. Zafar TA, Weaver CM, Zhao Y, Martin BR, Wastney ME. Nondigestible oligosaccharides increase calcium absorption and suppress bone resorption in ovariectomized rats. J Nutr. 2004;134:399–402.
104. Mitamura R, Hara H. Ingestion of difructose anhydride III partially restores calcium absorption impaired by vitamin D and estrogen deficiency in rats. Eur J Nutr. 2006;45:242–9. doi:10.1007/s00394-006-0592-0.

105. Scholz-Ahrens KE, Açil Y, Schrezenmeir J. Effect of oligofructose or dietary calcium on repeated calcium and phosphorus balances, bone mineralization and trabecular structure in ovariectomized rats*. Br J Nutr. 2002;88:365–77. doi:10.1079/BJN2002661.
106. Morohashi T, Sano T, Ohta A, Yamada S. True calcium absorption in the intestine is enhanced by fructooligosaccharide feeding in rats. J Nutr. 1998;128:1815–8.
107. Coudray C, Feillet-Coudray C, Tressol JC, Gueux E, Thien S, Jaffrelo L, et al. Stimulatory effect of inulin on intestinal absorption of calcium and magnesium in rats is modulated by dietary calcium intakes short and long-term balance studies. Eur J Nutr. 2005;44:293–302. doi:10.1007/s00394-004-0526-7.
108. Nzeusseu A, Dienst D, Haufroid V, Depresseux G, Devogelaer J-P, Manicourt D-H. Inulin and fructo-oligosaccharides differ in their ability to enhance the density of cancellous and cortical bone in the axial and peripheral skeleton of growing rats. Bone. 2006;38:394–9. doi:10.1016/j.bone.2005.09.006.
109. Shiga K, Nishimukai M, Tomita F, Hara H. Ingestion of difructose anhydride III, a non-digestible disaccharide, improves postgastrectomy osteopenia in rats. Scand J Gastroenterol. 2006;41:1165–73. doi:10.1080/00365520600575753.
110. Lobo AR, Filho JM, Alvares EP, Cocato ML, Colli C. Effects of dietary lipid composition and inulin-type fructans on mineral bioavailability in growing rats. Nutrition. 2009;25:216–25. doi:10.1016/j.nut.2008.08.002.
111. Takahara S, Morohashi T, Sano T, Ohta A, Yamada S, Sasa R. Fructooligosaccharide consumption enhances femoral bone volume and mineral concentrations in rats. J Nutr. 2000;130:1792–5.
112. Demigné C, Jacobs H, Moundras C, Davicco M-J, Horcajada M-N, Bernalier A, et al. Comparison of native or reformulated chicory fructans, or non-purified chicory, on rat cecal fermentation and mineral metabolism. Eur J Nutr. 2008;47:366–74. doi:10.1007/s00394-008-0736-5.
113. Mathey J, Puel C, Kati-Coulibaly S, Bennetau-Pelissero C, Davicco MJ, Lebecque P, et al. Fructooligosaccharides maximize bone-sparing effects of soy isoflavone-enriched diet in the ovariectomized rat. Calcif Tissue Int. 2004;75:169–79. doi:10.1007/s00223-004-0128-7.
114. van den Heuvel EGHM, Muys T, van Dokkum W, Schaafsma G. Oligofructose stimulates calcium absorption in adolescents. Am J Clin Nutr. 1999;69:544–8.
115. van den Heuvel EGHM, Schaafsma G, Muys T, van Dokkum W. Nondigestible oligosaccharides do not interfere with calcium and nonheme-iron absorption in young, healthy men. Am J Clin Nutr. 1998;67:445–51.
116. Whisner CM, Martin BR, Schoterman MHC, Nakatsu CH, McCabe LD, McCabe GP, et al. Galacto-oligosaccharides increase calcium absorption and gut bifidobacteria in young girls: a double-blind cross-over trial. Br J Nutr. 2013;110:1292–303. doi:10.1017/S000711451300055X.
117. van den Heuvel EGHM, Muijs T, Brouns F, Hendriks HFJ. Short-chain fructo-oligosaccharides improve magnesium absorption in adolescent girls with a low calcium intake. Nutr Res. 2009;29:229–37. doi:10.1016/j.nutres.2009.03.005.
118. Martin BR, Braun MM, Wigertz K, Bryant R, Zhao Y, Lee W, et al. Fructo-oligosaccharides and calcium absorption and retention in adolescent girls. J Am Coll Nutr. 2010;29:382–6.
119. Tahiri M, Tressol JC, Arnaud J, Bornet FRJ, Bouteloup-Demange C, Feillet-Coudray C, et al. Effect of short-chain fructooligosaccharides on intestinal calcium absorption and calcium status in postmenopausal women: a stable-isotope study. Am J Clin Nutr. 2003;77:449–57.
120. Yap KW, Mohamed S, Yazid AM, Maznah I, Meyer DM. Dose-response effects of inulin on the faecal short-chain fatty acids content and mineral absorption of formula-fed infants. Nutr Food Sci. 2005;35: 208–19. doi: 10.1080/1040 8399509527714
121. Mineo H, Hara H, Tomita F. Short-chain fatty acids enhance diffusional ca transport in the epithelium of the rat cecum and colon. Life Sci. 2001;69:517–26.
122. Heijnen AM, Brink EJ, Lemmens AG, Beynen AC. Ileal pH and apparent absorption of magnesium in rats fed on diets containing either lactose or lactulose. Br J Nutr. 1993;70:747–56.

123. Fanaro S, Marten B, Bagna R, Vigi V, Fabris C, Peña-Quintana L, et al. Galacto-oligosaccharides are bifidogenic and safe at weaning: a double-blind randomized multicenter study. J Pediatr Gastroenterol Nutr. 2009;48:82–8. doi:10.1097/MPG.0b013e31817b6dd2.
124. Kim YY (Kyung HUSR of K, Jang KH (Samcheok NUSR of K, Kang SA (Konkuk USR of K, Ha WK (Pasteur MC. L. HR of K, Lee EY (Kyung HUSR of K, Cho YH (Kyung HUSR of K, et al. The effect of chicory fructan fiber on calcium absorption and bone metabolism in Korean postmenopausal women. Nutr Sci. 2004.
125. Slevin MM, Allsopp PJ, Magee PJ, Bonham MP, Naughton VR, Strain JJ, et al. Supplementation with calcium and short-chain fructo-oligosaccharides affects markers of bone turnover but not bone mineral density in postmenopausal women. J Nutr. 2014;144:297–304. doi:10.3945/jn.113.188144.
126. Donohoe DR, Garge N, Zhang X, Sun W, O'Connell TM, Bunger MK, et al. The microbiome and butyrate regulate energy metabolism and autophagy in the mammalian colon. Cell Metab. 2011;13:517–26. doi:10.1016/j.cmet.2011.02.018.
127. Li C-J, Li RW, Elsasser TH. MicroRNA (miRNA) expression is regulated by butyrate-induced epigenetic modulation of gene expression in bovine cells. Genet Epigenet. 2010;3:23–32.
128. Henagan T, Navard A, Ye J. Sodium butyrate remodels whole genome nucleosome maps and attenuates high fat diet-induced mitochondrial dysfunction in skeletal muscle from C57BL6/J mice (1072.1). FASEB J. 2014;28:1072.1.
129. Raschka L, Daniel H. Mechanisms underlying the effects of inulin-type fructans on calcium absorption in the large intestine of rats. Bone. 2005;37:728–35. doi:10.1016/j.bone.2005.05.015.
130. Scheppach W, Bartram P, Richter A, Richter F, Liepold H, Dusel G, et al. Effect of short-chain fatty acids on the human colonic mucosa in vitro. JPEN J Parenter Enteral Nutr. n.d. 1992;16:43–8.
131. Fukushima A, Ohta A, Sakai K, Sakuma K. Expression of calbindin-D9k, VDR and Cdx-2 messenger RNA in the process by which fructooligosaccharides increase calcium absorption in rats. J Nutr Sci Vitaminol (Tokyo). 2005;51:426–32. doi:10.3177/jnsv.51.426.
132. Clark EM, Ness AR, Bishop NJ, Tobias JH. Association between bone mass and fractures in children: a prospective cohort study. J Bone Miner Res. 2006;21:1489–95. doi:10.1359/jbmr.060601.
133. Weaver CM, Gordon CM, Janz KF, Kalkwarf HJ, Lappe JM, Lewis R, et al. The National Osteoporosis Foundation's position statement on peak bone mass development and lifestyle factors: a systematic review and implementation recommendations. Osteoporos Int. 2016;27:1281–386. doi:10.1007/s00198-015-3440-3.
134. Scientific Report of the 2015 Dietary Guidelines Advisory Committee. n.d.. http://www.health.gov/dietaryguidelines/2015-scientific-report/. Accessed 23 July 2015.

Probiotics in Gut-Bone Signaling

Jonathan D. Schepper, Regina Irwin, Jun Kang, Kevin Dagenais, Tristan Lemon, Ally Shinouskis, Narayanan Parameswaran, and Laura R. McCabe

Introduction

Each year more than two million fractures occur because of osteoporosis [1]. Numerous therapies have been developed for the prevention and treatment of osteoporosis. As a first approach, patients are asked to make changes to their lifestyle (i.e., exercise, cessation of smoking) and diet (including vitamin D and calcium supplementation) [2]. For patients at a higher risk of fractures, pharmacologic treatments (drugs and biologics) are used to inhibit bone resorption or stimulate bone formation [3]. Despite the many treatment options, we have yet to stop the increase in osteoporosis fractures. This may be in part due to patient concerns about side effects (although rare) from many pharmaceutical/drug-based therapies [4]. Given that 67 million Americans are predicted to have low bone mass by 2020, it is important to continue to identify additional therapeutic approaches/targets for osteoporosis.

One therapeutic target receiving increasing attention is the intestinal microbiome, which is an important regulator of physiologic functions of many organs including bone. The intestinal microbiota accounts for 90% of the cells in our body

J.D. Schepper • R. Irwin • J. Kang • K. Dagenais • T. Lemon • A. Shinouskis
Department of Physiology, Michigan State University, East Lansing, MI, USA
e-mail: narap@msu.edu; mccabel@msu.edu

N. Parameswaran
Comparative Medicine and Integrative Biology Program, East Lansing, MI, USA

Department of Physiology, Michigan State University, East Lansing, MI, USA

L.R. McCabe
Department of Physiology and Department of Radiology, Biomedical Imaging Research Centre, Michigan State University, East Lansing, MI, USA
e-mail: mccabel@msu.edu

L.R. McCabe, N. Parameswaran (eds.), *Understanding the Gut-Bone Signaling Axis*, Advances in Experimental Medicine and Biology 1033,
DOI 10.1007/978-3-319-66653-2_11

and amounts to ~100 trillion microbes comprising ~1,000 species and 28 different phyla [5]. In addition to outnumbering host cell number, the gut microbiota also express 100-fold more genes compared to the human genome [5]. As the microbiome coevolve with us, changes in its composition can consequently influence our human health [6]. Dysbiosis (a microbial imbalance) is linked to disease and bone loss; however, more importantly, the reverse is also true: treatment with probiotics can beneficially modulate the gut microbiota to enhance health, including that of bone [7–10]. In this review we will focus on (1) probiotics (definition, history, nomenclature, types), (2) the overall effects of probiotics on bone health, and (3) mechanisms of probiotic prevention of bone pathologies.

Probiotics

Probiotic: Defined

The word "probiotic" is derived from the Latin word "pro" and the Greek word "bios" meaning "for life;" this contrasts with "antibiotic" meaning "against life" [11–20]. While "good for life" is a general definition of probiotics, the detailed definition of what constitutes a probiotic has been difficult to achieve and has changed over time. In the 1950s, Werner Kollath, a German scientist, used the word "probiotic" to be inclusive of all organic and inorganic supplements that restored the health of malnourished patients [11, 12, 19, 20]. Years later, probiotics were further defined as substances produced by one microorganism to promote growth of another microorganism [11, 12, 16, 18–26]. In the 1970s, Fujii and Cook described probiotics as compounds that build resistance to infection in the host but do not inhibit the growth of microorganisms in vitro [11, 18, 27]. In the 1980s and 1990s, there was a surge of different probiotic definitions. For example, in 1990 Parker defined probiotics as organisms or substances in feed supplements which contribute to intestinal microbial balance [11, 14, 18, 19, 22, 24, 28]. Parker's general definition was unsatisfactory to many since the word "substances" included chemical supplements such as antibiotics [18, 28]. Most researchers cited the definition of Fuller, who, in 1989, defined probiotics as live microbial feed supplements [11, 18, 19, 22, 24, 25]. Fuller's definition stressed the importance of live cells as an essential part of the effective probiotic [18]. His definition also stated that a probiotic or supplement will benefit the host by improving the intestinal microbial balance [11, 26]. Many thought this definition was not as applicable to humans as it was to animals [11]. Subsequently, in the early 1990s, the definition was broadened to include viable mono or mixed cultures of live microorganisms which, when given to humans or animals, benefit the host by improving the properties of the indigenous microflora [29]. In the late 1990s, Salminen offered the view of incorporating nonviable bacteria in the probiotic definition [11, 28]. Finally, in 2001, after consultation of international scientists working on behalf of the FAO/WHO (Food and Agricultural Organization/World Health Organization), probiotics were proposed to be defined "as live microorganisms that when administered in adequate amounts will confer a health benefit on the host" [11,

15, 19, 21, 24, 30, 31]. Misuse of the probiotic term became a major problem in the ensuing years. For this reason, the International Scientific Association for Probiotics and Prebiotics (ISAPP) organized a meeting of clinical and scientific experts on probiotics in October 2013 to reexamine the concept and definition of probiotics [31]. The ISAPP panel recommended that the definition of probiotic as defined by FAO/WHO in 2001 is broad enough to enable a wide range of products to be developed and at the same time sufficiently narrow to impose some core requirements [24, 31]. Thus, probiotics are currently remain defined as live microorganisms that when administered in adequate amounts will confer a health benefit on the host.

The History of Probiotic Discovery

Probiotic use can be traced back over 10,000 years ago [32]. During the Neolithic period of the Stone Age, animal domestication and husbandry developed [20]. Ancient oriental people, as well as Phrygian, Sarmatian, and Macedonian nomadic shepherds, drank milk from cows, sheep, goats, horses, and camels. Traditional Egyptian fermented milk products (Laban Rayeb and Laban Khad) were consumed as early as 7000 BCE [11, 19, 32]. Both iconographic and written evidence from 3000 to 2000 BCE indicate that Hindi, Egyptians, Greeks, and Romans all used fermented milk products [11]. Fermenting milk was also evident in the Middle and Far East of Asia and spread throughout eastern Europe and Russia by the Tartars, Huns, and Mongols during their land conquests [11]. Fermented products other than milk, such as beer, bread, wine, kefir, kumis, and cheese, were also consumed [32] since fermentation increased their long-term storage [11, 19, 20].

The ancient Ayurvedic texts, written between 400 and 200 BCE, linked a long and healthy life with the intake of milk and dairy products [20]. To store the milk, it was customary to use containers made from animal skins or stomachs [19, 20]. The containers were a source of bacteria, most likely ancestors of *Lactobacillus acidophilus* and *Lactobacillus bulgaricus*, which came into contact with the milk [20]. One Turkish legend describes a shepherd, traveling the hot desert, who forgot he had milk in a goatskin bag. When he checked, the milk had transformed into a thick, creamy, and tasty custard; this new product was referred to as yogurt [20]. For the Turkish people, yogurt was the elixir of life, as they believed that this food gave physical and inner well-being and could prolong life [20].

The modern history of probiotics begins in the late nineteenth to early twentieth century. Elie Metchnikoff (a Nobel laureate), as well as Theodor Escherich, studied microbial communities in feces and described the need for a complex intestine (microbe-wise) [33]. Metchnikoff was a Kharkov/Ukrainian scientist working at the Pasteur Institute [19, 20]. Pasteur had identified the microorganisms responsible for fermentation, but it was Metchnikoff who investigated the effects these microbes had on human health [20]. Metchnikoff associated the longevity of Bulgarian rural people (who had an average lifespan of 87 years) to their regular consumption of fermented dairy products such as yogurt [19, 20, 24, 34]. Metchnikoff described two bacteria types: one that leads to putrefying luminal contents and produces unhealthy waste products (NH3, H2S, amines) and another that ferments luminal contents and produces beneficial metabolic products (i.e., lactic acid) [35]. This was a key concept because probiotic bacteria secrete enzymes that are not produced by human intestinal cells. These enzymes can ferment nondigestible poly-carbohydrates (mainly dietary fiber) to produce energy for the bacteria as well as other factors such as short-chain fatty acids (SCFA) and lactic acid which benefits the intestinal epithelium [36]. Metchnikoff theorized that the production of lactic acid would prevent the toxic effects of putrefying microbes. This further led Metchnikoff to suggest that lactobacilli may benefit gastrointestinal metabolism and counteract illness and aging [11, 20, 24]; thus, he considered lactobacilli a probiotic [20, 25, 37]. Thanks to Metchnikoff, the dairy industry began in France and subsequently spread throughout Europe, using fermented milk obtained from *Bacillus bulgaricus*, *Streptococcus thermophiles*, and *Lactobacillus delbrueckii* [19].

About the same time that Metchnikoff was making his discoveries of lactic acid-producing bacteria, French pediatrician Dr. Henry Tissier observed that children with diarrhea had a low number of "Y"-shaped bacteria in their stools [19, 24, 26]. Healthy children had an abundance of these bacteria. In 1905, he isolated the bacteria, *Bacillus bifidus*, and linked its presence in children to those who were breastfed [33]. He suggested these bacteria could be administered to patients with diarrhea to help restore their healthy flora (eubiosis) and used it to recolonize the gut of children [14, 19–21, 27, 28, 33]. As the health benefits of milk-associated bacteria became better known, fermented dairy products were appearing around the world. For example, in 1935 a Japanese microbiologist, Dr. Shirota, isolated *Lactobacillus casei* and added it to a dairy drink that was eventually marketed. Today, food products containing probiotics are usually dairy, mainly due to the historical association of lactic acid bacteria with fermented milk [11, 20, 32].

Probiotic Nomenclature and Types

Probiotics are widely consumed and have a long history of safe use. Bacteria names are derived from descriptors of the bacteria (i.e., *Lactobacillus*, "lacto" meaning "milk" and "bacillus" meaning "rod-shaped"), a scientist's name (i.e., *Pasteurella*, found by Louis Pasteur), the place where found (i.e., *Legionella longbeachae*, found in Long Beach California), or an organization (i.e., *Legionella* and the American Legion). In addition to a general name, the bacteria are described based on a taxonomic/genetic hierarchy [39]. Based on this system, bacteria are divided into phylum, class, order, family, genus, species and subspecies, and/or strain (Fig. 1). With more than 23 bacteria phyla, it is easy to see the abundance of specific probiotics and the complexity of their names. Current evidence indicates that the beneficial effect of probiotics are strain specific [22]. It is also important to note that not all bacteria within a species act the same and/or can be regarded as a probiotic. Below, we discuss several of the most notable probiotics including lactic acid bacteria, *Bifidobacteria* and *Enterococcus* (also see Table 1).

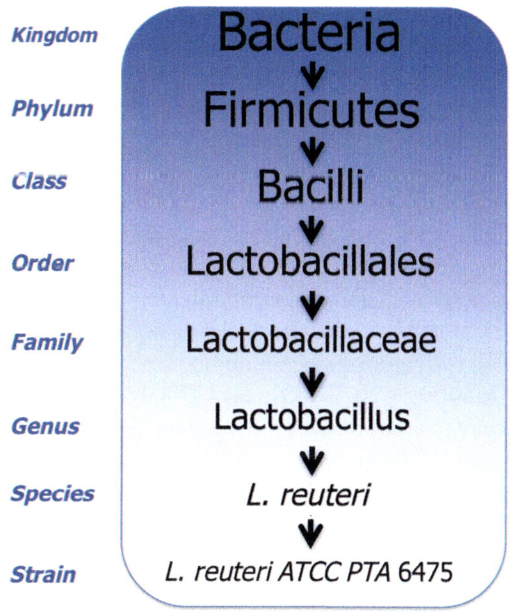

Fig. 1 Scientific nomenclature: an example of bacterial scientific nomenclature for the *Lactobacillus reuteri* ATCC PTA 6475 strain

Table 1 Common probiotic bacteria

Genus	Species	Genus	Species	Other
Lactobacillus	*acidophilus*	*Bifidobacterium*	*longum*	*Enterococcus faecalis*
	crispatus		bifidum	*Enterococcus faecium*
	johnsonii		*infantis*	*Lactococcus lactis*
	gasseri		*animalis*	*Escherichia coli (Nissle 1917)*
	casei		*adolescentis*	*Propionibacterium freudenreichii*
	rhamnosus		*lactis*	*Saccharomyces cerevisiae*
	reuteri		*breve*	*Streptococcus thermophilus*
	plantarum			*Bacillus cereus*
	fermentum			*Bacillus subtilis*
	salivarius			

Adapted from [40, 41]

Lactic Acid Bacteria/*Lactobacillales*

Lactic acid bacteria (also known as LAB) are one of the most important groups of bacteria/probiotics with health benefits that are thought to result in part from their production of lactic acid, their major fermentation product [11, 34, 42]. In general, they are gram-positive, acid-tolerant, asporogenous rods and cocci which are oxidase, catalase, and benzidine negative; they lack cytochromes, do not reduce nitrates to nitrite, are gelatinase negative, and are unable to utilize lactate [11, 38, 42]. Lactic acid bacteria obtained from fermented milk products have been used for centuries. Traditional fermented milk is a useful source of probiotics because it contains a complex composition of lactic acid bacterial species. In a recent study, 148 lactic acid bacterial strains were isolated from Kurut, a traditional naturally fermented yak milk from China [43]. Additional studies are evaluating these traditional fermented products as potential natural sources of probiotic bacteria [43].

Lactic acid bacteria, which consist of a diverse genera, are grouped as either homofermenters or heterofermenters based on the fermentation end product [38, 42]. Homofermenters produce lactic acid from glucose as a major product, and heterofermenters produce a number of products such as carbon dioxide, acetic acid, ethanol, as well as lactic acid [38, 42]. Homofermentive lactics include the genera *Streptococcus* which produces the L(+) lactate isomer and *Pediococcus* which produces DL lactate [42]. Heterofermentive lactics consist of the genus *Leuconostoc* which produce D(−) lactate and a subgroup of the genus *Lactobacillus*, the *Betabacteria* which produce DL lactate [42].

Lactobacilli are ubiquitous in nature and are usually found in carbohydrate-rich environments [11]. They also are a part of the normal flora in the intestinal tract of many animals. The genus *Lactobacillus* belongs to the phylum *Firmicutes*, class *Bacilli*, order *Lactobacillales*, and family *Lactobacillaceae* [11]. The most commonly isolated species are *Lactobacillus acidophilus*, *L. salivarius*, *L. casei*, *L. plantarum*, *L. fermentum*, *L. reuteri*, *L. rhamnosus*, *L. gasseri*, and *L. brevis* from human intestine [11]. Several of these, *Lactobacillus acidophilus*, *Lactobacillus rhamnosus*, *Lactobacillus casei*, and *Lactobacillus reuteri*, have been extensively studied and well documented [44].

Lactobacillus acidophilus, which was first isolated from children's feces by Ernst Moro in 1900, is capable of colonizing the human colon, has antimicrobial effects, and can be used to treat intestinal infections [26, 44]. *Lactobacillus rhamnosus GG* or *Lactobacillus GG* (LGG) is commonly used in dairy products marketed for infant and children's consumption. *Lactobacillus GG* was isolated from human feces in 1983 and is indigenous to the human intestinal flora, has a tolerance to low pH environment, and adheres to the gastrointestinal tract [44, 45]. LGG is effective in treating diarrhea [19, 46, 47]. *Lactobacillus gasseri* colonizes the gastrointestinal tract, oral cavity, and vagina in humans and is believed to contribute or potentiate probiotic activity in part by reducing fecal mutagenic enzymes as well as stimulate macrophages [44].

Bifidobacteria

Bifidobacteria are the predominant intestinal organism of breastfed infants. These bacteria are rod-shaped, non-gas producing, and anaerobic. Breast milk has been found to contain lactic acid bacteria as well as *Bifidobacteria*, both now included in formulas and foods targeted to preterm and full-term infants [43]. Bifidobacteria are generally characterized as gram-positive, nonspore-forming, nonmotile, and catalase-negative anaerobes [11]. Initially they had been assigned to the genera *Bacillus*, *Bacteroides*, *Nocardia*, *Lactobacillus*, and *Corynebacterium*, before being recognized as a separate genera in 1974 and included in the *Actinomycetaceae* family [11, 44]. This family consists of five genera: *Bifidobacterium*, *Propionibacterium*, *Miycobaccterium*, *Corynebacterium*, and *Brevibacterium* [11]. Currently there are 32 species in the genus *Bifidobacterium*, 12 are isolated from human sources, 15 from animal intestinal tracts or rumen, 3 from honeybees, and the other 2 are found in fermented milk and sewage [11, 38]. Species found in humans are *Bifidobacterium adolescentis*, *B. angulatum*, *B. bifidum*, *B. breve*, *B. catenulatum*, *B. dentium*, *B. infantis*, *B. longum*, and *B. pseudocatenulatum* [11, 44]. These probiotic species can induce immunoglobulins, improve food nutritional value by assimilation of substrates not metabolized by the host, and have potential anticarcinogenic activity and folic acid synthesis [44]. Specifically, *Bifidobacterium infantis* has been found to significantly improve symptoms in patients with irritable bowel disease [19].

Enterococcus

There are 37 species of *Enterococcus* which have been validated for use as probiotics [48]. Enterococci are singular, double- or short-chained gram-positive cocci [44]. These bacteria occur in many habitats such as soil, surface water, ocean water, sewage, on plants, and in the gastrointestinal tract of animals and humans, with *E. faecalis* being the most predominant [48]. Bacteria of the *Enterococcus* genus can also be used to treat diarrhea, irritable bowel syndrome, are considered to be an alternative for antibiotics, and are used for lowering cholesterol and immune regulation [44, 48].

Other Probiotics

Besides the human gastrointestinal tract, the gastrointestinal tracts of other animals such as pigs, rats, and poultry are also good sources of probiotics [43, 47]. Other probiotic strains have been discovered in marine and freshwater fish such as rainbow trout and shrimp [43] as well as in non-fermented foods such as meat and fruits [43]. *Lactobacillus* strains from brine of naturally fermented olives and from pickled juices have also demonstrated probiotic properties [43]. Other popular probiotics are *Streptococcus thermophilus*, *Lactococcus lactis* subsp. *lactis*, *Leuconostoc mesenteroides*, *Propionibacterium freudenreichii*, *Pediococcus acidilactici*, *Sporolactobacillus inulinus*, *Escherichia coli*, other bacteria of the *Bacillus* species, other lactic acid bacteria species, and *Saccharomyces cerevisiae* and *Saccharomyces boulardii* yeasts. Many popular probiotics are added to dairy products and can have favorable effects on human health [11, 19, 21, 22, 34, 44]. There is a selection criteria regarding probiotic strains used in such products. There are several components of this criteria: (a) the bacterium must be reported in the literature, (b) concrete proof of assistance to health must exist, (c) the bacterium must be able to colonize the gastrointestinal tract and have a regulatory role in microbial balance in that area, (d) the bacterium must be resistant to low pH values and bile salts in order to be able to sustain their viability, (e) the bacterium must possess natural antibiotic effect in order to prevent pathogen growth with their antimicrobial activity, (f) the bacterium must be safe to consume and show no antibiotic resistance, and (g) the bacterium must be suitable for commercialization [11, 22, 24, 30, 43, 44].

Commensal Bacteria

Through coevolution, humans not only tolerated the presence of the intestinal microbiota but also evolved to use the colonization of commensal microbes for immune development and function, intestinal barrier integrity, and overall health [49]. Commensal microbes comprise the resident bacteria that live on the human body and in the intestine amount to over 500 different strains including probiotic strains. The composition of intestinal microbes differs depending upon the intestinal region, with gradients existing both vertically and longitudinally (Fig. 2) [50]. Along the longitudinal axis, the number of microbiota increases distally with the greatest level in the colon ($\sim 10^{12}$). Along the vertical axis, certain bacteria are found in the upper mucus layer above the epithelium, while others prefer the lumen. Different microbes thrive in different regions because of the local environment, which is influenced by luminal dietary contents, bile, pH, mucus, other bacteria, etc. Several of the major probiotic strains that were originally isolated from humans include *Lactobacillus acidophilus*, bifidobacteria, several LAB strains [43], and *Lactobacillus rhamnosus GG* [44, 45]. In the intestine, the balance of beneficial bacteria with neutral or inflammatory bacteria is critical. Thus, intestinal dysbiosis (microbe imbalance) leads to a reduction in the beneficial commensal microbes and can contribute to disease [49]. Probiotic intake can help restore commensal microbe balance.

Probiotics and Bone Health

Probiotics Regulate the Gut-Bone Axis

Oral probiotics benefit the intestine as well as extraintestinal organs including the bone [8–10, 51, 52]. The bone is a dynamic organ that depends on a fine balance between the bone-forming osteoblasts and bone-resorbing osteoclasts. An imbalance in this process can lead to bone disease. Bone homeostasis can be regulated by hormones such as estrogen, parathyroid hormone, as well as by immune cells [53–55]. The gastrointestinal system also plays a key role in bone health, most notably by regulating absorption of minerals such as calcium, phosphorous, and magnesium as well as by being major producers of endocrine factors that signal to bone cells, such

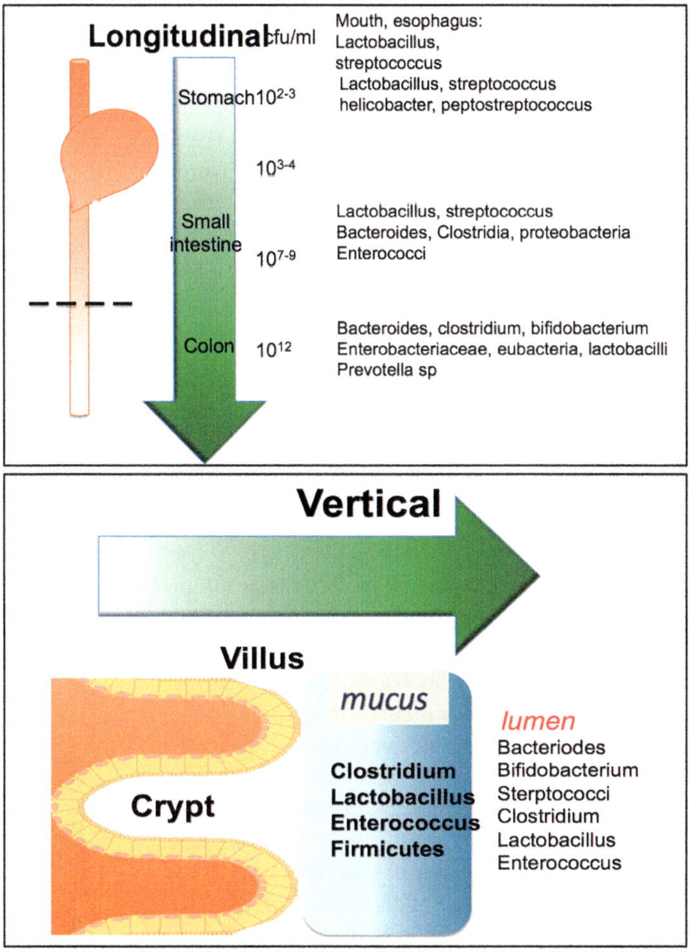

Fig. 2 Regional bacterial changes of the intestine. The intestine is a major source of commensal microbes containing more than 500 species. Along the longitudinal axis, the number of bacteria increases distally. Along the vertical axis, the majority of bacteria are in the lumen with some in the top mucus layer. Microbes colonize different environments based on a number of factors including pH and the nutrients available

as incretins and serotonin. Therefore, agents/conditions that influence intestinal physiology can impact bone health. Recent studies, including some from our lab, indicate that in addition to mineral absorption, the intestinal microbiota can be a critical player in regulating bone physiology [7, 8, 52, 56, 57]. Thus, we and others have examined the influence of probiotics on gut microbiome and how this modulates bone health. The effect of probiotics on the gut-bone axis is determined by a variety of factors. In this subsection we will discuss studies examining the effect of probiotics on bone during growth, aging, and menopause. In addition, we will discuss the role of sex in bone responses to probiotics as well as the safety of probiotics.

Probiotic Effects on Growth

Stability of the intestinal microbiota composition is a critical regulator of intestinal homeostasis throughout life, from newborn to adulthood. Increasing evidence also indicates that intestinal homeostasis plays a key role in the development of healthy strong bone during childhood and adolescence, which ultimately leads to a healthy adult skeleton [58]. By comparing microbiota from undernourished and healthy children from a Malawian birth cohort, Blanton et al. [59] demonstrated that the microbiota is causally related to childhood nutrition. More importantly, the microbiota effects were functionally transmittable to germ-free mice (mice lacking a microbiome). Specifically, germ-free mice whose intestines were populated with microbiota from the undernourished children displayed reduced growth, altered bone morphology, and metabolic dysfunction compared to mice populated with age-matched healthy microbiota [59]. Supplementation with two bacterial strains (*Ruminococcus gnavus* and *Clostridium symbiosum*) added to the microbiome from undernourished children ameliorated growth abnormalities in the mice, supporting a role for microbiome composition and by extension probiotics in growth regulation [59]. In support of these findings, Schwarzewr et al. [60] show that undernourished mice supplemented with the probiotic *Lactobacillus plantarum* are able to maintain normal growth rates. Specifically, undernutrition suppresses growth and bone growth parameters (femur length, cortical thickness, cortical bone fraction, and trabecular fraction of the femur), and these effects were prevented by *L. plantarum* treatment [60]. Importantly, and in agreement with Blanton et al. [59], the presence and/or composition of microbiota during development was shown to be important for regulating mouse growth rates. By comparing wild-type and germ-free mice, the group found that growth parameters were decreased in the germ-free mice which were 4% shorter and weighed less than the WT mice. This response was shown to be dependent on the IGF-1-IGF-1R axis (Fig. 3). Analysis of growth hormone (GH), IGF-1, and IGFBP-3 levels indicated a significant decrease in germ-free compared to wild-type mice 56 days after birth while on undernourished diet [60]. Supplementation with *L. plantarum* brought IGF-1 and IGFBP-3 back to wild-type levels, suggesting *L. plantarum* can recapitulate the beneficial effects of the microbiota on the IGF-1-IGF-1R axis [60]. Yan et al. [61] also demonstrate the important role of the gut microbiota in regulating IGF-1 expression, bone formation, and growth in mice. These effects cross animal species and are seen in *Drosophila* as well. Specifically, *Drosophila* display growth suppression in response to undernutrition or lack of a microbiome [62]. When germ-free flies are repopulated with probiotic lactobacilli strains, the flies regain their ability to grow at normal rates [62], and the IGF axis is restored [63]. In humans, Steenhout et al. examined the impact of probiotic-supplemented formulas on growth in both healthy and vulnerable populations [64]. They concluded that the probiotic *Bifidobacterium lactis* has a positive effect on growth in infants born to mothers with human immunodeficiency virus [64]. Taken together, these studies demonstrate that a healthy gut microbiome is important for bone growth during development.

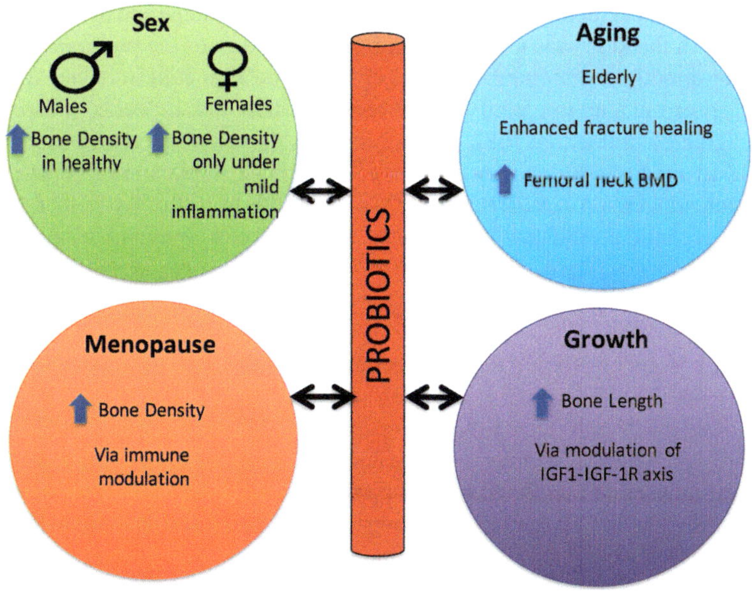

Fig. 3 Probiotics bone effects in different populations. Probiotics benefit bone health across differing populations. The host bone responses are dependent upon factors such as sex, aging, menopause, and growth

Probiotic Effects on Aging Bone

Aging is associated with many complications including osteoporosis. The use of probiotics to benefit longevity and health dates back to ancient Ayurvedic texts (400 and 200 BCE) [20]. Given this, it is surprising that only recently research has begun to focus on the critical role and mechanisms of microbiome/probiotic regulation of aging conditions, such as osteoporosis. While there currently are several ongoing studies examining probiotic effects on bone health in the elderly, only a few studies have been published to date. In one study, *Lactobacillus casei* Shirota was given to elderly male and female patients ($n = 417$); after 4 months of treatment these patients showed enhanced fracture healing (distal radius) compared to patients with placebo treatment (Fig. 3) [65]. In a similar study, 50 postmenopausal women with osteopenia (50–72 years of age) were randomly assigned to take either GeriLact (7 probiotic bacteria species) or a placebo for 6 months. The multispecies probiotic GeriLact significantly decreased biomarkers of bone resorption in comparison with the placebo group, though no significant changes in bone mineral density were observed during this period of treatment [66]. Interestingly, the probiotic treatment did significantly decreased serum levels of parathyroid hormone and the pro-inflammatory marker TNF-α [66]. Another study, which saw an effect on bone density, involved the treatment of osteoporotic males (64–67 years of age) with kefir fermented milk for 6 months. The group found a 5% increase in femoral neck bone mineral density measured by DEXA [67]. This study supports a benefit of probiotics on bone health, but it is important to recognize that only 24 subjects were studied and the contribution of calcium in the kefir was not separated from the effects of the probiotic bacteria. While not directly examining bone, a recent study by Han et al. screened a library of *C. elegans* mutants to identify bacterial metabolites that influence lifespan and reduce aging complications [68]. The polysaccharide colonic acid was found to be involved in mediating longevity and reducing aging complications, supporting a role for intestinal microbes in regulating lifespan and health. Taken together, ancient texts and recent data indicate the potential for probiotics to maintain bone health throughout life.

Probiotic Effects on Menopausal Osteoporosis

The natural loss of estrogen due to menopause is the most important risk factor for osteoporosis in women. Women, over the course of their lifetime, lose about 50% of their trabecular bone and 30% of their cortical bone; about half of the bone loss occurs during the first 10 years after menopause [69]. Recent studies have examined the influence of the microbiota and probiotic treatment during osteoporosis especially under conditions of estrogen deficiency in animal models. For example, while we previously noted that intact healthy female mice do not display a bone response to *L. reuteri*, we found that *L. reuteri* treatment can prevent ovariectomy-induced bone loss in mice, suggesting that lack of estrogen may influence responsiveness to *L. reuteri* effects on bone (Fig. 3) [52]. These findings were confirmed by others using similar or distinct probiotics [44, 48, 63]. In a recent study, Li et al. [70] demonstrated that microbiota is necessary for sex steroid deficiency-induced bone loss. Female wild-type and germ-free mice were given Lupron (ovarian sex steroid antagonist) to block the effect of estrogen in mice. While wild-type mice lost bone as expected, the germ-free mice did not lose bone, demonstrating that the microbiota may be essential for estrogen deficiency-induced bone loss [70]. While Lupron increased intestinal permeability in wild-type mice, it did not affect permeability in the germ-free mice. Supplementation of conventional mice with *Lactobacillus rhamnosus* GG (LGG) or VSL#3 reduced gut permeability and intestinal inflammation and protected mice against bone loss induced by ovariectomy-induced estrogen deficiency [70].

Probiotics have been proposed to function in multiple ways under estrogen-deficient conditions. One important mechanism is through the suppression of osteoclastogenesis, an event that is upregulated during estrogen deficiency/menopause. Our studies showed that *L. reuteri* can suppress OVX-induced increases in bone marrow CD4+ T lymphocytes, which are responsible for the overstimulation of osteoclasts (Fig. 5) [52]. In addition, we have also shown that a 3 kd fraction of the *L. reuteri* can inhibit osteoclastogenesis in vitro [52]. Similarly, Ohlson et al. showed that the probiotics could affect pro-inflammatory cytokines such as TNF-α and IL-1β, as well as increase osteoprotegerin levels, all of which will decrease osteoclastogenesis. Similar attenuation of bone loss was also demonstrated with soymilk that was supplemented with *L. paracasei* subsp. *paracasei* NTU101 or *L. plantarum*

NTU 102 in ovariectomized mice [8]. Narva et al. have also demonstrated a similar outcome with the use of fermented milk, valyl-prolyl-proline, and *Lactobacillus helveticus* LBK-16H in ovariectomized rats [71]. Finally, Rodrigues et al. showed that synbiotics, in this study a combination of prebiotics (Yacon flour) and probiotics (*Bifidobacterium longum*), increased bone mineral content in rats [51]. Together, these studies demonstrate an important role for oral probiotics in reversing estrogen deficiency-induced bone loss.

Influence of Sex on Probiotic Effectiveness

Sex hormones are known to play a critical role in regulating bone density [72]. For example, males have greater bone density than females mainly due to differences in cortical bone expansion and greater trabecular bone volume [73, 74]. In addition, studies indicate that some mouse models display gender differences in response to hormones, such as PTH, which regulate bone [75]. Similarly, in one of the earliest bone studies to identify sex-specific responses to probiotic use, our lab administered *Lactobacillus reuteri* ATCC PTA 6475 (*L. reuteri*) to healthy male and healthy female mice for 4 weeks [9]. *L. reuteri* increased bone volume fraction and bone mineral density in healthy male mice, and this was associated with a suppression of intestinal inflammation (Fig. 3) [9]. Surprisingly, these effects were not observed in female mice, demonstrating that *L. reuteri* treatment influences bone (and gut) in a sex-specific manner [9]. This is also consistent with studies that induce intestinal inflammation by infecting mice with *H. hepaticus*; in these studies, the pathogenic bacteria caused intestinal inflammation and bone loss in male mice but did not have a significant effect in female mice [76]. Taken together the findings suggest that female mice do not respond to either "bad" or "good" bacteria. In later studies, we identified that intact female mice can respond to probiotic (*L. reuteri*) treatment, but only when they are put into mild inflammatory state through dorsal surgical incision [7], supporting a potential role for inflammatory cells and estrogen in regulating female responses to luminal bacteria.

Probiotic Safetyc Throughout Life

The above studies indicate that probiotics hold great promise for supporting bone health. While generally regarded as safe (GRAS), there are some situations where probiotics need to be used cautiously. Patients with compromised immune systems, with significant intestinal barrier dysfunction, or with severe/critical illness may be susceptible to adverse effects such as sepsis, fungemia, and intestinal ischemia [77]; under these conditions the concern is that the load of intestinal bacteria, even though beneficial, could lead to inflammation and crossover into the blood system where immune cells may be compromised and unable to remove/kill the bacteria. Recent tolerability studies for one probiotic, *Lactobacillus rhamnosus* GG (LGG), are very positive. Children with Crohn's disease, which involves a barrier break, tolerate orally supplemented LGG and displayed a side effect profile comparable with placebo [78]. Similarly, elderly patients (66–80 years old) did not display serious adverse effects in response to probiotic (LGG) treatment [79]. Mild symptoms that can occur include bloating, gas, and nausea during the adaptation to probiotic ingestion. As with any new therapy, it is important to carry out these safety and tolerability studies.

Mechanisms of Probiotic Prevention of Bone Pathophysiology

Effect of Probiotics in Dysbiosis-Induced Bone Loss

Dysbiosis is caused by an imbalance of gut microbiota composition/function [80]. While primarily an ailment of the gut, dysbiosis can have systemic effects due to increased permeability of the intestinal mucosa [81]. This can result in bacterial products such as lipopolysaccharide to enter systemic circulation resulting in systemic and local tissue inflammation at distant sites including the bone (Fig. 4) [82, 83]. Our lab has shown that dysbiosis caused by an infectious *H. hepaticus* bacteria can induce gut inflammation as well as bone loss in male mice [84]. Long-term antibiotic treatment can also induce dysbiosis and has been shown to influence the bone. Specifically, male mice treated with antibiotics (ampicillin and neomycin) from 4 to 16 weeks of age display decreased bone strength and reduced B and T cell populations [85]. In a periodontal model of

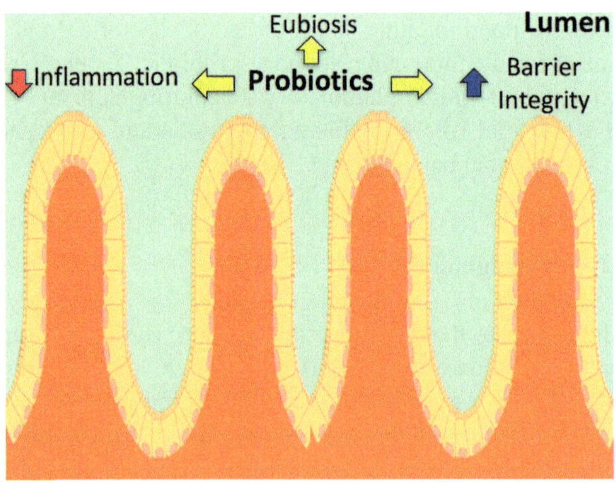

Fig. 4 Model of probiotic mechanistic signals regulating bone density. A disruption in gut microbiota homeostasis can lead to increased inflammation and gut permeability resulting in systemic organ inflammation, including within the bone. Prevention of local gut inflammation and permeability by promoting a healthy gut microflora (eubiosis) is one of the many ways probiotics can benefit bone health

dysbiosis, bone loss was observed [86]. Activation of nucleotide-binding oligomerization domain containing 1 (NOD1), a receptor for immune function in the gut, spared bone loss in these mice, indicating that it could have important effects in similar cases in humans [86].

Probiotic treatment can benefit dysbiosis and gut health through maintaining intestinal barrier function and thereby preventing toxins from entering systemic circulation [87–91]. In a study causing enteropathogenic *E.coli* (EPEC) induced dysbiosis, administration of probiotic *E.coli* Nissle 1917 increased specific claudin expression and prevented increases in intestinal permeability seen after infection with EPEC (Fig. 4) [92]. While pathogenic dysbiosis can damage the intestinal barrier, several studies have shown that this barrier can be rescued through the use of specific probiotics [93–96]. These studies suggest that several conditions linked with gut dysbiosis can be improved through the proper treatment with probiotics. Along with treating the intestinal permeability observed in dysbiosis, probiotics have also been shown to have positive effects on bone health in dysbiosis models. Periodontal disease characterized by dysbiosis of the healthy oral bacterial flora leading to increased inflammation and subsequent bone loss was prevented with probiotic administration. Using this model, mice with periodontitis that were treated with *Lactobacillus brevis* CD2 displayed decreased bone loss and lower expression of pro-inflammatory cytokines such as tumor necrosis factor, interleukin-1β, and interleukin-17A (Fig. 5). Similar studies in a rat model of periodontal disease indicate that probiotics (*Bacillus subtilis* and *Saccharomyces cerevisiae*) can decrease bone resorption, increased bone density, and decreased inflammation [97, 98]; dysbiosis was also prevented by probiotic treatment [97].

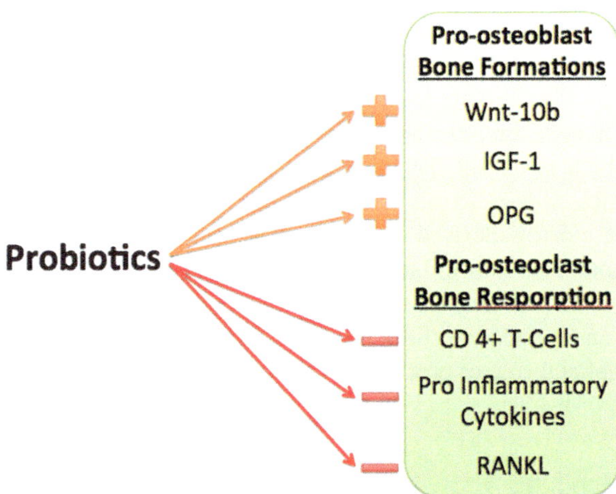

Fig. 5 Mechanism of probiotics beneficial bone affects. Probiotic treatment can modulate the differentiation and function of osteoblasts through changes in Wnt10b, insulin-like growth factor-1, and OPG as well as osteoclasts through modulation of CD4+ T-cells, pro-inflammatory cytokines, and RANKL)

Effect of Probiotics in IBD-Induced Bone Loss

Inflammatory bowel disease (IBD) can have detrimental effects on bone health by affecting the actions of osteoblasts and osteoclasts and promoting osteoporosis [99]. IBD is characterized by gut dysbiosis which generates an inflammatory response both locally and systemically, including within the bone marrow and bone [84]. Thus, IBD-induced intestinal inflammation is the primary pathology that leads to IBD-induced osteoporosis [99]. When the dysbiosis is recognized by the immune system, an inflammatory response occurs that includes the release of many pro-inflammatory cytokines such as TNF-α, IL-6, IL-11, and IL-17, as well as prostaglandin E_2 [100]. Cytokine expression is also elevated in bone [84, 101, 102]. The elevation of pro-inflammatory cytokines promotes osteoclast activity and also suppresses osteoblast activity; the latter occurs by decreasing maturation and increasing cell death. IBD also affects the RANK-RANKL-OPG pathway of bone metabolism and promotes excessive bone loss [103]. Prostaglandin E_2 promotes RANKL and inhibits OPG, which results in greater osteoclast activation. For a comprehensive review of how IBD affects bone, please refer to the chapter by Dr. Sylvester.

Recent studies have shown the protective effects of probiotics on IBD-induced gut inflammation and on bone. Administration of a commercially available probiotic VSL#3 in a mouse model of ulcerative colitis led to decreased gut permeability and aided in treatment of inflammatory symptoms (Fig. 5) [89]. Using other probiotics, such as *L. reuteri* (R2LC), in IL-10-deficient colitis models attenuated disease development, normalizing gut barrier function and reducing pro-inflammatory cytokines and histological disease score [104]. Consistent with these studies, DSS-induced colitis caused increases in gut permeability in female BALB/c mice which was prevented with treatment of *Bifidobacterium longum* CCM 7952 (Bl) [105]. Additional studies indicate that the modulation of toll-like receptor 9 (TLR9) is necessary for the beneficial effects of probiotics in ulcerative colitis treatment [106].

Although these studies did not look at the direct effect of probiotics on the bone, they do indicate that probiotics can have beneficial effects on IBD-induced gut inflammation), which is one of the main components of IBD-induced bone loss. However, probiotics appear to have differential effects on bone inflammation. Treatment of bone marrow-derived dendritic cells from mice with VSL#3 showed increases in both pro- and anti-inflammatory cytokine levels [107]. Taken together these studies show that probiotic treatment of IBD patients may be beneficial to correct the dysbiosis and reduce intestinal inflammation, but further studies are needed to solidify the beneficial role of probiotics.

Effect of Probiotics in Type 1 Diabetes-Induced Bone Loss

Type 1 diabetes is a chronic autoimmune disease characterized by destruction of insulin-producing pancreatic β-cells, resulting in the requirement for exogenous insulin to control blood glucose levels. The consequent metabolic dysregulation has many deleterious consequences including bone loss. T1D-induced osteoporosis is thought to result primarily from the dysregulation of osteoblastic activity. Given that probiotics benefit bone health, probiotic treatment in this model has been examined. This is based on early studies indicating a role for the gut microbiome in T1D development. One of the original studies in nonobese diabetic mice (NOD) showed that NOD mice lacking MyD88 protein (adaptor for multiple innate immune receptors that recognize bacterial stimuli) did not develop T1D [108]. This protection is dependent on the commensal microbes because germ-free MyD88-negative NOD mice develop severe diabetes, whereas bacterial colonization attenuates T1D [108]. Thus, commensal bacteria may be important to reduce disease susceptibility. Consistent with these findings, another group showed that early life antibiotics alter the gut microbiota and its metabolic capacities, intestinal gene expression, and T cell populations leading to accelerated T1D in NOD mice [109]. In addition, our lab has demonstrated that modulation of the gut microbiota with probiotic *L. reuteri 6475* can prevent streptazotocin (STZ)-induced T1D-mediated bone loss in mice. In this study, male (C57BL/6 14 weeks old) mice were given an STZ injection to induce type 1 diabetes which displayed a 35 % reduction in bone volume fraction 4 weeks postinjection [10]. Treatment with *L. reuteri* 6475 prevented this bone loss. This was further supported by trabecular bone data, which revealed that *L. reuteri* 6475 prevented the increase in trabecular spacing and reduction in trabecular number induced by T1D. STZ-induced T1D bone loss comes from reduced osteoblast activity, which was consistent with decreased osteocalcin (bone formation) serum markers and decreases in mineral apposition rate (MAR) compared to controls. *L. reuteri* 6475 prevented decreases in both osteocalcin and MAR suggesting that probiotics, specifically in this model, can have an anabolic effect on the bone [10]. Additionally, part of T1D's bone pathology is an increase in bone marrow adiposity, indicating an altered lineage commitment of bone marrow stromal cells toward the adipocyte over osteoblast lineage [110, 111]. In this study, consistent with benefiting bone health, *L. reuteri* 6475-treated T1D mice did not display increases in adipocyte number [86]. Furthermore Wnt10b signaling which in mesenchymal precursor cells stimulates osteoblastogenesis and inhibits adipogenesis was decreased in T1D mouse bone (Fig. 5). Treatment with probiotic *L. reuteri* 6475 fully restored whole bone Wnt10b gene expression back to normal levels [86]. These findings suggest that probiotic use can prevent T1D bone loss by modulation of expression of Wnt10b in bone.

Conclusions

There are many studies supporting the role for the microbiome in the regulation of bone heath. Direct supplementation of beneficial probiotic bacteria can affect bone health by regulating aspects of gut such as preventing dysbiosis and/or increases in gut permeability and inflammation. However, more research is needed to understand the signaling pathways that link the gut microbiome to bone. Future studies should focus on identifying mechanisms in which probiotics/microbiome are able to regulate osteoblast/clast activities. These studies are important for developing future treatments for osteoporosis.

References

1. National Osteoporosis Foundation. https://www.nof.org/
2. Kanis JA, McCloskey EV, Johansson H, Cooper C, Rizzoli R, Reginster JY. European guidance for the diagnosis and management of osteoporosis in postmenopausal women. Osteoporos Int. 2013;24:23–57.
3. Collins FL, Rios-Arce ND, Schepper JD, Parameswaran N, McCabe LR. The potential of probiotics as a therapy for osteoporosis. Microbiol Spectrum. 2017. 5(4).
4. Khosla S, Shane E. A crisis in the treatment of osteoporosis. J Bone Miner Res. 2016;31:1485–7.
5. Fukuda S, Ohno H. Gut microbiome and metabolic diseases. Semin Immunopathol. 2014;36:103–14.
6. Ley R, Turnbaugh P, Klein S, Gordon J. Human Gut microbes associated with obesity. Nature. 2006;444:1022–3.
7. Collins FL, Irwin R, Bierhalter H, Schepper J, Britton RA, Parameswaran N, McCabe LR. Lactobacillus reuteri 6475 increases bone density in intact females only under an inflammatory setting. PLoS One. 2016;11:e0153180.
8. Ohlsson C, Engdahl C, Fåk F, Andersson A, Windahl SH, Farman HH, Movérare-Skrtic S, Islander U, Sjögren K. Probiotics protect mice from ovariectomy-induced cortical bone loss. PLoS One. 2014;9:e92368.
9. McCabe LR, Irwin R, Schaefer L, Britton RA. Probiotic use decreases intestinal inflammation and increases bone density in healthy male but not female mice. J Cell Physiol. 2013;228:1793–8.
10. Zhang J, Motyl KJ, Irwin R, MacDougald OA, Britton RA, McCabe LR. Loss of bone and Wnt10b expression in male type 1 diabetic mice is blocked by the probiotic lactobacillus reuteri. Endocrinology. 2015;156:3169–82.
11. Azizpour K, Bahrambeygi S, Mahmoodpour S, Azizpour A. History and basic of probiotics. Res J Biol Sci. 2009;4:409–26.
12. Hamilton-Miller JMT, Gibson GR, Bruck W. Some insights into the derivation and early uses of the word "probiotic". Br J Nutr. 2003;90:845.
13. Rijkers GT, de Vos WM, Brummer R-J, Morelli L, Corthier G, Marteau P. Health benefits and health claims of probiotics: bridging science and marketing. Br J Nutr. 2011;106:1291–6.
14. Fuller R. Probiotics. J Appl Bacteriol Suppl. 1986;63:1S–7S.
15. FAO. Probiotics in food. Food Nutr Pap. 2001;85:71.
16. Kechagia M, Basoulis D, Konstantopoulou S, Dimitriadi D, Gyftopoulou K, Skarmoutsou N, Fakiri EM. Health benefits of probiotics: a review. ISRN Nutr. 2013;2013:481651.
17. Morelli L, Capurso L. FAO/WHO guidelines on probiotics 10 years later. J Clin Gastroenterol. 2012;46:10–1.
18. O'Sullivan MG, Thornton G, O'Sullivan GC, Collins JK. Probiotic bacteria: myth or reality? Trends Food Sci Technol. 1992;3:309–14.
19. Gogineni VK. Probiotics: history and evolution. J Anc Dis Prev Remedies. 2013;1:1–7.
20. Gasbarrini G, Bonvicini F, Gramenzi A. Probiotics history. J Clin Gastroenterol. 2016;50:S116–9.
21. Senok AC, Ismaeel AY, Botta GA. Probiotics: facts and myths. Clin Microbiol Infect. 2005;11:958–66.
22. Fuller R. Probiotics in man and animals. J Appl Bacteriol. 1989;66:365–78.

23. Lilly DM, Stillwell RH. Probiotics: growth-promoting factors produced by microorganisms. Science. (80-. 2017;147:747–8.
24. Anukam K, Reid G. Probiotics: 100 years (1907–2007) after Elie Metchnikoff's observation. Commun Curr Res Educ Top trends Appl Microbiol. 2007;2:466–74.
25. Guarner F, Khan AG, Garisch J, Eliakim R, Gangl A, Krabshuis J, Thomson A, Lemair T. Probiotics and prebiotics. Probiotics prebiotics-World Gastroenterol Organ Glob Guidel. 2011:1–28.
26. McFarland LV. From yaks to yogurt: the history, development, and current use of probiotics. Clin Infect Dis. 2015;60:S85–90.
27. Fujii A, Bush JH, Shores KE, Johnson RG, Garascia RJ, Cook ES. Probiotics: antistaphylococcal activity of 4-aminocyclohexanecarboxylic acid, aminobenzoic acid, and their derivatives and structure? Activity relationships. J Pharm Sci. 1977;66:844–8.
28. Salminen S, Ouwehand A, Benno Y, Lee YK. Probiotics: how should they be defined? Trends Food Sci Technol. 1999;10:107–10.
29. Huis in't Veld JHJ, Havenaar R. Probiotics and health in man and animal. J Chem Technol Biotechnol. 1991;51:562–7.
30. FAO, WHO. Guidelines for the evaluation of probiotics in food. 2002:1–11.
31. Hill C, Guarner F, Reid G, Gibson GR, Merenstein DJ, Pot B, Morelli L, Canani RB, Flint HJ, Salminen S, Calder PC, Sanders ME. Expert consensus document: the International Scientific Association for Probiotics and Prebiotics consensus statement on the scope and appropriate use of the term probiotic. Nat Rev Gastroenterol Hepatol. 2014;11:9.
32. Ozen M, Dinleyici EC. The history of probiotics: the untold story. Benef Microbes. 2015;6:159–65.
33. Calatayud GA, Suárez JE. A new contribution to the history of probiotics. Benef Microbes. 2017;8:323–5.
34. Masood MI, Qadir MI, Shirazi JH, Khan IU. Beneficial effects of lactic acid bacteria on human beings. Crit Rev Microbiol. 2011;37:91–8.
35. Brüssow H. Microbiota and healthy ageing: observational and nutritional intervention studies. Microb Biotechnol. 2013;6:326–34.
36. Lino PA, Martins MAP, De S e Silva ME, De Abreu MHNG. Anxiolytics, sedatives, and hypnotics prescribed by dentists in Brazil in 2010. Biomed Res Int. 2017;2017:1–5.
37. Metchnikoff E. The prolongation of life; optimistic studies. New York: G.P. Putnam's Sons; 1908.
38. Soccol CR, Vandenberghe LP d S, Spier MR, Medeiros ABP, Yamaguishi CT, Lindner JDD, Pandey A, Thomaz-Soccol V. The potential of probiotics: a review. Food Technol Biotechnol. 2010;48:413–34.
39. Wiley J. Bergey's manual of systematics of Archaea and bacteria. Bergey's Man Trust. 2001.
40. Babel W, Endo I, Enfors S-O, Fiechter A, Hoare M, Hu W-S, Mattiasson B, Nielsen J, Schügerl K, Stephanopoulos G, Von Stockar U, Tsao GT, Ulber R, Wandrey C, Zhong J-J, De Vrese M, Schrezenmeir J. Probiotics, prebiotics, and synbiotics. Adv Biochem Engin/Biotechnol. 2008;111:1–66.
41. Ouwehand AC, Salminen S, Isolauri E. Probiotics: an overview of beneficial effects. Antonie Van Leeuwenhoek. 2002;82:279–89.
42. Carr FJ, Chill D, Maida N. The lactic acid bacteria: a literature survey. Crit Rev Microbiol. 2002;28:281–370.
43. Fontana L, Bermudez-Brito M, Plaza-Diaz J, Munoz-Quezada S, Gil A. Sources, isolation, characterisation and evaluation of probiotics. Br J Nutr. 2013;109(Suppl):S35–50.
44. Yerlikaya O. Starter cultures used in probiotic dairy product preparation and popular probiotic dairy drinks. Food Sci Technol. 2014;34:221–9.
45. Gorbach SL. Probiotics and gastrointestinal health. Am J Gastroenterol. 2000;95:2–4.
46. Tuohy KM, Probert HM, Smejkal CW, Gibson GR. Using probiotics and prebiotics to improve gut health. Drug Discov Today. 2003;8:692–700.
47. Williams NT. Probiotics. Am J Heal Pharm. 2010;67:449–58.

48. Franz CMAP, Huch M, Abriouel H, Holzapfel W, Gálvez A. Enterococci as probiotics and their implications in food safety. Int J Food Microbiol. 2011;151:125–40.
49. Lin L, Zhang J. Role of intestinal microbiota and metabolites on gut homeostasis and human diseases. BMC Immunol. 2017;18:2.
50. Sekirov I, Russell SL, Antunes LCM, Finlay BB. Gut microbiota in health and disease. Physiol Rev. 2010;90:859–904.
51. Rodrigues FC, Castro ASB, Rodrigues VC, Fernandes SA, Fontes EAF, de Oliveira TT, Martino HSD, de Luces Fortes Ferreira CL. Yacon flour and Bifidobacterium longum modulate bone health in rats. J Med Food. 2012;15:664–70.
52. Britton RA, Irwin R, Quach D, Schaefer L, Zhang J, Lee T, Parameswaran N, McCabe LR. Probiotic *L. reuteri* treatment prevents bone loss in a menopausal ovariectomized mouse model. J Cell Physiol. 2014;229:1822–30.
53. Cenci S, Weitzmann MN, Roggia C, Namba N, Novack D, Woodring J, Pacifici R. Estrogen deficiency induces bone loss by enhancing T-cell production of TNF-alpha. J Clin Invest. 2000;106:1229–37.
54. Hock JM, Gera I, Fonseca J, Raisz LG. Human parathyroid hormone-(l–34) increases bone mass in ovariectomized and orchidectomized rats*. Endocrinology. 1988;122:2899–904.
55. Yamada C. Role of incretins in the regulation of bone metabolism. Nihon Rinsho. 2011;69:842–7.
56. Ohlsson C, Sjögren K. Effects of the gut microbiota on bone mass. Trends Endocrinol Metab. 2015;26:69–74.
57. Parvaneh K, Ebrahimi M, Sabran MR, Karimi G, Hwei ANM, Abdul-Majeed S, Ahmad Z, Ibrahim Z, Jamaluddin R. Probiotics (*Bifidobacterium longum*) increase bone mass density and upregulate *Sparc* and *Bmp-2* genes in rats with bone loss resulting from ovariectomy. Biomed Res Int. 2015;2015:1–10.
58. Sommer F, Bäckhed F. The gut microbiota—masters of host development and physiology. Nat Rev Microbiol. 2013;11:227–38.
59. Blanton LV, Charbonneau MR, Salih T, Barratt MJ, Venkatesh S, Ilkaveya O, Subramanian S, Manary MJ, Trehan I, Jorgensen JM, Fan Y-M, Henrissat B, Leyn SA, Rodionov DA, Osterman AL, Maleta KM, Newgard CB, Ashorn P, Dewey KG, Gordon JI. Gut bacteria that prevent growth impairments transmitted by microbiota from malnourished children. Science. 2016; doi:10.1126/science.aad3311.
60. Schwarzer M, Makki K, Storelli G, Machuca-Gayet I, Srutkova D, Hermanova P, Martino ME, Balmand S, Hudcovic T, Heddi A, Rieusset J, Kozakova H, Vidal H, Leulier F. Lactobacillus plantarum strain maintains growth of infant mice during chronic undernutrition. Science. (80-.). 2016;351:854.
61. Yan J, Herzog JW, Tsang K, Brennan CA, Bower MA, Garrett WS, Sartor BR, Aliprantis AO, Charles JF. Gut microbiota induce IGF-1 and promote bone formation and growth. Proc Natl Acad Sci U S A. 2016;113:E7554–63.
62. Storelli G, Defaye A, Erkosar B, Hols P, Royet J, Leulier F. Lactobacillus plantarum promotes drosophila systemic growth by modulating hormonal signals through TOR-dependent nutrient sensing. Cell Metab. 2011;14:403–14.
63. Hyun S. Body size regulation and insulin-like growth factor signaling. Cell Mol Life Sci. 2013;70:2351–65.
64. Steenhout PG, Rochat F, Hager C. The effect of bifidobacterium lactis on the growth of infants: a pooled analysis of randomized controlled studies. Ann Nutr Metab. 2009;55:334–40.
65. Lei M, Hua L-M, Wang D-W. The effect of probiotic treatment on elderly patients with distal radius fracture: a prospective double-blind, placebo-controlled randomised clinical trial. Benef Microbes. 2016;7:631–7.
66. Jafarnejad S, Djafarian K, Fazeli MR, Yekaninejad MS, Rostamian A, Keshavarz SA. Effects of a multispecies probiotic supplement on bone health in osteopenic postmenopausal women: a randomized, double-blind, controlled trial. J Am Coll Nutr. 2017;19:1–10.
67. Tu M-Y, Chen H-L, Tung Y-T, Kao C-C, Hu F-C, Chen C-M. Short-term effects of kefir-fermented milk consumption on bone mineral density and bone metabolism in a randomized clinical trial of osteoporotic patients. PLoS One. 2015;10:e0144231.

68. Han B, Sivaramakrishnan P, Lin C-C, Neve L, He J, Tay L, Wei R, Sowa J, Sizovs A, Du G, Wang J, Herman C, Wang M. Microbial genetic composition tunes hosts longevity. Cell. 2017;169:1249–62.
69. Finkelstein JS, Brockwell SE, Mehta V, Greendale GA, Sowers MR, Ettinger B, Lo JC, Johnston JM, Cauley JA, Danielson ME, Neer RM. Bone mineral density changes during the menopause transition in a multiethnic cohort of women. J Clin Endocrinol Metab. 2008;93:861–8.
70. Li J-YJ, Chassaing B, Tyagi AMA, Vaccaro C, Luo T, Adams J, Darby TM, Weitzmann MN, Mulle JG, Gewirtz AT, Jones RMRRM, Pacifici R. Sex steroid deficiency-associated bone loss is microbiota dependent and prevented by probiotics. J Clin Invest. 2016;126:1–15.
71. Narva M, Rissanen J, Halleen J, Vapaatalo H, Väänänen K, Korpela R. Effects of bioactive peptide, Valyl-Prolyl-Proline (VPP), and lactobacillus helveticus fermented milk containing VPP on bone loss in ovariectomized rats. Ann Nutr Metab. 2007;51:65–74.
72. Manolagas SC, O'Brien CA, Almeida M. The role of estrogen and androgen receptors in bone health and disease. Nat Rev Endocrinol. 2013;9:699–712.
73. Vanderschueren D, Laurent MR, Claessens F, Gielen E, Lagerquist MK, Vandenput L, Börjesson AE, Ohlsson C. Sex steroid actions in male bone. Endocr Rev. 2014;35:906–60.
74. Nieves JW, Formica C, Ruffing J, Zion M, Garrett P, Lindsay R, Cosman F. Males have larger skeletal size and bone mass than females, despite comparable body size. J Bone Miner Res. 2004;20:529–35.
75. Wang Y, Sakata T, Elalieh HZ, Munson SJ, Burghardt A, Majumdar S, Halloran BP, Bikle DD. Gender differences in the response of CD-1 mouse bone to parathyroid hormone: potential role of IGF-I. J Endocrinol. 2006;189:279–87.
76. Harris L, Senagore P, Young VB, McCabe LR. Inflammatory bowel disease causes reversible suppression of osteoblast and chondrocyte function in mice. Am J Physiol Gastrointest Liver Physiol. 2009;296:G1020.
77. Fijan S. Microorganisms with claimed probiotic properties: an overview of recent literature. Int J Environ Res Public Health. 2014;11:4745–67.
78. Bousvaros A, Guandalini S, Baldassano RN, Botelho C, Evans J, Ferry GD, Goldin B, Hartigan L, Kugathasan S, Levy J, Murray KF, Oliva-Hemker M, Rosh JR, Tolia V, Zholudev A, Vanderhoof JA, Hibberd PL A randomized, double-blind trial of lactobacillus GG versus placebo in addition to standard maintenance therapy for children with Crohn's disease. Inflamm Bowel Dis. 2005;11:833.
79. Hibberd PL, Kleimola L, Fiorino A-M, Botelho C, Haverkamp M, Andreyeva I, Poutsiaka D, Fraser C, Solano-Aguilar G, Snydman DR. No evidence of harms of probiotic lactobacillus rhamnosus GG ATCC 53103 in healthy elderly-a phase I open label study to assess safety, tolerability and cytokine responses. PLoS One. 2014;9:e113456.
80. Li M, Liang P, Li Z, Wang Y, Zhang G, Gao H, Wen S, Tang L. Fecal microbiota transplantation and bacterial consortium transplantation have comparable effects on the re-establishment of mucosal barrier function in mice with intestinal dysbiosis. Front Microbiol. 2015;6:692.
81. Tremellen K, Pearce K. Dysbiosis of Gut Microbiota (DOGMA) – a novel theory for the development of polycystic ovarian syndrome. Med Hypotheses. 2012;79:104–12. doi:10.1016/j.mehy.2012.04.016.
82. Cani PD, Neyrinck AM, Fava F, Knauf C, Burcelin RG, Tuohy KM, Gibson GR, Delzenne NM. Selective increases of bifidobacteria in gut microflora improve high-fat-diet-induced diabetes in mice through a mechanism associated with endotoxaemia. Diabetologia. 2007;50:2374–83.
83. Cani PD, Amar J, Iglesias MA, Poggi M, Knauf C, Bastelica D, Neyrinck AM, Fava F, Tuohy KM, Chabo C, Waget A, Delmée E, Cousin B, Sulpice T, Chamontin B, Ferrières J, Tanti J-F, Gibson GR, Casteilla L, Delzenne NM, Alessi MC, Burcelin R. Metabolic endotoxemia initiates obesity and insulin resistance. Diabetes. 2007;56:1761.
84. Irwin R, Lee T, Young VB, Parameswaran N, McCabe LR. Colitis induced bone loss is gender dependent and associated with increased inflammation. Inflamm Bowel Dis. 2013;19:1586.
85. Guss JD, Horsfield MW, Fontenele FF, Sandoval TN, Luna M, Apoorva F, Lima SF, Bicalho RC, Singh A, Ley RE, van der Meulen MC, Goldring SR, Hernandez CJ. Alterations to

the gut microbiome impair bone strength and tissue material properties. J Bone Miner Res. 2017;32:1343.
86. Chaves de Souza JA, Frasnelli SCT, Curylofo-Zotti F d A, Ávila-Campos MJ, Spolidório LC, Zamboni DS, Graves DT, Rossa C. *NOD1* in the modulation of host-microbe interactions and inflammatory bone resorption in the periodontal disease model. Immunology. 2016;149:374–85.
87. Rosenfeldt V, Benfeldt E, Valerius NH, Pærregaard A, Michaelsen KF. Effect of probiotics on gastrointestinal symptoms and small intestinal permeability in children with atopic dermatitis. J Pediatr. 2004;145:612–6.
88. Stratiki Z, Costalos C, Sevastiadou S, Kastanidou O, Skouroliakou M, Giakoumatou A, Petrohilou V. The effect of a bifidobacter supplemented bovine milk on intestinal permeability of preterm infants. Early Hum Dev. 2007;83:575–9.
89. Madsen K, Cornish A, Soper P, McKaigney C, Jijon H, Yachimec C, Doyle J, Jewell L, De Simone C. Probiotic bacteria enhance murine and human intestinal epithelial barrier function. Gastroenterology. 2001;121:580–91.
90. Zareie M, Johnson-Henry K, Jury J, Yang P-C, Ngan B-Y, McKay DM, Soderholm JD, Perdue MH, Sherman PM. Probiotics prevent bacterial translocation and improve intestinal barrier function in rats following chronic psychological stress. Gut. 2006;55:1553–60.
91. Bron PA, Kleerebezem M, Brummer R-J, Cani PD, Mercenier A, MacDonald TT, Garcia-Ródenas CL, Wells JM. Can probiotics modulate human disease by impacting intestinal barrier function? Br J Nutr. 2017;117:93–107.
92. Zyrek AA, Cichon C, Helms S, Enders C, Sonnenborn U, Schmidt MA. Molecular mechanisms underlying the probiotic effects of Escherichia coli Nissle 1917 involve ZO-2 and PKC? Redistribution resulting in tight junction and epithelial barrier repair. Cell Microbiol. 2007;9:804–16.
93. Anderson RC, Cookson AL, McNabb WC, Kelly WJ, Roy NC. Lactobacillus plantarum DSM 2648 is a potential probiotic that enhances intestinal barrier function. FEMS Microbiol Lett. 2010;309.:no-no:184.
94. Resta–Lenert S, Barrett KE. Probiotics and commensals reverse TNF-α– and IFN-γ–induced dysfunction in human intestinal epithelial cells. Gastroenterology. 2006;130:731–46.
95. Qin H, Zhang Z, Hang X, Jiang Y. L. plantarum prevents enteroinvasive Escherichia coli-induced tight junction proteins changes in intestinal epithelial cells. BMC Microbiol. 2009;9:63.
96. Moorthy G, Murali MR, Devaraj SN. Lactobacilli facilitate maintenance of intestinal membrane integrity during Shigella dysenteriae 1 infection in rats. Nutrition. 2009;25:350–8.
97. Messora MR, Oliveira LFF, Foureaux RC, Taba M, Zangerônimo MG, Furlaneto F a C, Pereira LJ. Probiotic therapy reduces periodontal tissue destruction and improves the intestinal morphology in rats with ligature-induced periodontitis. J Periodontol. 2013;84:1818–26.
98. Garcia VG, Knoll LR, Longo M, Novaes VCN, Assem NZ, Ervolino E, de Toledo BEC, Theodoro LH. Effect of the probiotic *Saccharomyces cerevisiae* on ligature-induced periodontitis in rats. J Periodontal Res. 2016;51:26–37.
99. Armour KE, Van'T Hof RJ, Grabowski PS, Reid DM, Ralston SH. Evidence for a pathogenic role of nitric oxide in inflammation-induced osteoporosis. J Bone Miner Res. 1999;14:2137–42.
100. Ott SJ, Plamondon S, Hart A, Begun A, Rehman A, Kamm MA, Schreiber S. Dynamics of the mucosa-associated flora in ulcerative colitis patients during remission and clinical relapse. J Clin Microbiol. 2008;46:3510–3.
101. Harris L, Senagore P, Young VB, McCabe LR. Inflammatory bowel disease causes reversible suppression of osteoblast and chondrocyte function in mice. Am J Physiol – Gastrointest Liver Physiol. 2009;296
102. Irwin R, Raehtz S, Parameswaran N, McCabe LR. Intestinal inflammation without weight loss decreases bone density and growth. Am J Physiol – Regul Integr Comp Physiol. 2016;311:R1149.

103. Boyce BF, Xing L. The RANKL/RANK/OPG pathway. Curr Osteoporos Rep. 2007;5:98–104.
104. Madsen KL, Doyle JS, Jewell LD, Tavernini MM, Fedorak RN. Lactobacillus species prevents colitis in interleukin 10 gene-deficient mice. Gastroenterology. 1999;116:1107–14.
105. Srutkova D, Schwarzer M, Hudcovic T, Zakostelska Z, Drab V, Spanova A, Rittich B, Kozakova H, Schabussova I. Bifidobacterium longum CCM 7952 promotes epithelial barrier function and prevents acute DSS-induced colitis in strictly strain-specific manner. PLoS One. 2015;10:e0134050.
106. Rachmilewitz D, Katakura K, Karmeli F, Hayashi T, Reinus C, Rudensky B, Akira S, Takeda K, Lee J, Takabayashi K, Raz E. Toll-like receptor 9 signaling mediates the anti-inflammatory effects of probiotics in murine experimental colitis. Gastroenterology. 2004;126:520–8.
107. Mariman R, Kremer B, Koning F, Nagelkerken L. The probiotic mixture VSL#3 mediates both pro- and anti-inflammatory responses in bone marrow-derived dendritic cells from C57BL/6 and BALB/c mice. Br J Nutr. 2014;112:1088–97.
108. Wen L, Ley RE, Volchkov PY, Stranges PB, Avanesyan L, Stonebraker AC, Hu C, Wong FS, Szot GL, Bluestone JA, Gordon JI, Chervonsky AV. Innate immunity and intestinal microbiota in the development of type 1 diabetes. Nature. 2008;455:1109–13.
109. Livanos AE, Greiner TU, Vangay P, Pathmasiri W, Stewart D, McRitchie S, Li H, Chung J, Sohn J, Kim S, Gao Z, Barber C, Kim J, Ng S, Rogers AB, Sumner S, Zhang X-S, Cadwell K, Knights D, Alekseyenko A, Backhed F, Blaser MJ. Antibiotic-mediated gut microbiome perturbation accelerates development of type 1 diabetes in mice. Nat Microbiol. 2016;1:16140.
110. Botolin S, Faugere M-C, Malluche H, Orth M, Meyer R, McCabe LR. Increased bone adiposity and peroxisomal proliferator-activated receptor-? 2 Expression in type I diabetic mice. Endocrinology. 2005;146:3622–31.
111. Motyl KJ, Raetz M, Tekalur SA, Schwartz RC, McCabe LR. CCAAT/enhancer binding protein ?-deficiency enhances type 1 diabetic bone phenotype by increasing marrow adiposity and bone resorption. AJP Regul Integr Comp Physiol. 2011;300:R1250–60.

Bone Mechanical Function and the Gut Microbiota

C.J. Hernandez

Introduction

The primary function of bone is to resist physical forces. Bones provide a stiff structure that supports the body and provides leverage points for muscle insertions to enable locomotion. Additionally, bones provide mechanical protection to vital organs. Failure of bone to serve its mechanical function leads to fracture. For this reason, the mechanical performance of bone is perhaps the most clinically relevant effect of gut-bone signaling.

Impaired bone mechanical performance can result in a clinical fracture during activities of daily living. Fractures that occur from modest loads such as a fall from standing height or loads associated with rising from a chair are known as fragility fractures. Fragility fractures are often the result of osteoporosis, a condition in which bone mass and density are reduced. A key clinical assay used to diagnose osteoporosis is bone mineral density (BMD). Individuals with low bone mineral density are more likely to experience a fragility fracture. Whether or not an individual develops low bone mineral density characteristic of osteoporosis is determined by the amount of bone accrued during growth and maturation as well as rates of bone loss later in life [1, 2]. While bone mineral density is a useful indicator of the risk of fragility fracture, there are aspects of fracture risk that are not explained by BMD. For example, risk of fragility fracture exceeds what is expected from

C.J. Hernandez (✉)
Sibley School of Mechanical and Aerospace Engineering, Cornell University, 355 Upson Hall, Ithaca, NY, USA

Meinig School of Biomedical Engineering, Cornell University, Ithaca, NY, USA

Hospital for Special Surgery, New York, NY, USA
e-mail: cjh275@cornell.edu

L.R. McCabe, N. Parameswaran (eds.), *Understanding the Gut-Bone Signaling Axis*, Advances in Experimental Medicine and Biology 1033,
DOI 10.1007/978-3-319-66653-2_12

Table 1 The terms used to describe mechanical performance at the whole bone are shown along with the corresponding tissue-level terms

Whole bone mechanical properties (units)	Tissue mechanical properties (units)
Load (N)	Stress (N/mm^2)
Displacement (mm)	Strain (mm/mm)
Stiffness (N/mm)	Elastic modulus (N/mm^2)
Ultimate load (N)	Ultimate stress (N/mm^2)

BMD in a number of clinical conditions including advanced age [3], diabetes and obesity [4, 5], and inflammatory bowel disease [6, 7].

Over the last 10 years, improvements in high-throughput sequencing have led to rapid improvements in our understanding of the human microbiome and its effect on disease. The human microbiome consists of the microbial species and associated molecular products that reside on the surfaces of the human body. The microbiome of an individual consists of over 1,000 distinct microbial species including bacteria, archaea, and single-celled eukaryotes. Alterations in the constituents and metabolic activity of the human microbiome have been associated with a host of chronic diseases including obesity, diabetes [8], and cardiovascular disease [9]. Interestingly, many of the chronic conditions associated with an altered gut microbiome are also risk factors for fragility fracture (Table 1) [10], suggesting a potential link between the gut microbiome and bone fragility. An association between the gut microbiome and bone would suggest that the microbiome could be used as a biomarker of fracture risk or even a therapeutic target to prevent fractures.

The goal of this chapter is to summarize how gut-bone signaling may influence the ability of bone to resist physical forces and avoid fragility fracture. I first review bone biomechanics and then discuss links between the microbiome and bone.

Mechanical Properties of Bone

This section has two parts: a conceptual review of biomechanics targeted to readers with little engineering background, followed by a more technical review of the aspects of bone quality that determine whole bone strength and fracture risk.

Conceptual Review of Biomechanics: Mechanical Stress and Strain and Material Failure

Most readers are familiar with mechanical failure in man-made devices. Mechanical components of automobiles, buildings, and even office furniture fail from time to time and must be replaced. Engineers prevent premature failure in man-made devices by applying well-established concepts that describe how forces are distributed throughout the structure and how the material from which the structure is made resists failure. The same physical mechanisms that cause failure of man-made devices also apply to the mechanical failure of bones. Here I review two concepts that are fundamental to understanding the biomechanical performance of bone in relation to fragility fractures: (1) The mechanical performance of the whole bone is determined by bone morphology and density as well as the mechanical performance of bone tissue material itself and (2) mechanical failure of a whole bone is not always the result of impaired bone tissue strength but can also occur as a result of changes in tissue mechanical properties other than strength.

The key aspects of mechanical function of a whole bone are stiffness and strength. Stiffness is the relationship between the applied load and the resulting deformation. Stiffness is the same parameter that is described in introductory physics textbooks as a spring constant. The strength of a whole bone is the maximum load that can be applied before failure (measured as the ultimate load; see Table 1). The ultimate load expresses the functional strength of the bone but often doesn't tell the complete story regarding fragility fracture. For example, a whole bone may be weak due to unusual morphology, poor internal structure, impaired bone tissue material properties, or a combination of all three. It is often useful to understand the causes for variation whole bone strength by separating the effects of morphology from the effects of tissue material properties. Tissue material properties are analogous to whole bone mechanical properties (Table 1). The concentration of force at a point within the bone structure is known as the tissue stress. Stress is expressed as force per unit cross-sectional area (the same units as pressure). The amount of tissue deformation at a location within the whole bone is expressed as strain. Strain is the amount of deformation relative to the initial shape (a unitless parameter, sometimes expressed as mm/mm or as a percent). The relationship between tissue stress and strain is the elastic modulus. The elastic modulus is the stiffness of the tissue and is independent of bone morphology. The strength of a material is expressed as the ultimate stress. When a whole bone is loaded, the forces applied to the bone are distributed throughout the structure creating a complex distribution of tissue stress and strain.

The distribution of tissue stress and strain within a whole bone is influenced by external morphology as well as internal structure. A simple example of the effects of internal structure on tissue stress and strain is the effect of cross-sectional geometry on resistance to bending in a long bone such as a femur. When a long bone is submitted to bending loads (a common loading mode), tissue stress and strain are distributed in a pattern determined by cross-sectional geometry. The key geometric factor describing how a bone resists bending is the second moment of area, also known as the "moment of inertia" in the engineering literature. The moment of inertia is directly proportional to whole bone stiffness and strength in bending. Changes in cross-sectional geometry can have a large effect on whole bone stiffness and strength even when the amount of material in the structure is not changed (Table 2). Although the cross-section of the diaphysis of a long bone is used to illustrate the effect of geometry, similar effects of internal geometry are present in

Table 2 The effect of cross-sectional geometry on the whole bone stiffness and strength in bending is shown. All three of the example have the same cross-sectional area but differ in terms of inner and outer radius. The bone with the larger outer radius has a strength 1.6 times greater even though the cross-sectional area is identical

Parameter	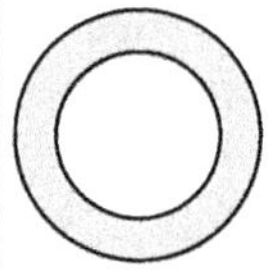	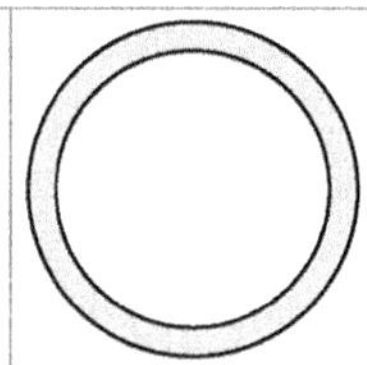	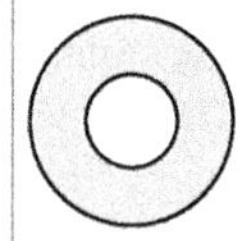
Cross-sectional area	1	1	1
Outer radius	1	1.30	0.83
Whole bone bending stiffness (flexural rigidity)	1	2.10	0.58
Whole bone bending strength (onset of failure)	1	1.60	0.69

regions of the skeleton with both cancellous and cortical bone [11, 12]. Readers interested in more detailed mathematical discussions of the moment of inertia are referred to prior work [13].

Understanding the differences between whole bone mechanical function and tissue material properties is necessary to understand the underlying causes of increased bone fragility. A clinical parameter often associated with whole bone strength is bone mineral density. Bone mineral density, measured using dual energy x-ray absorptiometry or quantitative computed tomography, is a single numerical value that summarizes the size and density of a whole bone but provides no information about internal geometry or tissue material properties. It is therefore possible that a bone can display high bone mineral density, but may still be prone to fragility fracture due to impaired internal geometry or tissue strength. Understanding the underlying cause is important because some interventions are more effective at improving internal geometry, while others may be more effective at improving tissue strength.

Mechanical failure of a whole bone is commonly attributed to excessive mechanical loads such as those from a fall, but not all fractures occur during a single loading event. For example, as many as 60% of vertebral fractures occur without a single memorable loading event and are either spontaneous or incidental [14]. The idea that a structure can fail through processes other than a single overload is not commonly discussed in the bone literature [15]. However, most readers are familiar with the fact that mechanical failure is not always caused by a single event: a load carrying part of an automobile may undergo mechanical failure following thousands of cycles of low-magnitude loading during regular use. Clearly such a component did not fail because it had insufficient strength (it survived many previous loads of the same magnitude). Tissue strength, measured as the ultimate stress, describes tissue failure under a single load (Table 1) but provides little information about failure that is caused by multiple loads. Material properties that describe tissue failure but are distinct from tissue strength include fracture toughness (resistance to failure caused by rapid crack growth) and fatigue strength (resistance to failure caused by thousands to millions of cycles of loads well below the ultimate stress) [16]. Fracture toughness and fatigue strength are associated with tissue brittleness. Fracture toughness, in particular, has recently been identified as a tissue material property that can greatly influence fracture risk and is modified in a number of different clinical conditions/regions of the skeleton [17–19]. It is possible that bone tissue can display a high bone strength (resistance to failure under a single load) and low fracture toughness (resistance to rapid crack growth) [20]. The vast majority of research in bone mechanics has examined whole bone and tissue strength, but relatively little is known about whole bone failure caused by inadequate fracture toughness or fatigue strength. It is likely that future advancements in fracture prevention will involve improving material properties other than strength [15].

Hierarchical Nature of Bone Mechanical Performance

Bone is a hierarchical material, a term that means that bone structure is made up of smaller, divisible parts. Whole bone mechanical function aggregates effect of the mechanical performance of all of these smaller parts, each acting at a length scale ranging from nanometers to centimeters (Fig. 1, Table 3). A clinical condition that influences whole bone mechanical function does so by changing one or more of these traits. Here I review the characteristics of bone at each length scale.

Whole bone size and density have long been associated with resistance to fracture. Whole bone density is measured in vivo using dual energy x-ray absorptiometry and/or quantitative computed tomography. Bone mineral density is strongly correlated to whole bone strength as measured in the laboratory [12, 21, 22]. However, there are aspects of bone geometry and internal structure that are not well described by bone mineral density but nevertheless influence whole bone mechanical performance. Two aspects of bone internal structure that influence whole bone mechanical performance are the relative amounts of cortical and cancellous bone and variability in cancellous bone density. Cortical bone is stiffer and stronger than cancellous bone and is therefore much more influential in determining whole bone strength, although cancellous bone still plays an important role, especially in bones such as the vertebrae that have relatively little cortical bone tissue [23–25]. Variation in the density of cancellous bone within a whole bone can also influence mechanical performance [26], and recent studies have shown that increased heterogeneity of bone density at the millimeter scale influences the stiffness and strength of vertebrae [27, 28].

Bone tissue at the scale of 3–5 mm is the next smallest hierarchical scale of bone. Examination of bone mechanical properties at the scale of 3–5 mm is popular because at this scale: (1) It is possible to dissect specimens of uniform size for experimental analysis, thereby separating the effects of whole bone morphology and internal architecture; (2) the size scale is large enough so that the microstructure can be summarized by a single number (e.g., bone volume fraction); and (3) the scale is similar to the resolution of common clinical imaging modalities such as a

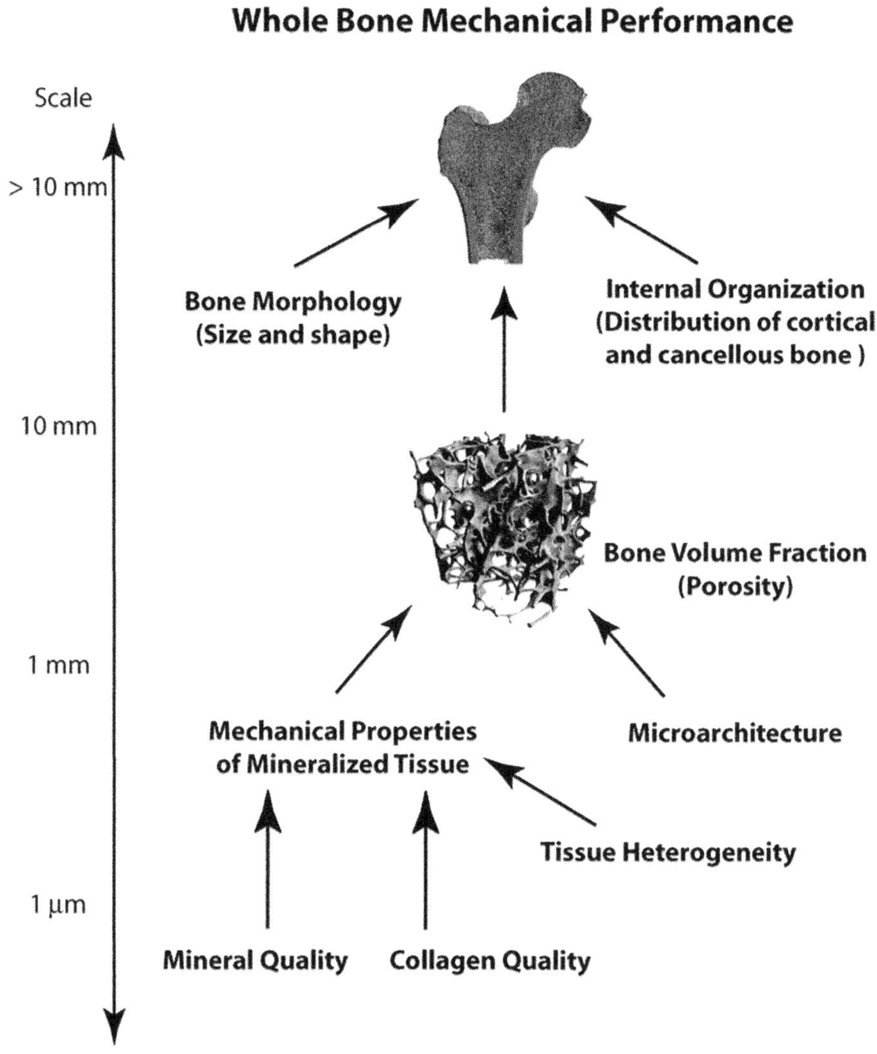

Fig. 1 The mechanical performance of a whole bone is determined by factors affecting bone at multiple length scales (Adapted from Hernandez and Keaveny [53] with permission)

quantitative computed tomography and can therefore be directly applied to finite element models to estimate whole bone stiffness and strength [29]. Bone volume fraction is the single most important contributor to elastic modulus and strength at the scale of 3–5 mm [30]. Average tissue degree of mineralization can also influence elastic modulus and strength by the following empirical relationship [31]:

$$E(\text{GPa})(\text{BV}/\text{TV})^{2.58}\,\alpha^{2.74},$$

$$\sigma_{\text{Ult}}(\text{MPa})(\text{BV}/\text{TV})^{1.92}\,\alpha^{2.79},$$

Table 3 Characteristics of bone that influence bone mechanical performance are shown

Scale (m)	Bone characteristics
$>10^{-3}$	Whole bone morphology (size and shape)
	Bone density distribution
	Relative proportion of cortical/cancellous bone
10^{-6}–10^{-3}	Bone volume fraction
	Microarchitecture
	Average degree of mineralization
10^{-9}–10^{-6}	Remodeling cavity number, size, and distribution
	Tissue heterogeneity
$<10^{-9}$	Collagen structure and cross-linking
	Mineral type and crystal alignment
	Non-collagenous proteins
	Matrix-bound water

Adapted from Hernandez and Keaveny [53]

where E is the elastic modulus, σ_{Ult} is the ultimate stress in compression, BV/TV is the bone volume fraction, and α is the degree of mineralization (ash mass/total mass).

Microarchitecture also influences bone mechanical performance, but the effects are complicated. The mechanical properties of bone at the millimeter scale are anisotropic, a term that means that tissue material properties depend on the orientation of loads. The anisotropy of bone tissue mechanical properties is primarily due to the fact that microstructures (trabeculae in cancellous bone and Haversian canals in cortical bone) are preferentially oriented in the same direction as stresses caused by habitual loading in vivo [32, 33]. The mechanical properties of bone tissue at the millimeter scale are different in directions transverse to the microstructural orientation. However, if we consider only the mechanical properties oriented with the microstructure, detailed measures of microarchitecture (trabecular thickness, trabecular separation, etc. [34]) explain little of the variation in bone stiffness and strength beyond what can be inferred from bone volume fraction (see above) [30].

At the scale of 10–100 s micrometers, bone tissue mechanical performance is influenced by microscale porosity and tissue heterogeneity. Pores in bone tissue caused by bone remodeling such as resorption cavities (also referred to as resorption lacunae or osteoclast lacunae, in humans typically 30 μm in depth and 80,000 μm^2 in area [35]) can cause localized stress concentrations in bone tissue. Extensive examination of the effects of resorption cavities on cancellous bone, however, has failed to observe an effect of resorption cavities on elastic modulus, ultimate stress, or fatigue life at the scale of 3–5 mm [19, 36, 37]. Microscale porosity in cortical bone has recently become a topic of great interest because it may impair the mechanical properties of cortical bone in ways that are not apparent from clinical measures of bone mineral density [38, 39]. Heterogeneity of bone tissue at the scale of 10–100 μm is also believed to influence the mechanical properties of bone tissue [40]. Heterogeneity in bone tissue at this scale is a result of bone remodeling (see below). Tissue heterogeneity at the scale of 10–100 μm results from slight variations among prior locations of bone remodeling (bone structural units) and can alter the elastic modulus of cancellous bone at the scale of 3–5 mm [41–43]. The effect of tissue heterogeneity on other mechanical properties of bone such as ultimate stress remains poorly understood.

Mechanically relevant changes in bone tissue also occur at the scale of micrometers or nanometers and include alterations in mineral content and quality, the relative amounts of collagen and non-collagenous proteins and matrix-bound water. Characteristics of bone mineral at scales below 10 μm that influence bone tissue mechanical properties include the mineral-to-matrix ratio and crystallinity. Crystallinity has a profound effect on bone tissue elastic modulus and strength [44, 45]. Collagen is the primary organic component of bone tissue, but there are also a number of non-collagenous proteins including osteocalcin and osteopontin that may also influence tissue mechanical performance. Bone tissue that is deficient in osteocalcin or osteopontin is more brittle and therefore more likely to fail [18]. Type I collagen in bone is synthesized by osteoblasts but can undergo posttranslational modifications resulting in cross-links with neighboring molecules. Collagen cross-links may be enzymatic or nonenzymatic. Nonenzymantic cross-links such as pentosidine can make the collagen network more rigid resulting in a bone tissue that is more brittle [46, 47]. In addition, matrix-bound water has recently been shown to have a profound effect on bone mechanical properties including tissue strength and brittleness [48–51].

The characteristics of bone that have been discussed so far have been studied for some time, but there is still much we do not know about how these characteristics contribute to fragility fracture. First, most of our understanding of bone mechanical performance is based on evaluation of elastic modulus and strength. While these two tissue material properties are extremely important, mechanical failure can also occur through mechanisms dominated by other material properties such as fracture toughness [17] or fatigue strength [16]. A characteristic of bone may improve one mechanical property while impairing another [15]. For example, increased bone tissue degree of mineralization is associated with increased elastic modulus and ultimate stress (see above) but has also been associated with impaired bone tissue brittleness [52] and increased risk of tissue failure through mechanisms dominated by fracture toughness or fatigue strength. Second, there is much we do not yet understand about micro- and nanoscale properties of bone tissue and their effect on whole bone mechanical performance. It is possible for a characteristic of bone to be extremely influential in determining mechanical performance of bone tissue at the scale of 100 μm but have no net effects at larger scales, either because the characteristic is too small to influence failure at larger scales or because of compensatory changes in other characteristics of bone [53]. This limitation in our understanding is compounded by the fact that changes in tissue fracture toughness and fatigue strength often originate at scales smaller than 100 μm.

Lastly, it is not always clear how some characteristics of bone tissue are modified in vivo. In some cases, genetic deficiencies can impair bone tissue mechanical properties. However, after a bone is initially formed, there are only two ways that bone structure can be modified: bone remodeling and bone modeling. Bone remodeling is a process in which bone resorption occurs at a location, followed soon after by bone formation that replaces some or all of the resorbed bone tissue. One function of bone remodeling is to remove older bone tissue and replace it with new tissue [54]. Imbalances in the volume of bone resorbed and formed at each remodeling event can cause net changes in bone volume that, over time, can lead to considerable changes in whole bone density. Bone modeling involves either formation or resorption at a location. Bone remodeling and modeling cause changes in local bone volume that can, over time, lead to changes in internal structure and whole bone morphology.

Bone remodeling and modeling may also regulate the material properties of bone tissue at the micro- and nanoscale. After the organic component of bone tissue is synthesized, it begins to mineralize, starting with a rapid primary phase of mineralization followed by an extended secondary phase of mineralization that lasts months or even years [55–57]. Hence, there are small differences in tissue degree of mineralization among regions of bone tissue depending on the time at which the bone was formed [56]. As mentioned above, tissue degree of mineralization can influence bone tissue elastic modulus, ultimate stress, and potentially fracture toughness. In addition to alterations in tissue degree of mineralization, bone tissue accumulates advanced glycation end products (AGEs) over time [47]. The rate of accumulation of AGEs in bone is slow but not yet well understood and may be influenced by systemic factors such as increased serum glucose [47]. The accumulation of AGEs is associated with the formation of collagen cross-links which can promote tissue brittleness [18, 46]. Both secondary mineralization and collagen cross-linking processes occur over time without active cell activity and suggest that the mechanical properties of bone tissue are influenced by the length of time the tissue has been present in the body, a parameter referred to as "tissue age." Differences in tissue age among bone structural units (regions of bone tissue formed during a single remodeling event, 20–100 μm in characteristic size; see above) result in tissue heterogeneity that can influence bone mechanical performance. To date, the only proposed process for altering bone tissue mechanical properties is to remove and replace tissue through bone remodeling. Bone remodeling is a surface-based process, however, and bone tissue is only remodeled if it is sufficiently close to a bone surface that is accessible to osteoclasts and osteoblasts. Such regions include the surfaces of a trabecula and the surfaces of Haversian or Volkmann's canals or a newly forming osteon/Haversian canal (a "cutting cone"). The fact that bone remodeling only occurs at surfaces limits the ability to turn over bone tissue; hence, there are some regions of bone tissue that never remodel during an individual's lifespan [56]. Additionally, the fact that bone remodeling is constrained to surfaces means that changes in bone tissue material properties that are caused by aging are very difficult to reverse; computational models suggest that increases in tissue age associated with antiresorptive treatments may remain for more than a decade after treatment is suspended [58]. Hence, impairment of bone morphology and bone density is readily reversed by bone remodeling and modeling, but changes in tissue material properties, once established, are not easily modified.

Mechanisms Linking the Microbiome to Bone Strength

The human microbiome is the community of microbial species and their molecular products that reside on the human body. The vast majority of commensal microbes in the human body are in the gut. Changes in the gut microbiota have been associated with inflammatory bowel diseases [59], obesity [8], metabolic disease [60], malnutrition [61], neurological disorders [62], cancer [63], and cardiovascular disease [9]. The effects of the microbiome on physiology have been studied since the advent of antibiotics but have traditionally been difficult to study because the vast majority of species within the microbiome are not readily grown under laboratory conditions (so-called unculturable organisms) [64]. Advancements in high-throughput sequencing in the past decade have made it possible to identify many of these organisms, resulting in increased interest in the role of the gut microbiome in health. This chapter includes a conceptual review of the microbiome for nonexperts in the topic followed by a more technical review of the effects of the microbiome on bone. The review of the effects of the gut microbiome on bone concentrates on the potential effect on bone mechanical performance, and the interested reader is referred to other chapters in this book for alterations in bone cell physiology.

Conceptual Review: What Is the Gut Microbiota?

The gut microbiota of an individual consists of over 1,000 distinct species of bacteria, archaea, viruses, and single-celled eukaryotes. The current approach for characterizing the constituents of the gut microbiota is through sequencing of the 16S rRNA gene. The resulting analysis provides the relative abundance of prokaryotic organisms present and can be analyzed at different phylogenetic levels (phyla, class, etc.). Here we review two concepts that are fundamental for understanding the microbiome: (1) The microbiome is a complex network of interdependent species and (2) the organisms within the microbiome maintain a relatively robust dynamic equilibrium that resists the introduction of new species.

Each species within the microbiome interacts with the others in a complex network that involves competition for resources, predatory and parasitic interactions, as well as symbiotic-like interdependencies [65]. The interactions among organisms within the gut flora make it difficult to identify a single species as the cause of a disease process. Even when a single organism is associated with a disease-like phenotype, one cannot ignore the possibility that the identified organism is not the cause of the disease but is simply part of an interdependent network with other organisms that are the direct cause [66]. Additionally, most of what we know about the contents of the microbiome is based on phylogenetic characterization of the species that are present; this tells us what organisms are present but does not describe how the organisms participate in the community. The functional capacity met by each species within the interacting network may be more important to understanding the relationship of the microbiome to disease [67, 68].

The network of species within a microbiome helps a healthy gut flora resist the introduction of new species [68]. A new organism introduced into the community must compete with established organisms and the network of interdependent species. In order to survive and populate the community, the newly introduced organism must have some competitive advantage (uses a different food resource, generates an antibiotic to hinder competitors, etc.). The ability of the commensal flora to limit the introduction of new species is called resistance and is useful to prevent colonization by pathogenic species. However, resistance in the microbial community can also make it difficult to repair an unhealthy gut flora. The commensal population of a mature gut flora fluctuates daily based on host diet and activity patterns [67, 69]. Although the gut microbiota fluctuates, the microbial community within the gut is robust to transient disruptions such as short-term alterations in diet, a trait called resilience. Extreme stimuli such as antibiotic treatment can cause long-lasting changes to some components of the gut flora even if the majority of the microbiota returns to its initial state [67, 70]. Even when changes in microbial composition occur, they may simply involve replacement of a species with another organism that fills the same functional role. The ability of the microbiome to recover function after a disruption is known as functional redundancy [68].

Mechanisms Linking the Gut Microbiota to Bone

The association between the gut microbiome and organs distant from the gut is not immediately obvious. I have found it useful to classify mechanisms that link gut microbiota to bone into three groups [10]: (1) regulation of nutritional absorption, (2) stimulation of the immune system, and (3) translocation of microbes or microbe-associated molecular patterns (MAMPs) across the endothelial barrier (Fig. 2).

The gut microbiota can have a profound influence on nutritional absorption in the gut. Low-dose antibiotics have long been known to increase rates of animal growth and do so primarily by altering the population of the gut flora in a way that enhances caloric absorption [71–73]. The gut microbiota also synthesize a number of vitamins useful to the body including pyridoxal phosphate, thiamine (B1), niacin (B3), pantothenic acid (B5), biotin (B7), cobalamin (B12), folate, vitamin K, and tetrahydrofolate [74, 75]. Vitamins synthesized by the gut flora are absorbed at the gut endothelium and distributed through the systemic circulation.

The gut microbiota interacts directly with the host immune system. The microbiota may come into direct contact with immune cells, in particular dendritic cell processes that extend into the mucosa. Additionally metabolites released from the

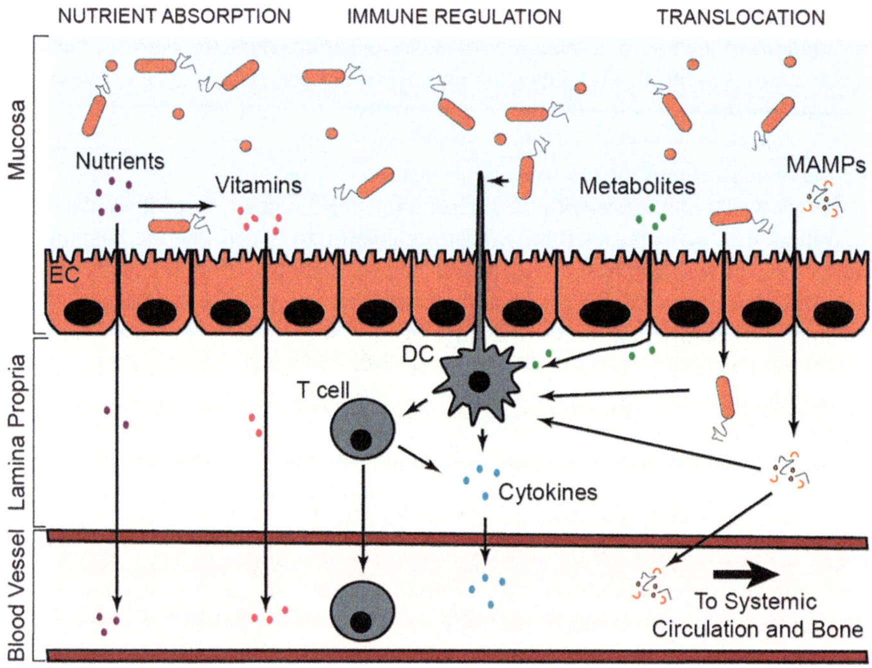

Fig. 2 Proposed mechanisms linking the gut microbiome to bone are illustrated. Mechanisms are divided into microbiome-derived regulation of nutrients and vitamins, microbiome-derived regulation of the host immune system, and translocation of microbes and microbe-associated molecular pattern (MAMP) molecules. *Abbreviations*: *DC* dendritic Cell, *EC* endothelial cell, *MAMPs* microbe-associated molecular patterns (From Hernandez et al. [10] used with permission)

gut microbiota may pass the endothelial barrier and initiate responses in resident dendrites and T cells [76]. Short chain fatty acids are a type of well-recognized bacterial product known to stimulate the immune system [76–80]. The immune cells may then release inflammatory or anti-inflammatory molecules that enter the systemic circulation and eventually reach bone where they can alter bone cell physiology. Additionally, immune cells simulated by gut flora may migrate to the bone marrow and directly regulate bone cells through expression of RANKL or other molecules [81, 82].

Lastly, microbes from the gut flora may migrate through the gut endothelial barrier, a process known as bacterial translocation. Live bacteria that penetrate the endothelial barrier are exposed to a broader array of immune cells and trigger different immune responses than bacteria within the intestinal lumen. Additionally, some live bacteria populate the epithelial as well as lymphoid tissue near the epithelium [83]. Live bacteria may be phagocytosed by immune cells and transported to regions of the body distant from the gut [84]. Additionally, microbe-associated molecular patterns (MAMPs) including flagellin, peptidoglycan, and lipopolysaccharide may translocate across the endothelium and into the systemic circulation where they can be transported to regions such as bone and activate innate immune receptors in local cell populations. Many bone cells and their precursors express receptors that are sensitive to bacterial products including TLR2 (which responds to peptidoglycan), TLR4 (which responds to lipopolysaccharide), and TLR5 (which responds to flagellin) [85–90].

The Gut Microbiota Influences Bone Morphology and Density

To date there are no clinical data directly associating the microbiome with bone disease. However, alterations in the gut microbiota have been observed in a number of conditions that are risk factors for fracture including advanced age [91] and clinical indices of frailty in the elderly [92] (see also Table 2). As there are no clinical data to report on the topic, the discussion here will concentrate on preclinical studies in mice.

Studies in mice demonstrate that the microbiome can influence bone morphology and density. Mice receiving gut microbiota from humans with poor nutritional history experience impaired bone growth, an effect that is mediated to some degree by treatment with probiotics [93, 94]. Additionally, the gut microbiota can influence rates of bone loss following depletion of sex hormones; germ-free mice are protected from bone loss following chemically induced ovariectomy [95], while the application of antibiotics or probiotics to rodents has also been shown to mediate bone loss following ovariectomy [95–99]. Germ-free mice (never exposed to a live microbe; see description below) have been reported to have increased bone mass [100] as well as decreased bone mass [94] as compared to mice with a normal microbiome. Disruption of the gut microbiota through antibiotic treatment can also lead to changes in bone morphology and density, but again it is not clear if the effect promotes denser bones or less dense bones [101–104].

The reasons for the conflicting findings in studies with germ-free and antibiotic-treated mice are not yet understood but may be due to the fact that the studies varied in terms of mouse genetic background, sex, and age at which the bone was examined (see below). Differences in genetic background can alter the microbiome by changing host-microbe interactions. It is possible that the genetic background of one mouse leads to a microbiome that promotes increased bone mass, while in another mouse, the microbiome promotes impaired bone mass. Modification of the microbiome may therefore result in very different findings in different mouse strains [10]. Sex hormones may also influence the contents of the microbiome and lead to differential effects when the microbiota are altered. Animal age is another factor that has been poorly controlled in prior work. Prior reports overwhelmingly examined bone structure in growing mice (less than 12 weeks of age). Rates of bone growth in young animals are rapid, and a condition such as altered microbiome may slow or delay bone growth without changing the final bone phenotype achieved at skeletal maturity. Hence, it is likely that some of the conflicting findings regarding the effects of the microbiome are due to slight variations in rates of bone growth. Another factor that may contribute to differences among studies is differences in imaging modality. Many studies have relied on mouse DXA, which is a low-resolution imaging approach and may not detect important differences in bone among mouse strains.

In addition to differences in methodologies, there are a few key components missing from recent studies of the microbiome and bone. First, few of the studies listed above provided an analysis of the constituents of the gut flora in experimental animals, and those that did report microbiota provided only the most rudimentary assessment of bone phenotype. As a result, detailed assessment of the morphologic changes in bone associated with alteration in the gut flora cannot yet be linked to microbial phyla or species. Lastly, perhaps the greatest limitation of these studies is that none of them report bone mechanical properties.

Challenges in Understanding the Effects of the Gut Microbiota on Bone

Further study of the effects of the microbiome on bone will require experimental investigations in live animals. Animal studies of the microbiome, however, can be sensitive to factors rarely controlled or reported in the bone literature including animal chow (including vendor and lot number), bedding in animal cages, housing facility (specific pathogen-free v. conventional), and the vendor supplying experimental animals (or even the housing facility in the vendor's network). Here I discuss some approaches that must be considered when examining the effects of the microbiome on bone.

The gut microbial community is shaped by environmental exposure and diet. Animals raised in specific pathogen-free facility may be exposed to very different microbes than animals raised in conventional housing. Subtle differences in the microbiota that influences phenotype may occur when mice are raised under specific pathogen-free conditions as compared to conventional housing [105]. Additionally, it is not uncommon for there to be slight differences in gut microbiota from cage to cage within the same room. However, variation of the microbiota within a cage is rare because the gut flora are readily transferred among cohoused mice through coprophagy. Diet also influences the microbiome and should be controlled throughout an experiment. However, commercial animal chow is sometimes modified by vendors in response to changes in their suppliers, leading to differences in diet over time. The bedding within animal cages may also influence the microbiome. Some beddings are ingested by mice (corncob bedding) leading to changes in the gut environment and the microbiota. Lastly, there are recognized differences in the microbiome among animal suppliers. Some organisms known to regulate host-microbe interactions are endemic in the facilities of some vendors but not others [106]. Even different cohorts of mice received from the same vendor may harbor distinct gut flora that can greatly influence the effects of treatments [107]. As a result of the environmental effects on the microbiome, it is recommended that investigators studying the effect of the microbiome on bone raise experimental and control groups in the same facility, with the same chow, bedding, and housing, preferably at the same time to limit random variations in the microbiota that could influence results.

As mentioned above, the microbiome of an individual is relatively robust to transient disruptions. As a result, one cannot simply add or subtract individual species from an established gut flora. There are only a few established experimental approaches for altering the gut flora in a controlled manner. Available approaches include the use of germ-free animals, transfer of gut flora into germ-free animals, chronic treatment with oral antibiotics, and genetic models [108, 109]. Germ-free animals are raised in sterile incubators to ensure that the animals never experience a live microbe. Germ-free animals are therefore an extreme example of the effects of the microbiome on a phenotype. However, a major limitation of the use of germ-free animals is that failure to experience live microbes impairs development of the immune system. Germ-free animals are useful, however, because the absence of the microbiota makes it possible to populate the animals with an entirely new commensal population, either with select microbes or with an entire microbial community from a donor. A limitation of transferring the gut flora to a germ-free mouse is that not all microbial species from a donor can survive in a new host, and as a result, the constituents of the commensal flora may change as the microbial community adjusts to the new host. Changes in the gut flora over time may occur faster than alterations in bone, making it difficult to attribute changes in bone to the contents of the gut flora. Chronic oral antibiotics are another means of regulating the gut flora. Antibiotics with poor oral bioavailability can be used so that their effect is limited to the gut flora. Antibiotic treatment decimates some populations of the gut microbiota and enriches the population of others. Antibiotic treatment can be applied to animals with established commensal flora and active immune systems, thereby avoiding some of the limitations of germ-free mice, but the investigator has relatively little control of the final contents of the gut flora after treatment. Lastly, differences in genetic background can greatly influence the contents of the gut microbiota by changing host-microbe interactions. Profound differences in the gut microbiota have been observed among inbred mouse strains [110]. The use of genetic models is perhaps the most natural manner of modulating the gut microbiota but also provides the least control over the resulting changes in the gut microbial population.

Conclusions

Over the last 40 years, the field of bone and mineral research has developed robust experimental approaches for controlling factors that regulate bone metabolism and bone structure such as circulating hormones and mechanical force. The microbiome presents a new and challenging factor to control in experiments. A recent study found that the effect of alterations in the microbiome on bone remodeling depends on the duration of the stimulus [101], highlighting the complexity of the microbiome and the need for preclinical studies of bone that revisit some of the factors that are assumed to be consistent in most bone research (diet, animal vendor source, etc.). Additionally, although the primary function of bone in the body is mechanical, a major limitation to our understanding of the role of the microbiome on bone disease is the lack of studies relating the mechanical performance of whole bones or bone tissue to alterations in the gut microbiome.

Clinical studies relating the microbiome to bone morphology and bone mineral density achieved in adulthood have not yet been reported. Large studies of human populations have been quite useful for identifying the effects of the microbiome on other diseases such as obesity [66], and it is likely that such large studies will provide information regarding bone mineral density and other risk factors for fracture. The ability of the microbiome to regulate bone in humans may have considerable clinical significance because, while there are effective treatments for osteoporosis, these treatments cannot be applied indefinitely [111]. Hence, there limitations to our ability to reverse established osteoporosis and a need for preventive strategies. The degree to which the microbiome can be modulated to prevent the development of osteoporosis remains to be seen (Table 4).

Table 4 Alterations in the gut microbiota have been associated with many of the factors that alter bone mass, bone mineral density (BMD), and fracture risk

Contributor to osteoporosis	Reported alterations in gut microbiota
Poor acquisition of bone mass during growth leading to low BMD in adulthood	Absence of gut microbiota associated with altered bone mass in mice [94, 100, 101]
Alterations in circulating sex hormones	Chemically induced estrogen depletion does not result in bone loss in germ-free animals [95]
	Probiotic treatment reduces ovariectomy associated bone loss [95, 96]
Diet/nutrition	Gut microbiota regulate production/absorption of vitamins [74]
Aging	Gut microbiota composition is correlated with indices of frailty in the elderly (Barthel index, functional independence measures) [92, 112]
Obesity/diabetes	Gut microbiota influence caloric intake and the development of obesity [8, 66]
Gastrointestinal disease	Inflammatory bowel disease is related to the microbiome and leads to osteopenia independent of its effects on nutrition [6–8]

Adapted from Hernandez et al. [10], used with permission

References

1. Hernandez CJ, Beaupre GS, Carter DR. A theoretical analysis of the relative influences of peak BMD, age-related bone loss and menopause on the development of osteoporosis. Osteoporos Int. 2003;14(10):843–7.
2. Marcus R, Dempster DW, Bouxsein ML. The nature of osteoporosis. In: Marcus R, Feldman D, Dempster DW, Luckey M, Cauley JA, editors. Osteoporosis. 1. 4th ed. Waltham: Academic; 2013. p. 21–30.
3. Hui SL, Slemenda CW, Johnston CC Jr. Age and bone mass as predictors of fracture in a prospective study. J Clin Invest. 1988;81(6):1804–9.
4. Vestergaard P. Discrepancies in bone mineral density and fracture risk in patients with type 1 and type 2 diabetes – a meta-analysis. Osteoporos Int. 2007;18(4):427–44.
5. Karim L, Bouxsein ML. Effect of type 2 diabetes-related non-enzymatic glycation on bone biomechanical properties. Bone. 2016;82:21–7.
6. Ali T, Lam D, Bronze MS, Humphrey MB. Osteoporosis in inflammatory bowel disease. Am J Med. 2009;122(7):599–604.
7. Byrne FR, Morony S, Warmington K, Geng Z, Brown HL, Flores SA, et al. CD4+CD45RBHi T cell transfer induced colitis in mice is accompanied by osteopenia which is treatable with recombinant human osteoprotegerin. Gut. 2005;54(1):78–86.
8. Turnbaugh PJ, Ley RE, Mahowald MA, Magrini V, Mardis ER, Gordon JI. An obesity-associated gut microbiome with increased capacity for energy harvest. Nature. 2006;444(7122):1027–31.
9. Tang WHW, Hazen SL. The contributory role of gut microbiota in cardiovascular disease. J Clin Invest. 2014;124(10):4204–11.
10. Hernandez CJ, Guss JD, Luna M, Goldring SR. Links between the microbiome and bone. J Bone Miner Res. 2016;31(9):1638–46.
11. Whealan KM, Kwak SD, Tedrow JR, Inoue K, Snyder BD. Noninvasive imaging predicts failure load of the spine with simulated osteolytic defects. J Bone Joint Surg Am. 2000;82(9):1240–51.
12. Crawford RP, Keaveny TM. Relationship between axial and bending behaviors of the human thoracolumbar vertebra. Spine. 2004;29(20):2248–55.
13. van der Meulen MC, Jepsen KJ, Mikic B. Understanding bone strength: size isn't everything. Bone. 2001;29(2):101–4.
14. Cooper C, Atkinson EJ, O'Fallon WM, Melton LJ. Incidence of clinically diagnosed vertebral fractures: a population-based study in Rochester, Minnesota, 1985–1989. J Bone Miner Res. 1992;7:221–7.
15. Hernandez CJ, van der Meulen MC. Understanding bone strength is not enough. J Bone Miner Res. 2017;32(6):1157–62. doi:10.1002/jbmr.3078.
16. Dowling NE. Mechanical behaviour of materials. 3rd ed. Upper Saddle River: Pearson Prentice Hall; 2007.
17. Ettinger B, Burr DB, Ritchie RO. Proposed pathogenesis for atypical femoral fractures: lessons from material research. Bone. 2013;55(2):495–500.
18. Poundarik AA, Diab T, Sroga GE, Ural A, Boskey AL, Gundberg CM, et al. Dilatational band formation in bone. Proc Natl Acad Sci U S A. 2012;109(47):19178–83.
19. Torres AM, Matheny JB, Keaveny TM, Taylor D, Rimnac CM, Hernandez CJ. Material heterogeneity in cancellous bone promotes deformation recovery after mechanical failure. Proc Natl Acad Sci U S A. 2016;113(11):2892–7.
20. Ritchie RO. The conflicts between strength and toughness. Nat Mater. 2011;10(11):817--22. doi:10.1038/nmat3115.
21. Courtney AC, Wachtel EF, Myers ER, Hayes WC. Age-related reductions in the strength of the femur tested in a fall loading configuration. Trans Orthop Res Soc. 1994;19:188.

22. Moro M, Hecker AT, Bouxsein ML, Myers ER. Failure load of thoracic vertebrae correlates with lumbar bone-mineral density measured by Dxa. Calcif Tissue Int. 1995;56(3):206–9.
23. Lotz JC, Cheal EJ, Hayes WC. Stress distributions within the proximal femur during gait and falls: implications for osteoporotic fracture. Osteoporos Int. 1995;5(4):252–61.
24. Van Rietbergen B, Huiskes R, Eckstein F, Ruegsegger P. Trabecular bone tissue strains in the healthy and osteoporotic human femur. J Bone Miner Res. 2003;18(10):1781–8.
25. Eswaran SK, Gupta A, Adams MF, Keaveny TM. Cortical and trabecular load sharing in the human vertebral body. J Bone Miner Res. 2006;21(2):307–14.
26. Crawford RP, Cann CE, Keaveny TM. Finite element models predict in vitro vertebral body compressive strength better than quantitative computed tomography. Bone. 2003;33(4):744–50.
27. Hussein AI, Jackman TM, Morgan SR, Barest GD, Morgan EF. The intravertebral distribution of bone density: correspondence to intervertebral disc health and implications for vertebral strength. Osteoporos Int. 2013;24(12):3021–30.
28. Hussein AI, Morgan EF. The effect of intravertebral heterogeneity in microstructure on vertebral strength and failure patterns. Osteoporos Int. 2013;24(3):979–89.
29. Keaveny TM. Biomechanical computed tomography-noninvasive bone strength analysis using clinical computed tomography scans. Ann N Y Acad Sci. 2010;1192:57–65.
30. Hernandez CJ. How can bone turnover modify bone strength independent of bone mass? Bone. 2008;42(6):1014–20.
31. Hernandez CJ, Beaupre GS, Keller TS, Carter DR. The influence of bone volume fraction and ash fraction on bone strength and modulus. Bone. 2001;29(1):74–8.
32. Bartel DL, Davy DT, Keaveny TM. Orthopaedic biomechanics: mechanics and design in musculoskeletal systems. Upper Saddle River: Prentice Hall; 2006.
33. Karim L, Hussein AI, Morgan EF, Bouxsein ML. The mechanical behavior of bone. In: Marcus R, Feldman D, Dempster DW, Luckey M, Cauley JA, editors. Osteoporosis. 4th ed. San Diego: Academic; 2013. p. 431–52.
34. Bouxsein ML, Boyd SK, Christiansen BA, Guldberg RE, Jepsen KJ, Muller R. Guidelines for assessment of bone microstructure in rodents using micro-computed tomography. J Bone Miner Res. 2010;25(7):1468–86.
35. Goff MG, Slyfield CR, Kummari SR, Tkachenko EV, Fischer SE, Yi YH, et al. Three-dimensional characterization of resorption cavity size and location in human vertebral trabecular bone. Bone. 2012;51(1):28–37.
36. Lambers FM, Bouman AR, Tkachenko EV, Keaveny TM, Hernandez CJ. The effects of tensile-compressive loading mode and microarchitecture on microdamage in human vertebral cancellous bone. J Biomech. 2014;47(15):3605–12.
37. Goff MG, Lambers FM, Nguyen TM, Sung J, Rimnac CM, Hernandez CJ. Fatigue-induced microdamage in cancellous bone occurs distant from resorption cavities and trabecular surfaces. Bone. 2015;79:8–14.
38. Nishiyama KK, Macdonald HM, Buie HR, Hanley DA, Boyd SK. Postmenopausal women with osteopenia have higher cortical porosity and thinner cortices at the distal radius and tibia than women with normal aBMD: an in vivo HR-pQCT study. J Bone Miner Res. 2010;25(4):882–90.
39. Patsch JM, Burghardt AJ, Yap SP, Baum T, Schwartz AV, Joseph GB, et al. Increased cortical porosity in type 2 diabetic postmenopausal women with fragility fractures. J Bone Miner Res. 2013;28(2):313–24.
40. Lloyd AA, Wang ZX, Donnelly E. Multiscale contribution of bone tissue material property heterogeneity to trabecular bone mechanical behavior. J Biomech Eng. 2015;137(1):0108011–18.
41. van der Linden JC, Birkenhager-Frenkel DH, Verhaar JA, Weinans H. Trabecular bone's mechanical properties are affected by its non-uniform mineral distribution. J Biomech. 2001;34(12):1573–80.

42. Jaasma MJ, Bayraktar HH, Niebur GL, Keaveny TM. The effects of intraspecimen variations in tissue modulus on the apparent mechanical properties of trabecular bone. Trans Orthop Res Soc; 2001; San Francisco.
43. Bourne BC, van der Meulen MC. Finite element models predict cancellous apparent modulus when tissue modulus is scaled from specimen CT-attenuation. J Biomech. 2004;37(5):613–21.
44. Yerramshetty JS, Akkus O. The associations between mineral crystallinity and the mechanical properties of human cortical bone. Bone. 2008;42(3):476–82.
45. Bi X, Patil CA, Lynch CC, Pharr GM, Mahadevan-Jansen A, Nyman JS. Raman and mechanical properties correlate at whole bone- and tissue-levels in a genetic mouse model. J Biomech. 2011;44(2):297–303.
46. Hernandez CJ, Tang SY, Baumbach BM, Hwu PB, Sakkee AN, van der Ham F, et al. Trabecular microfracture and the influence of pyridinium and non-enzymatic glycation-mediated collagen cross-links. Bone. 2005;37(6):825–32.
47. Vashishth D. The role of the collagen matrix in skeletal fragility. Curr Osteoporos Rep. 2007;5(2):62–6.
48. Horch RA, Gochberg DF, Nyman JS, Does MD. Non-invasive predictors of human cortical bone mechanical properties: T(2)-discriminated H NMR compared with high resolution x-ray. PLoS One. 2011;6(1):e16359.
49. Granke M, Does MD, Nyman JS. The role of water compartments in the material properties of cortical bone. Calcif Tissue Int. 2015;97(3):292–307.
50. Granke M, Makowski AJ, Uppuganti S, Does MD, Nyman JS. Identifying novel clinical surrogates to assess human bone fracture toughness. J Bone Miner Res. 2015;30(7):1290–300.
51. Manhard MK, Uppuganti S, Granke M, Gochberg DF, Nyman JS, Does MD. MRI-derived bound and pore water concentrations as predictors of fracture resistance. Bone. 2016;87:1–10.
52. Currey JD. Bones: structure and mechanics. Princeton: Princeton University Press; 2002.
53. Hernandez CJ, Keaveny TM. A biomechanical perspective on bone quality. Bone. 2006;39(6):1173–81.
54. Parfitt AM. Misconceptions (2): turnover is always higher in cancellous than in cortical bone. Bone. 2002;30(6):807–9.
55. Hernandez CJ, Beaupré GS, Marcus R, Carter DR. A theoretical analysis of the contributions of remodeling space, mineralization and bone balance to changes in bone mineral density during alendronate treatment. Bone. 2001;29(6):511–6.
56. Akkus O, Polyakova-Akkus A, Adar F, Schaffler MB. Aging of microstructural compartments in human compact bone. J Bone Miner Res. 2003;18(6):1012–9.
57. Fuchs RK, Allen MR, Ruppel ME, Diab T, Phipps RJ, Miller LM, et al. In situ examination of the time-course for secondary mineralization of Haversian bone using synchrotron Fourier transform infrared microspectroscopy. Matrix Biol. 2008;27(1):34–41.
58. Hernandez CJ, Lopez HK, Lane JM. Theoretical consideration of the effect of drug holidays on BMD and tissue age. Osteoporos Int. 2014;25(5):1577–84.
59. Kostic AD, Xavier RJ, Gevers D. The microbiome in inflammatory bowel disease: current status and the future ahead. Gastroenterology. 2014;146(6):1489–99.
60. Chassaing B, Aitken JD, Gewirtz AT, Vijay-Kumar M. Gut microbiota drives metabolic disease in immunologically altered mice. Adv Immunol. 2012;116:93–112.
61. Kane AV, Dinh DM, Ward HD. Childhood malnutrition and the intestinal microbiome malnutrition and the microbiome. Pediatr Res. 2015;77(0):256–62.
62. Wang Y, Kasper LH. The role of microbiome in central nervous system disorders. Brain Behav Immun. 2014;38:1–12.
63. Schwabe RF, Jobin C. The microbiome and cancer. Nat Rev Cancer. 2013;13(11):800–12.
64. Stewart EJ. Growing unculturable bacteria. J Bacteriol. 2012;194(16):4151–60.
65. Chaffron S, Rehrauer H, Pernthaler J, von Mering C. A global network of coexisting microbes from environmental and whole-genome sequence data. Genome Res. 2010;20(7):947–59.
66. Goodrich JK, Waters JL, Poole AC, Sutter JL, Koren O, Blekhman R, et al. Human genetics shape the gut microbiome. Cell. 2014;159(4):789–99.

67. Lozupone CA, Stombaugh JI, Gordon JI, Jansson JK, Knight R. Diversity, stability and resilience of the human gut microbiota. Nature. 2012;489(7415):220–30.
68. Moya A, Ferrer M. Functional redundancy-induced stability of gut microbiota subjected to disturbance. Trends Microbiol. 2016;24(5):402–13.
69. David LA, Materna AC, Friedman J, Campos-Baptista MI, Blackburn MC, Perrotta A, et al. Host lifestyle affects human microbiota on daily timescales. Genome Biol. 2014;15(7):R89.
70. Dethlefsen L, Relman DA. Incomplete recovery and individualized responses of the human distal gut microbiota to repeated antibiotic perturbation. Proc Natl Acad Sci U S A. 2011;108(Suppl 1):4554–61.
71. Jukes TH, Williams WL. Nutritional effects of antibiotics. Pharmacol Rev. 1953;5(4):381–420.
72. Ross E, Yacowitz H. Effect of penicillin on growth and bone ash of chicks fed different levels of vitamin-D and phosphorus. Poult Sci. 1954;33(2):262–5.
73. Rusoff LL, Fussell JM, Hyde CE, Crown RM, Gall LS. Parenteral administration of aureomycin to young calves with a note on mode of action. J Dairy Sci. 1954;37(5):488–97.
74. Kau AL, Ahern PP, Griffin NW, Goodman AL, Gordon JI. Human nutrition, the gut microbiome and the immune system. Nature. 2011;474(7351):327–36.
75. Yatsunenko T, Rey FE, Manary MJ, Trehan I, Dominguez-Bello MG, Contreras M, et al. Human gut microbiome viewed across age and geography. Nature. 2012;486(7402):222–7.
76. Belkaid Y, Hand TW. Role of the microbiota in immunity and inflammation. Cell. 2014;157(1):121–41.
77. Arpaia N, Campbell C, Fan X, Dikiy S, van der Veeken J, de Roos P, et al. Metabolites produced by commensal bacteria promote peripheral regulatory T-cell generation. Nature. 2013;504(7480):451–5.
78. Furusawa Y, Obata Y, Fukuda S, Endo TA, Nakato G, Takahashi D, et al. Commensal microbe-derived butyrate induces the differentiation of colonic regulatory T cells. Nature. 2013;504(7480):446–50.
79. Smith PM, Howitt MR, Panikov N, Michaud M, Gallini CA, Bohlooly YM, et al. The microbial metabolites, short-chain fatty acids, regulate colonic Treg cell homeostasis. Science. 2013;341(6145):569–73.
80. Schroeder BO, Backhed F. Signals from the gut microbiota to distant organs in physiology and disease. Nat Med. 2016;22(10):1079–89.
81. Li Y, Toraldo G, Li A, Yang X, Zhang H, Qian WP, et al. B cells and T cells are critical for the preservation of bone homeostasis and attainment of peak bone mass in vivo. Blood. 2007;109(9):3839–48.
82. Pacifici R. T cells, osteoblasts, and osteocytes: interacting lineages key for the bone anabolic and catabolic activities of parathyroid hormone. Ann N Y Acad Sci. 2016;1364:11–24.
83. Fung TC, Artis D, Sonnenberg GF. Anatomical localization of commensal bacteria in immune cell homeostasis and disease. Immunol Rev. 2014;260(1):35–49.
84. Amar J, Chabo C, Waget A, Klopp P, Vachoux C, Bermudez-Humaran LG, et al. Intestinal mucosal adherence and translocation of commensal bacteria at the early onset of type 2 diabetes: molecular mechanisms and probiotic treatment. EMBO Mol Med. 2011;3(9):559–72.
85. Takami M, Kim N, Rho J, Choi Y. Stimulation by toll-like receptors inhibits osteoclast differentiation. J Immunol. 2002;169(3):1516–23.
86. Zou W, Bar-Shavit Z. Dual modulation of osteoclast differentiation by lipopolysaccharide. J Bone Miner Res. 2002;17(7):1211–8.
87. Itoh K, Udagawa N, Kobayashi K, Suda K, Li X, Takami M, et al. Lipopolysaccharide promotes the survival of osteoclasts via Toll-like receptor 4, but cytokine production of osteoclasts in response to lipopolysaccharide is different from that of macrophages. J Immunol. 2003;170(7):3688–95.
88. Chamberlain ND, Vila OM, Volin MV, Volkov S, Pope RM, Swedler W, et al. TLR5, a novel and unidentified inflammatory mediator in rheumatoid arthritis that correlates with disease activity score and joint TNF-alpha levels. J Immunol. 2012;189(1):475–83.

89. Kim SJ, Chen Z, Chamberlain ND, Essani AB, Volin MV, Amin MA, et al. Ligation of TLR5 promotes myeloid cell infiltration and differentiation into mature osteoclasts in rheumatoid arthritis and experimental arthritis. J Immunol. 2014;193(8):3902–13.
90. Kassem A, Henning P, Kindlund B, Lindholm C, Lerner UH. TLR5, a novel mediator of innate immunity-induced osteoclastogenesis and bone loss. FASEB J. 2015;29(11):4449–60.
91. Mariat D, Firmesse O, Levenez F, Guimaraes V, Sokol H, Dore J, et al. The firmicutes/bacteroidetes ratio of the human microbiota changes with age. BMC Microbiol. 2009;9:123.
92. Jackson M, Jeffery IB, Beaumont M, Bell JT, Clark AG, Ley RE, et al. Signatures of early frailty in the gut microbiota. Genome Med. 2016;8(1):8.
93. Blanton LV, Charbonneau MR, Salih T, Barratt MJ, Venkatesh S, Ilkaveya O, et al. Gut bacteria that prevent growth impairments transmitted by microbiota from malnourished children. Science. 2016;351:6275.
94. Schwarzer M, Makki K, Storelli G, Machuca-Gayet I, Srutkova D, Hermanova P, et al. Lactobacillus plantarum strain maintains growth of infant mice during chronic undernutrition. Science. 2016;351(6275):854–7.
95. Li JY, Chassaing B, Tyagi AM, Vaccaro C, Luo T, Adams J, et al. Sex steroid deficiency-associated bone loss is microbiota dependent and prevented by probiotics. J Clin Invest. 2016;126:2049.
96. Britton RA, Irwin R, Quach D, Schaefer L, Zhang J, Lee T, et al. Probiotic L. reuteri treatment prevents bone loss in a menopausal ovariectomized mouse model. J Cell Physiol. 2014;229(11):1822–30.
97. Ohlsson C, Engdahl C, Fak F, Andersson A, Windahl SH, Farman HH, et al. Probiotics protect mice from ovariectomy-induced cortical bone loss. PLoS One. 2014;9(3):e92368.
98. Pytlik M, Folwarczna J, Janiec W. Effects of doxycycline on mechanical properties of bones in rats with ovariectomy-induced osteopenia. Calcif Tissue Int. 2004;75(3):225–30.
99. Williams S, Wakisaka A, Zeng QQ, Barnes J, Martin G, Wechter WJ, et al. Minocycline prevents the decrease in bone mineral density and trabecular bone in ovariectomized aged rats. Bone. 1996;19(6):637–44.
100. Sjogren K, Engdahl C, Henning P, Lerner UH, Tremaroli V, Lagerquist MK, et al. The gut microbiota regulates bone mass in mice. J Bone Miner Res. 2012;27(6):1357–67.
101. Yan J, Herzog JW, Tsang K, Brennan CA, Bower MA, Garrett WS, et al. Gut microbiota induce IGF-1 and promote bone formation and growth. Proc Natl Acad Sci U S A. 2016;22:113.
102. Cho I, Yamanishi S, Cox L, Methe BA, Zavadil J, Li K, et al. Antibiotics in early life alter the murine colonic microbiome and adiposity. Nature. 2012;488(7413):621–6.
103. Cox LM, Yamanishi S, Sohn J, Alekseyenko AV, Leung JM, Cho I, et al. Altering the intestinal microbiota during a critical developmental window has lasting metabolic consequences. Cell. 2014;158(4):705–21.
104. Nobel YR, Cox LM, Kirigin FF, Bokulich NA, Yamanishi S, Teitler I, et al. Metabolic and metagenomic outcomes from early-life pulsed antibiotic treatment. Nat Commun. 2015;6:7486.
105. Wen L, Ley RE, Volchkov PY, Stranges PB, Avanesyan L, Stonebraker AC, et al. Innate immunity and intestinal microbiota in the development of type 1 diabetes. Nature. 2008;455(7216):1109–13.
106. Ivanov II, Atarashi K, Manel N, Brodie EL, Shima T, Karaoz U, et al. Induction of intestinal Th17 cells by segmented filamentous bacteria. Cell. 2009;139(3):485–98.
107. Servick K. Mouse microbes may make scientific studies harder to replicate Science News. 2016. Available from: http://www.sciencemag.org/news/2016/08/mouse-microbes-may-make-scientific-studies-harder-replicate.
108. Laukens D, Brinkman BM, Raes J, De Vos M, Vandenabeele P. Heterogeneity of the gut microbiome in mice: guidelines for optimizing experimental design. FEMS Microbiol Rev. 2016;40(1):117–32.

109. Ericsson AC, Franklin CL. Manipulating the gut microbiota: methods and challenges. ILAR J. 2015;56(2):205–17.
110. O'Connor A, Quizon PM, Albright JE, Lin FT, Bennett BJ. Responsiveness of cardiometabolic-related microbiota to diet is influenced by host genetics. Mamm Genome. 2014;25(11–12):583–99.
111. Cosman F, de Beur SJ, LeBoff MS, Lewiecki EM, Tanner B, Randall S, et al. Clinician's guide to prevention and treatment of osteoporosis. Osteoporos Int. 2014;25(10):2359–81.
112. Claesson MJ, Jeffery IB, Conde S, Power SE, O'Connor EM, Cusack S, et al. Gut microbiota composition correlates with diet and health in the elderly. Nature. 2012;488(7410):178–84.

Index

L.R. McCabe, N. Parameswaran (eds.), *Understanding the Gut-Bone Signaling Axis*, Advances in Experimental Medicine and Biology 1033,
DOI 10.1007/978-3-319-66653-2

Printed by Printforce, the Netherlands